introduction to circuit analysis

intr

a

JOHN D. RYDER

PROFESSOR OF ELECTRICAL ENGINEERING
MICHIGAN STATE UNIVERSITY

PRENTICE-HALL, INC.
ENGLEWOOD CLIFFS, NEW JERSEY

Library of Congress Cataloging in Publication Data

RYDER, JOHN DOUGLAS.
Introduction to circuit analysis.

Includes bibliographical references.
1. Electric networks. 2. Electric circuits.
I. Title.
TK454.2.R88 621.319'2 72-8125
ISBN 0-13-481101-1

Introduction to Circuit Analysis / John D. Ryder

PRINTED IN THE UNITED STATES OF AMERICA

10 9 8 7 6 5 4 3 2 1

PRENTICE-HALL INTERNATIONAL, INC., LONDON
PRENTICE-HALL OF AUSTRALIA, PTY. LTD., SYDNEY
PRENTICE-HALL OF CANADA, LTD., TORONTO
PRENTICE-HALL OF INDIA PRIVATE LIMITED, NEW DELHI
PRENTICE-HALL OF JAPAN, INC., TOKYO

contents

two Circuit Responses: Natural and Transient 32

three Circuit Solution by the Laplace Transform 60

four Steady-State Response; Phasors; Power 92

five Nonlinear Elements; Rectifier Circuits 156

preface

This book is intended to provide a concise and cohesive treatment of those circuit topics which will be needed in other electrical engineering courses and to lay a foundation for later study of more advanced circuit theory. Building on the basic laws of electricity and stressing that all electrical science is based on concepts or models, a logical path of progression is followed. Complex frequency is introduced at an early stage, frequent use is made of equations written for the s domain, and the Laplace transform is made a mechanism for circuit solution rather than an uneasy experience in mathematical rigor. Pole-zero concepts are brought in at appropriate points and supported by mathematical analysis, and transfer functions are utilized. The hope is to prepare the student for modern circuit topics and methods of analysis.

Network models are stressed and carried to active circuit models and to controlled sources; resonance and coupled circuits provide opportunity for use of pole-zero concepts, and the transformation of impedances is developed as an exercise in circuit design to meet performance specifications.

Because of the increasing importance of operational amplifiers, or "pieces of wire with gain," feedback and the gain element are studied in a separate chapter as circuit topics; this is followed by chapters on passive and active *RC* filters. The former area is briefly approached along the historical constant-k path, and the reasons for modern design methods are pointed out as the discussion moves into Butterworth and Chebyshev designs. The chapter on active *RC* filters follows modern development and brings this subject to the undergraduate level. It illustrates the coupling of the gain element with circuit analysis, again developed by finding circuit arrangements to meet specified criteria.

The student is assumed to have a grounding in integral calculus, with some knowledge of the physical basis of electricity. Every effort has been made to keep the book to a manageable and teachable size, but Chapters 10 and 11 could be omitted in a basic course.

J. D. Ryder

introduction to circuit analysis

the fundamental electrical variables

one

1.1 Introduction

No one has ever seen an electron or an electric current. We think of electricity in terms of *charge*, *current*, *voltage*, and *electric or magnetic fields*, but these are only mental structures or concepts to help us understand electricity, since the phenomena associated with it are not directly observable. Even when we see an electric arc we do not see the current—we see only the visible light from the ionized air and incandescent carbon and metal particles in the path of the current.

It follows that experimental study of electricity is usually by inference and indirect methods so that the investigator aids his thought processes with the use of physical or mathematical *models*. These models are often simplified circuits, assembled in the laboratory, or are represented by mathematical equations which the investigator believes will predict performance equivalent to the action of the true electrical variables, whatever those variables may be in physical reality.

Examples of often-employed models include that of the *electron* as a particle having charge and mass, that of current as a *flow rate* of charged particles, and the concept of *flux lines* in electric or magnetic *fields*. Another model is that of the *hole* for the positive charge, which appears at a location in a semiconductor crystal from which a negative electron is absent. We infer that an electron, as a particle, has struck the television screen when we see emission of light from the screen, but we do not see the electron. In fact, we have no understanding of the actual form of an electron.

Because it was necessary to predict results from intangible indications of the presence of electricity, considerable mystery has always surrounded electrical forces, and this stimulated the research interests of early scientists, many of whom were mathematicians. It logically followed that mathematics became the tool and language used in the analysis of electrical phenomena, and this led to the postulation of basic laws in mathematical form, preceding useful applications of electricity. For instance, Faraday presented his law of electromagnetic induction some years before the development of the electromagnetic generator, and Maxwell mathematically predicted the possibility of radio transmission over twenty years before Hertz demonstrated the existence of radio waves in his laboratory. Electrical engineering has thus been set apart as more abstract and mathematical than some of the other engineering fields, whose early developments were fostered largely by direct observation.

At first study, electrical science and engineering may seem overly mathematical, but it should be remembered that their laws are solidly based on experimental evidence. Actually, the employment of mathematics by the early researchers was a fortunate circumstance. It was found that many of the phenomena of electricity could be related by mathematical equations, and solutions for these equations have led imaginative engineers to the development of the practical power, control, communications, and computational systems which surround us.

1.2 Models and concepts

We use the words *model* and *concept* interchangeably, because it is somewhat arbitrary as to whether our physical system should be studied with a simplified set of idealized components in a circuit model or represented in a conceptual model by a set of mathematical equations which predict the response of the system to external stimuli.

For instance, we may employ a model composed of a battery and a resistor to represent an electrical generator, or we may represent the performance of that generator by an equation, as we do when we replace the generator with a *Thévénin-equivalent* model. The use of mathematical models is often preferred over physical models or circuits assembled in the laboratory, since the mathematical models are easier to modify and can be used directly with the computer.

A model will be considered as being isolated from the remainder of its circuit or system and as producing a response believed closely analogous to the response produced by the actual physical circuit being represented. We usually expect idealized theoretical performance from the elements of models; that is, they will be expected to perform in accordance with the theoretical laws developed in the next several sections. In practice such ideal performance is not always obtained, and methods of handling such situations will be developed.

We start with the study of the basic electrical variables: charge, current, voltage, and power. To these we then add knowledge concerning the responses of our *passive elements*—inductance, capacitance, and resistance—when excited by our *active energy sources*, represented by generators, transistors, or vacuum tubes. We combine these elements and sources into useful *circuits* or *networks*, which are then simulated through simplified models known as *equivalent circuits*. When excited by the sources, these equivalent circuits yield responses at their terminals, as do the more complicated networks from which the models are derived. Overall system performance can be analyzed in a relatively simple manner by use of the model, as compared to the level of complexity in the analysis when working with the actual system. Analysis is speeded and our basic understanding of circuit performance is much improved by use of the equivalent circuit.

1.3 Basic system variables

Electrical systems are assembled by the connection of individual circuit elements in desired configurations. The *passive* or non-energy-generating elements, such as resistances, inductances, and capacitances, serve as energy reservoirs or sinks. *Active* elements introduce electrical energy into the system; this energy must be converted from some other form. The battery converts chemical energy; the rotating generator converts mechanical shaft energy to electrical energy; the thermocouple converts thermal energy to an electrical form.

These electrical circuit elements can be studied in terms of a pair of complementary variables: $y(t)$ as a flow rate or response *through* the element and $x(t)$ as a pressure variable applied *across* the element. For such determinations an element must have a pair of terminals for access. A terminal pair is called a *port*; it is used as a location for connection of driving forces, for extraction of energy or signals, and for measurement. Individual elements can be represented as *one-port devices*, as in Fig. 1.1(a).

Passive and active elements may be connected at their terminals into desired arrangements called *circuits* or *networks*. Conductors may be brought out from chosen connections in the network and assembled in pairs as ports of the network. A very common result appears at (b) in Fig. 1.1, where two ports are shown. The box then represents a *two-port network*, a needed concept because one port may be used for connection of a driving or input source, with the second port serving for connection of a load or output.

The number of ports that can be connected to a network may be increased without limit, resulting in n-port networks, but here our studies will concern only one-port and two-port forms.

The value of this approach may be illustrated by consideration of the complicated circuit in Fig. 1.1(c). This electronic microcircuit has designated terminals 1,8 as an *input port* and terminals 5,8 as an *output port*. We may be interested only in the overall function of the circuit, and if we can devise methods of predicting what happens at port 5,8 in response to a given electrical input at 1,8, then we do not really need to know the exact internal connections and so may represent the complicated circuit of (c) by the much simpler two-port box of (b).

By use of an appropriate instrument connected at the ports we can obtain indications proportional to the *across* variable, known as *voltage*, and the *through* effect, which is conceived as a flow rate of electrical charge

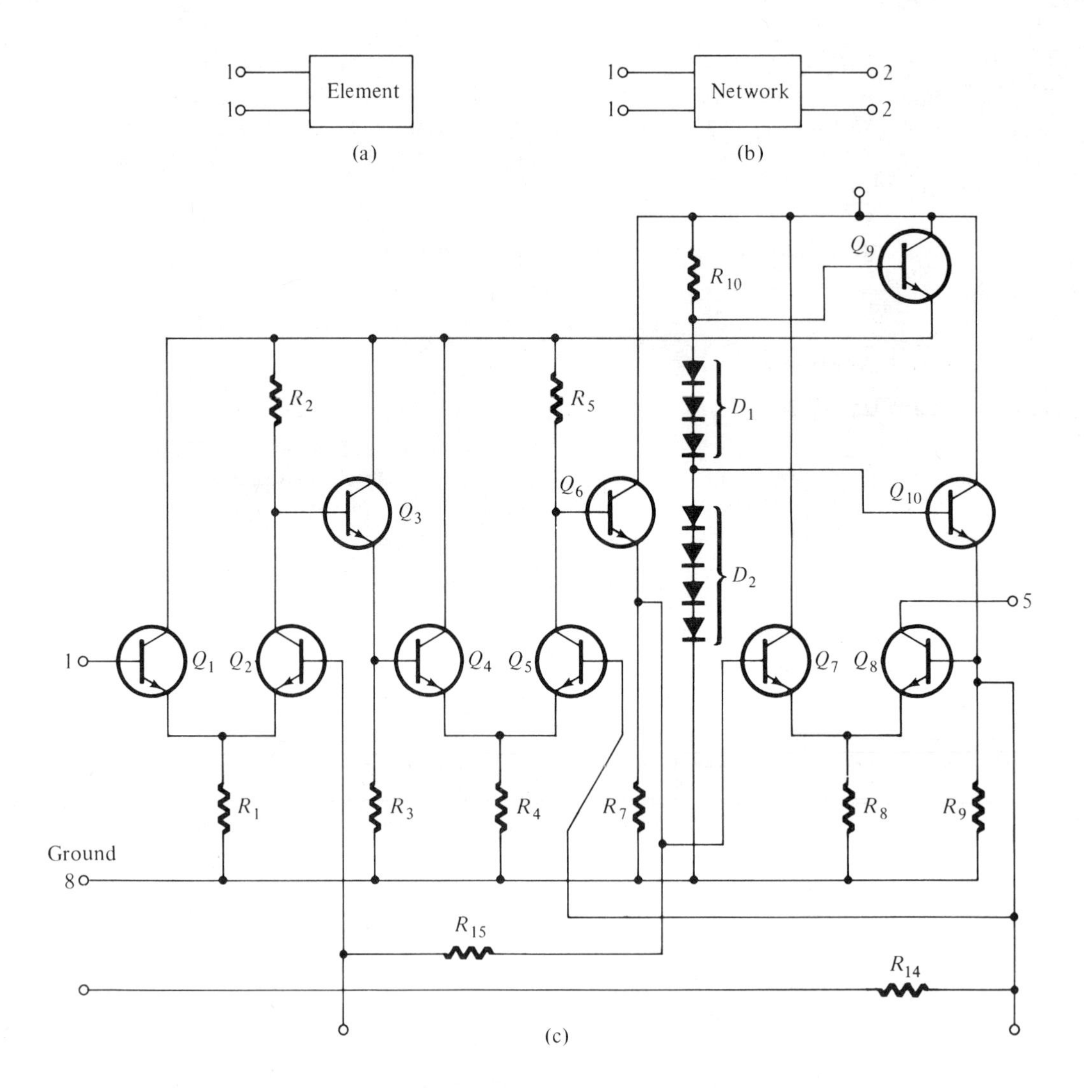

FIGURE 1.1

(a) One-port; (b) two-port; (c) electrical system.

or a *current* into or out of the port. By providing the instrument with a calibrated scale, the indications can give information in the real number system, i.e., positive or negative deflection and the magnitude of the deflection. Manipulation of the resultant numbers allows us to develop constraint relations which apply between the through and across variables at each port. Overall performance of the system of (c) in Fig. 1.1 can thus be measured in terms of the responses at the input and output ports and compared with desired performance.

We shall be concerned with the electrical across and through variables known as voltage and current. In general, these variables and the power variable p *will be time-dependent* and will be written as lowercase v, i, and p for simplicity in our equations. Average values of time-dependent functions will be designated by uppercase letters, as V, I, P.

1.4 Electrical charge

The earliest recorded observations of electrical phenomena were those of Thales of Greece, about the sixth century B.C., and were manifestations of electrical charges in rubbed amber (*elektron* in Greek). The positive and negative charge concept was not introduced until the eighteenth century and was then based on observation of the opposite attractive effects produced by frictional electrification of glass and of a resin such as amber.

The positive and negative charge concept was derived from the observation that there are charges which repel and charges which develop attractive forces. Coulomb (1736–1806) gave us the electrostatic force law, which predicts the force on a point charge q_2 when in the presence of a point charge q_1. An equal force is exerted on q_1 by the presence of q_2. The law states, in the MKSA system of units (Section 1.8), that

$$\mathbf{f}_e = \frac{q_1 q_2}{4\pi\varepsilon r^2}\mathbf{u}_{12} \qquad \text{N (newtons)} \qquad (1.1)$$

The charges q_1 and q_2 are measured in units of charge known as *coulombs* (C), r is the distance between q_1 and q_2 in meters (m), and $\mathbf{u}_{12}$ is a unit vector in the direction from q_1 to q_2. The charges carry their own signs, and if they are of like sign, then $\mathbf{f}_e$ is positive and outward from q_1 along the unit vector. If the charges are of unlike sign, then $\mathbf{f}_e$ is negative and the force is inward toward q_1 along the unit vector.

The constant of proportionality is

$$\varepsilon = \varepsilon_v \varepsilon_r$$

known as the *permittivity*. The constant ε_v is the *space permittivity* $= 10^7/4\pi c^2 \cong 8.85 \times 10^{-12}$ farad/meter (F/m), where c is the space velocity of light $\cong 3 \times 10^8$ meters/second (m/s). The term ε_r is called the *relative permittivity* and depends on the medium in which the force acts; for space, $\varepsilon_r = 1$.

The smallest integral unit of charge is that carried by the *electron*, one of the atomic building blocks. The electronic charge magnitude is given the symbol e and represents -1.602×10^{-19} C, or, in other words, the combined charge of 6.249×10^{18} electrons is needed to equal 1 C of negative charge. In discussing the electron, we often think of it as a particle, with a mass m_o of 9.109×10^{-31} kilogram (kg). However, a concept of an electron having wave properties is useful in the explanation of other phenomena.

EXAMPLE. If an electron is placed in vacuum at a distance of 1 centimeter (cm) from a point P having a positive charge of 1×10^{-4} C [100 microcoulombs (μC)], find the force on the electron.

Using Coulomb's law,

$$f_e = \frac{q_1 q_2}{4\pi\varepsilon r^2} = \frac{1 \times 10^{-4}(-1.602 \times 10^{-19})}{4\pi \times 8.85 \times 10^{-12} \times 0.01^2}$$

$$= -1.44 \times 10^{-9} \text{ N}$$

Because the force is negative, the electron is accelerated toward P.

1.5 Potential difference, the across variable

Coulomb's experimental law shows that electric charges interact at a distance. We conceive of charges establishing a region of influence about themselves, and when a charge experiences a force due to the presence of other charges, we say that it is in an *electric field*.

Newton's second law states that $\mathbf{f} = M\mathbf{g}$ in a gravitational system, and we define the strength of the gravitational field as the force per unit mass, or

$$\mathbf{g} = \frac{\mathbf{f}}{M} \qquad \text{N/kg}$$

By analogy with the gravitational field, we define the *electric field intensity* $\mathscr{E}$ as having a magnitude and direction determined by *the force per unit positive charge*; that is,

$$\mathscr{E} = \frac{\mathbf{f}_e}{q} \qquad \text{N/C} \tag{1.2}$$

Now consider Fig. 1.2(a) in which we show two plane electrodes A and B. The battery C separates the positive and negative charges, employing

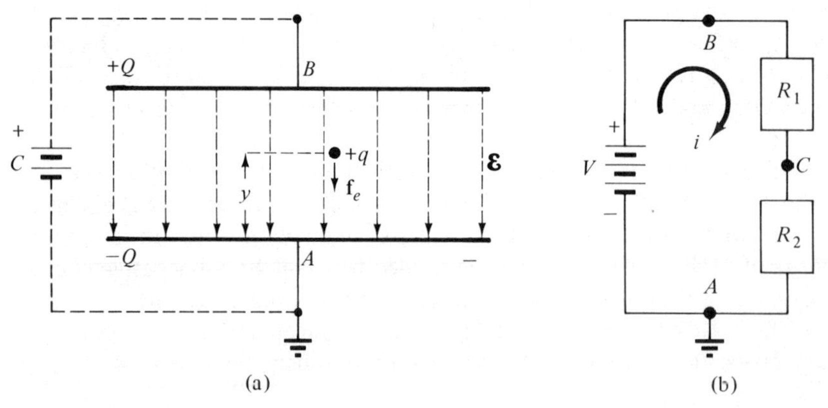

FIGURE 1.2

(a) Positive charge in an electric field; (b) current i in a circuit.

its internal chemical energy to perform the requisite work in overcoming the electric forces, and places a charge of $+Q$ on electrode A and $-Q$ on electrode B. An electric field $\mathscr{E}$ is then present between these planes of charge, as indicated by the dashed arrows. If we place a small positive test charge q at a position y in the space between the electrodes, we find that it is attracted toward the negative electrode A. The force acting on the charge is

$$\mathbf{f}_e = q\mathscr{E} \tag{1.3}$$

and is directed downward. The direction of the force acting on a positive charge is the defined positive direction of the electric field $\mathscr{E}$, indicated by the dashed arrows.

The gravitational potential at a point is defined as the work done per unit mass in bringing the mass from infinity to the point. Since infinity is not always a convenient starting point, we speak more often of differences of potential in which we mean the work done in moving the mass from a reference level to the point. Again by analogy, we define an *electric potential* as the work done per unit positive charge in moving the charge from infinity to the designated point in an electric field. As a more usual and practical measure, we define the *electric potential difference* as the work done per unit positive charge in moving the charge from a chosen reference point to the designated point in the electric field.

Due to its electrochemical action, the battery source C in Fig. 1.2 is able to separate positive and negative charges and place a charge of $+Q$ on the upper electrode B and a charge of $-Q$ on the lower electrode A. The work done by the source per unit charge moved is the electric potential difference established between the electrodes A and B, designated $\mathbf{v}_{BA}$. This potential difference can be determined by calculating the work done on a small test charge $+q$ placed on A and to be moved to B in the field existing between the charges $+Q$ and $-Q$ on the electrodes. The forced motion will be upwards in the negative direction of the electric field and so the work per unit charge q is

$$\frac{W}{q} = -\frac{1}{q}\int_A^B \mathbf{f}_e \, dy = -\frac{1}{q}\int_A^B q\mathscr{E} \, dy = \mathbf{v}_{BA}$$

and so

$$\mathbf{v}_{BA} = -\int_A^B \mathscr{E} \, dy \tag{1.4}$$

With the electrode areas large with respect to the spacing d, the intensity $\mathscr{E}$ will be uniform in the y direction in the region remote from the electrode edges. Then the potential difference between B and A due to charges $+Q$ and $-Q$ on the electrodes is

$$\mathbf{v}_{BA} = \mathscr{E} d \tag{1.5}$$

with B positive to A. The units of $\mathbf{v}_{BA}$ are joules per coulomb of charge (J/C), given the name *volt* (V) for convenience and in memory of the early electrical researcher Volta (1745–1827). Field intensity $\mathscr{E}$ is then expressible in *volts per meter.*

A battery uses chemical energy to move charges against its internal electric field. A rotating generator raises positive charges internally from a negative to a positive terminal and against the electric field by conversion of rotating shaft energy. These are common sources. We also have sinks or energy lossers, known as resistances. Referring to Fig. 1.2(b), the resistances at R_1 and R_2 cause a loss of charge energy, and point C is at a lower potential than is point B, or, in other words, C is negative to B. If we consider the reference plane or *ground* at A, then we may employ a double-subscript notation on potentials, the second subscript indicating the reference electrode; i.e., $\mathbf{v}_{BA} = +15$ V, $\mathbf{v}_{CA} = +10$ V, and $\mathbf{v}_{CB} = -5$ V.

Potential difference, often referred to only as potential, has a sign and a magnitude and must be stated with a point of reference given or implied. Herein we shall use a plus sign to indicate the point of positive potential.

1.6 Current, the through variable

An electric current is defined as the net rate of flow of electrical charge through a surface; with charge q in coulombs and time t in seconds, the current is

$$i = \frac{dq}{dt} \qquad \text{A (amperes)} \qquad (1.6)$$

A net rate of movement of 1 coulomb per second (C/s) represents a current of 1 A; the ampere was named for the French physicist André Ampère (1775–1836).

The valence electrons of most atoms in metallic conductors are able to drift in the space between the atoms of the crystal lattice; for the good conductors of Group I of the Periodic Table there is one such free electron per atom. Under conditions of equilibrium the thermal energies of these electrons produce motion in random directions, and there is no net movement of charge. However, application of an electric potential and an electric field along a conductor adds a directed drift component onto the otherwise random motions of the charges, and a current of electrons results.

In certain semiconductors the free charges are called *holes* and represent lattice sites from which electrons are missing. The hole charge or site is attractive to negative electrons and acts as a positive charge. The holes or vacancies can migrate, and conduction is then by positive charges.

Current may therefore be due to motion of either positive or negative charges, moving through the reference surface in both directions, as in Fig. 1.3. A negative charge moving to the left across the surface is equivalent to a positive charge moving to the right. The net current due to both positive and negative charges is then to the right.

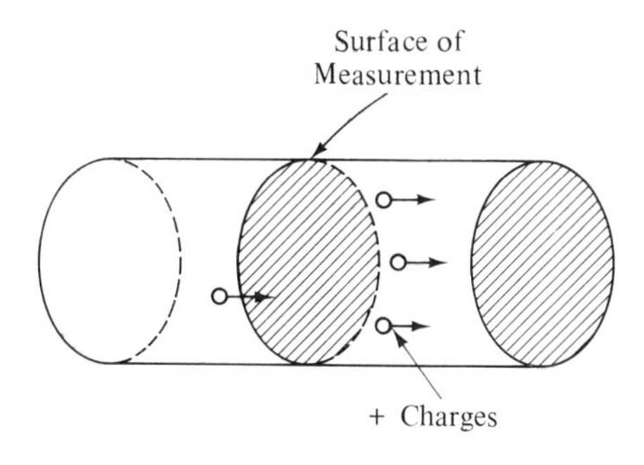

FIGURE 1.3

Motion of charge in a conductor.

1.7 Power

Electric potential difference between two points is the work required to move a unit charge between those points in an electric field. With work shown by w, as a function of time, then for a differential charge dq we have

$$v = \frac{dw}{dq} \tag{1.7}$$

Multiplying by $i = dq/dt$,

$$\frac{dw}{dq}\frac{dq}{dt} = vi = \frac{dw}{dt} \qquad \text{J/s} \tag{1.8}$$

But dw/dt represents the rate of change of energy and this is power, p, so

$$p = \frac{dw}{dt} = vi \qquad \text{W (watts)} \tag{1.9}$$

since a joule per second is called a *watt*. Since v and i are functions of time, p is also a function of time and is *instantaneous power*.

Energy is given by the integral equation

$$W = \int p\,dt = \int vi\,dt \tag{1.10}$$

Power is the result of the product of the potential *across* a one-port element and the current *through* that one-port. With current as a movement of positive charge, current *from* the positive element terminal implies that work has been done to raise the positive charges to that terminal; the one-port is an electrical energy source. If the current is directed *into* the positive element terminal, the charges lose energy in dropping from the positive terminal to the negative terminal; the element is then an energy sink.

1.8

The MKSA system and electrical units

Early in electrical system development it was realized that electricity and electrical science would override national boundaries, and therefore the International Electrotechnical Commission was created by international cooperation to achieve unity and standardization. The work of this commission has now been incorporated in a broader set of recommendations, based on the metric system and known as the International System of Units (*Système International* in French, and therefore SI), which covers the related fields of physics, mechanics, and electricity. This system, now generally used in electrical engineering, employs the rationalized MKSA units, founded on the meter, kilogram, second, and ampere for the basic units.

These basic units are defined in terms of various physical measurements, with the exception of the kilogram, which is represented by a particular platinum-iridium cylinder in a vault at Sèvres, France. The meter is equal to 1,650,763.73 wavelengths in vacuum of the light emitted in a particular line of the krypton-86 spectrum. The second is referred to the motion of the earth, and the ampere is derived in terms of a measurement of magnetic forces.

Electrical engineering honors its early researchers by utilizing their names for various derived units; thus we have the ampere, the volt, the coulomb, the ohm (Ω), the farad, and the henry (H). As indicated, abbreviations for these units are usually obtained from the initial letter of the name and when so derived are capitalized. Thus 10 A indicates a current of 10 amperes; 10 g (lowercase letter) shows a mass of 10 grams since no person is being honored.

Frequently it is necessary to work with quantities which are very large or very small multiples of the defined units. For instance, a charge may be stated as 1.5×10^{-6} C, but this quantity is more conveniently stated as 1.5 μC. The standard metric prefixes for designation of often-used powers of 10 are given in Table 1.1.

TABLE 1.1 **STANDARD MAGNITUDE PREFIXES**

Multiplier	*Prefix*	*Symbol*	*Multiplier*	*Prefix*	*Symbol*
10^{12}	tera	T	10^{-2}	centi	c
10^{9}	giga	G	10^{-3}	milli	m
10^{6}	mega	M	10^{-6}	micro	μ
10^{3}	kilo	k	10^{-9}	nano	n
10^{2}	hecto	h	10^{-12}	pico	p
			10^{-15}	femto	f
			10^{-18}	atto	a

It was originally proposed that prefixes for magnitude multipliers greater than unity be abbreviated as uppercase letters and that symbols for magnitudes less than unity be shown as lowercase letters. However, K was entrenched as the abbreviation for Kelvin temperature and H for the henry of inductance, and so k and h are employed as exceptions.

1.9 Circuit elements

Systems for control and conversion of electrical signals and energy are designed by interconnection of one-port elements. Two means of connection are possible, and when one-port elements carry the same current, they are said to be in *series*, while elements subjected to a common voltage are said to be in *parallel*; these connections are shown in Fig. 1.4.

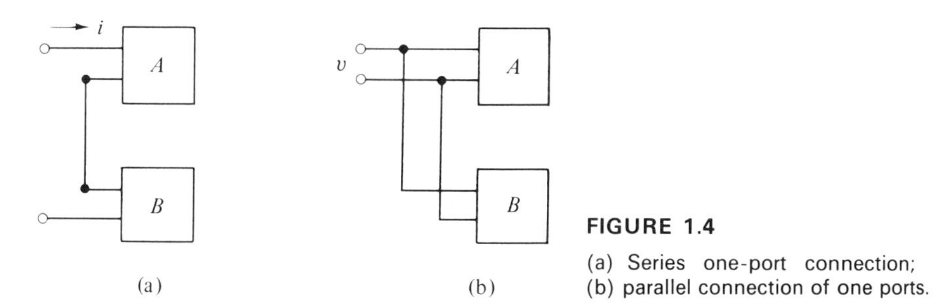

FIGURE 1.4
(a) Series one-port connection; (b) parallel connection of one ports.

A one-port element is described, or modeled, by a relation between its across-through variables, or, in electrical terms, the voltage-current relation. This v-i characteristic may be obtained analytically, or it may be determined from measurements made at the port and expressed mathematically or graphically. Using the v-i relation for each element and applying elementary field theory, we can derive energy relations. We can study the performance of combinations of elements into circuits and networks by the methods of circuit theory, to be developed later.

Here, our elements will be assumed to be ideal—or, in other words, the electric and magnetic fields are isolated and confined to the indicated elements, and charge conduction occurs only over prescribed paths. Later we shall consider the effects of less ideal elements.

1.10 Resistance

Charges moving through a metallic material are subject to collisions with the atoms present in the material. These collisions are not elastic and result in energy transfer from the charges to the atoms. In Fig. 1.5(a) the arrow represents the assumed direction of positive charges into the one-port

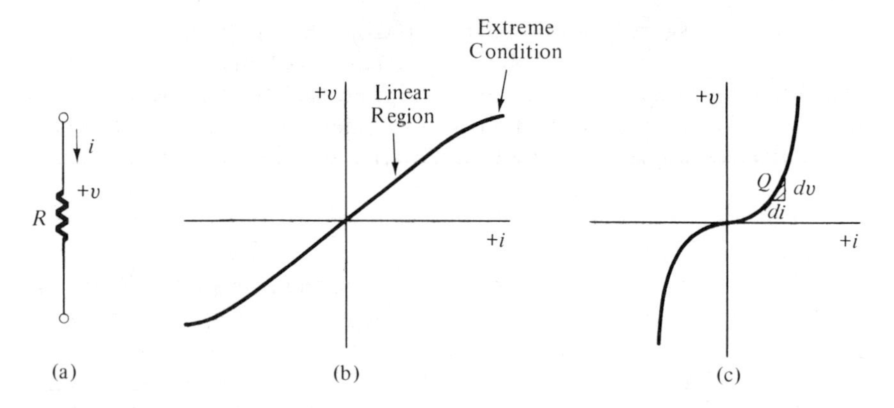

FIGURE 1.5
(a) Circuit symbol for a resistance; (b) volt-ampere relation for a resistor, with a linear region; (c) nonlinear resistor.

element. In passing along the element, the charges lose energy; since potential difference is a measure of that energy, the voltage at the current-entering end of the one-port is higher than at the lower or reference end. Therefore v is positive to reference at the current-entering end, as marked. The volt-ampere relation which expresses this loss of potential between terminals is Ohm's law, stated as

$$v = Ri \qquad (1.11)$$

where R is the *resistance*. Thus

$$R = \frac{v}{i} \qquad \Omega \qquad (1.12)$$

relates the across and through measurements of the one-port. The units of resistance R are volts per ampere, given the name *ohms*.

If plotted, Eq. 1.11 would be a straight-line relation between v and i, as in Fig. 1.5(b). Laboratory measurement of practical resistances may show departure from linearity at extreme current values; however, we shall adopt the linear plot to represent our *ideal* resistance element R. This linear relation requires that R be independent of voltage and current, and we shall also assume that it is independent of time.

Since power $p = vi$, we have for a resistance R

$$p = vi = Ri \times i = Ri^2 \qquad \text{W}$$

Assuming that the current is zero prior to $t = 0$, the energy supplied to a resistance R is

$$W = \int Ri^2\, dt \qquad \text{J} \qquad (1.13)$$

where i must be expressed as a function of time. The input energy is positive due to i^2, and it is finite for finite time. The integral represents a continuing summation of energy in the resistance R, for which we use the symbol W. Joule showed the equivalence of this energy to heat. There is no circuit means by which this energy may be returned to the current source, and failure to remove the heat by conduction, convection, or thermal radiation will result in a rising temperature in any practical resistance element—and its possible destruction.

A current of positive charge enters a resistance at the high-potential end and delivers energy to the one-port resistance element. The positive terminals of ammeter A and voltmeter V would be connected at the current-entering end for measurements of the through and across variables consistent with this reasoning.

The resistance of a material of uniform cross section, a wire, for example, is proportional to its length and inversely proportional to the cross-sectional area of conduction S, or

$$R = \frac{\rho L}{S} \qquad \Omega \tag{1.14}$$

where L and S are measured in meter units. The constant ρ is a function of the material and is called the *resistivity*, measured in *ohm-meters* (Ω-m).

The *conductance* of a material of uniform cross section is the reciprocal of the resistance and is written as

$$G = \frac{1}{R} = \frac{\sigma S}{L} \qquad \text{mhos} \tag{1.15}$$

The unit of conductance, as reciprocal ohms, is given the name *mho*, or ohm spelled backward. The constant of the material, σ, is the *conductivity*, measured in *mhos per meter*.

EXAMPLE. Hard-drawn copper has a resistivity $\rho = 1.77 \times 10^{-8}\ \Omega$-m at 20°C. A copper wire 2.053 mm in diameter (no. 12 AWG) is to be used, and we wish to find its resistance per meter.

The cross-sectional area is

$$S = \pi r^2 = \pi(1.026 \times 10^{-3})^2 = 3.31 \times 10^{-6}\ \text{m}^2$$

Then

$$R = \frac{\rho L}{S} = \frac{1.77 \times 10^{-8} \times 1}{3.31 \times 10^{-6}} = 0.00535\ \Omega/\text{m}$$

It is of interest to determine the average velocity of the charges in such a wire when carrying a current of 1 A. The density of atoms in copper is $8.4 \times 10^{28}/\text{m}^3$, and since the valence of copper is unity, it is assumed that each atom contributes one free electron for conduction. The volume of the wire, per meter of length, is

$$\text{Vol.} = S \times 1 = 3.31 \times 10^{-6}\ \text{m}^3$$

and the number of free charges, per meter of length, is

$$n = 8.4 \times 10^{28} \times 3.31 \times 10^{-6} = 2.78 \times 10^{23}$$

One ampere requires the movement of 6.249×10^{18} charges/s, and we can find the velocity as

$$\begin{aligned} \text{Velocity} &= \frac{\text{charges needed per second}}{\text{charges available per meter}} \\ &= \frac{6.249 \times 10^{18}}{2.78 \times 10^{23}}\ \text{m/s} \\ &= 2.24 \times 10^{-5}\ \text{m/s} = 0.0224\ \text{mm/s} \end{aligned}$$

Thus the velocity of the individual charges is small in a good conductor, and the conduction process in a solid is frequently referred to as a *drift* of charge.

In English units, conductor area is measured in *circular mils*, as the area of a circle of 1-mil, 0.001-in., diameter. The area of a round wire in circular mils is equal to the square of its diameter in mils. The resistivity of a material is then stated in *ohm-circular mils per foot*.

The resistance of conductors of variable cross section must be determined by incremental methods; that is,

$$dR = \frac{\rho\, dL}{S(L)} \tag{1.16}$$

and the area S must be expressed as the appropriate function of length.

Since the atoms of a conductor become more agitated as the temperature rises, more collisions between moving charges and atoms may be expected, and the energy loss in collisions increases. For good conductors (copper, silver, lead, nickel, etc.), this effect is linear with temperature over usual temperature ranges, and the resistance is expressed by

$$R_2 = R_1[1 + \alpha(T_2 - T_1)] \qquad \Omega \tag{1.17}$$

where the temperatures are in degrees centigrade and α is known as the *temperature coefficient of resistance*, given in ohms per ohm per degree centigrade. Equation 1.17 does not apply for temperatures near absolute zero.

Certain alloys, such as manganin, have been developed to provide near-zero values of temperature coefficient over the usual ambient temperature ranges. These alloys are useful in the manufacture of accurate constant resistors for measuring equipment.

Other materials, including the semiconductors germanium and silicon, show negative values of α. In these materials the number of free charges increases with temperature, and the conductivity improves with the increasing number of free charges.

Resistor elements are made in wire-wound or molded carbon forms or with a metal or carbon film deposited on a glass or ceramic base. They are rated in terms of the maximum power that they can dissipate without exceeding the limits in which their resistance remains essentially constant. Wire-wound forms are the most stable, but film types can also be manufactured to high standards of accuracy.

1.11 Capacitance

A capacitance is formed when two conducting electrodes, usually planes, are separated by an insulating medium, known as a *dielectric*. A typical schematic arrangement appears in Fig. 1.6. In a similar arrangement in Section 1.5 we showed that when a voltage was applied between the plates

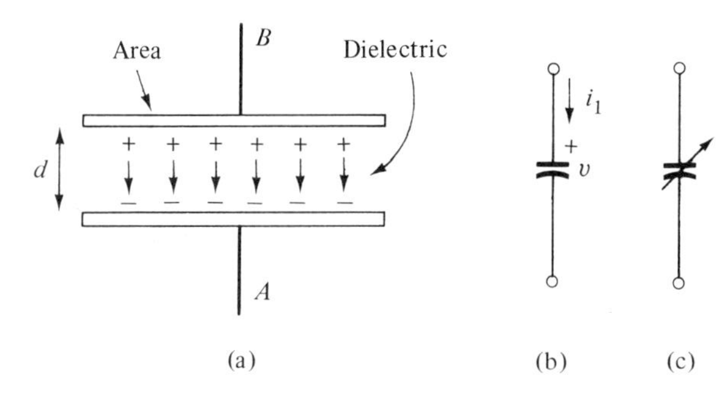

FIGURE 1.6

(a) Capacitance; (b) IEEE symbol for capacitance; (c) IEEE symbol for a variable capacitance.

a positive charge appeared on the plate at the positive terminal and a negative charge appeared on the plate at the negative terminal. The charge held is proportional to the voltage applied between the plates:

$$q = Cv \qquad \text{C} \tag{1.18}$$

The proportionality constant C is the measure of the *capacitance*, and

$$C = \frac{q}{v} \qquad \text{F} \tag{1.19}$$

in units of coulombs per volt, given the name *farad* in memory of Faraday's early work. Measured as the charge accumulated per volt applied, we shall here consider C as constant, independent of voltage, current, or time, although in practice we have some capacitors in which C varies with applied voltage.

When the applied voltage is varied, the charge is also varied, and a pair of observers stationed at A and B in the conductors to the capacitor electrodes would note charge passing from or to the capacitance; such charge movement represents a current to the capacitor. By differentiation of Eq. 1.18 we find the observed current to be

$$\frac{dq}{dt} = i = \frac{d(Cv)}{dt} = C\frac{dv}{dt} \quad \text{A} \tag{1.20}$$

with C constant. This equation represents a fundamental relation for a capacitance, linearly relating current and rate of change of voltage across the capacitance, in accordance with the symbolism of Fig. 1.6(b).

Integration of Eq. 1.20, with an initial charge Q_o on the capacitance at $t = 0$, gives

$$\begin{aligned} v &= \frac{1}{C}\int_0^T i\,dt + \frac{Q_o}{C} \\ &= \frac{1}{C}\int_0^T i\,dt + V_o \quad \text{V} \end{aligned} \tag{1.21}$$

where V_o appears as the voltage across the capacitance at $t = 0$, as a result of the initial charge Q_o. Equations 1.20 and 1.21 are the basic v and i, or through and across, relations for the capacitance one-port element. Because of the effect of the integral in Eq. 1.21, it is not possible for the voltage across a capacitance to change instantaneously; the needed infinite current is not physically obtainable.

We can determine the energy present in a capacitance as

$$W = \int_0^T vi\,dt = \int_0^V vC\frac{dv}{dt}\,dt = \int_0^V vC\,dv = \frac{Cv^2}{2} \quad \text{J} \tag{1.22}$$

The energy increases with the square of the voltage; the energy is stored and is all returned to the circuit if the final value of v is zero, since the upper limit of the integral is then zero. There is no loss of energy in an ideal capacitance. From physical evidence it can be shown that the energy is stored in the electric field in the dielectric.

For a parallel electrode capacitance, as in Fig. 1.6, the value of the capacitance can be calculated as

$$C = \frac{\varepsilon_v \varepsilon_r S}{d} \quad \text{F} \tag{1.23}$$

where area S and separation d are in meter units. The constant ε_v has been defined as $\cong 8.85 \times 10^{-12}$, and the relative permittivity, ε_r, is given the name *dielectric constant*, with values for various materials available in tables. For air, $\varepsilon_r = 1$.

EXAMPLE. A parallel-plate capacitance has an effective area $S = 25\ \text{cm}^2$, a plate separation $d = 0.5$ mm, and an air dielectric. The capacitance can be found as

$$C = \frac{\varepsilon_v \varepsilon_r S}{d} \qquad \text{F}$$

$$= \frac{8.85 \times 10^{-12} \times 1 \times 25 \times 10^{-4}}{0.5 \times 10^{-3}}$$

$$= 44.2 \times 10^{-12}\ \text{F} = 44.2\ \text{pF}$$

The farad is a large unit, and practical capacitances are usually in the microfarad or picofarad range.

As circuit elements, capacitances are manufactured in several forms. Large capacitances are made as a sandwich of metal foil electrodes with a dielectric such as mica, paper, or plastic. Small capacitances in the picofarad range employ ceramic dielectrics. In some large capacitances the electrodes are of aluminum foil with a very thin layer of aluminum oxide on one surface as the dielectric. The foils are separated by paper saturated with an electrolyte for current conduction—thus the name *electrolytic capacitors*.

1.12 Inductance

A current-carrying straight wire surrounds itself with a concentric magnetic field, modeled conceptually in terms of *lines of magnetic flux*. If the wire is wound into a helical coil as in Fig. 1.7(a), the flux created by each turn concentrates along the axis of the coil and links with additional turns of wire; the flux passes inside the coil and then around the outside of the helix. The

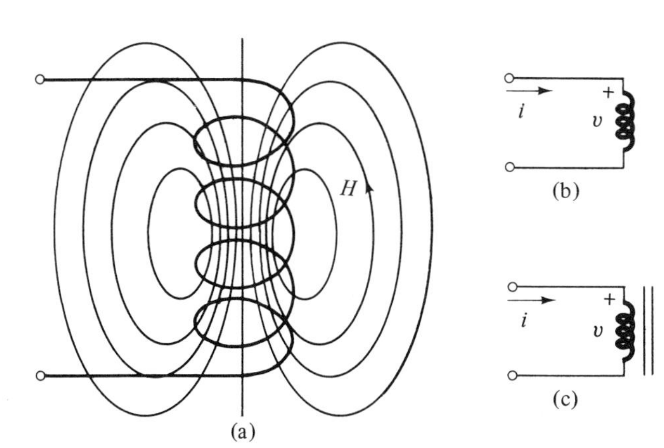

FIGURE 1.7

(a) Flux passing through a helical coil; (b) circuit representation of an inductance; (c) same, with an iron magnetic path.

flux linkages are the product $N\Phi = \lambda$, where $N =$ turns of the coil and $\Phi =$ the flux in webers (Wb) linking those turns. The linkages are greatly increased over the linkages present around the straight wire, where each flux line links the current only once.

The original concept of flux probably came from the patterns obtained when iron filings aligned themselves in curvilinear lines around the poles of a bar magnet. The opposite ends of the helical coil are analogous to the poles of the magnet.

Faraday, in 1831, showed that a *change in the flux linkages* of a coil produced a voltage in the coil, as indicated in Fig. 1.7(b) or (c). Faraday's law states that

$$v = \frac{d(N\Phi)}{dt} \qquad \text{V} \tag{1.24}$$

If the number of turns is constant,

$$v = N\frac{d\Phi}{dt} \qquad \text{V} \tag{1.25}$$

The voltage developed by the change in flux linkages has a polarity such as to oppose the change in current producing it. A current i, increasing in time in Fig. 1.7(b) or (c) induces a voltage $+v$ as shown, thereby opposing a source which tends to increase i; this is implied by the symbolism of the figure. Should i decrease with time, then di/dt becomes negative and the voltage reverses in polarity, thus tending to maintain the current.

If Φ and I are linearly related, as is true when the coil is surrounded by air or nonmagnetic material, then Eq. 1.24 can also be written

$$v = \frac{d(Li)}{dt} = L\frac{di}{dt} \tag{1.26}$$

The constant of proportionality between the voltage and the rate of change of the current is L, the *inductance*, measured in *henrys*. A coil which develops an induced voltage of 1 V when the current changes at a rate of 1 A/s has an inductance of 1 H. We shall here consider the L value of our ideal inductance as independent of voltage, current, or time.

By integration of Eq. 1.26 from $t = 0$ to $t = T$ we obtain a second fundamental relation for inductance, namely

$$i = \frac{1}{L}\int_0^T v\,dt + I_o \tag{1.27}$$

Current I_o is the value of any current in the inductor at $t = 0$. Because of the integration of v, the current will be delayed, or will follow the application of the potential. As a result of the integration a sudden change of current through an inductor is not possible, since the needed infinite voltage is not physically obtainable.

Since energy is the integral of instantaneous power, and since $v = L\,di/dt$, we can write

$$W = \int_0^T vi\,dt = \int_0^T L\frac{di}{dt}i\,dt = \int_0^{i_o} Li\,di$$

$$W = \frac{Li_o^2}{2} \qquad \text{J} \tag{1.28}$$

where i_o is the current at $t = T$. Equation 1.28 states the energy present in the magnetic field of an inductor with a current i_o. As the value of i_o is changed the stored energy changes, and if i_o is reduced to zero the integral becomes one with zero upper and lower limits and the magnetic energy becomes zero. The energy of the inductance has been returned to the circuit, and therefore an ideal inductance (without resistance) does not lose energy.

1.13 Inductors with iron in the magnetic circuit

The magnetic flux may be considered as the through quantity in a magnetic field, and a magnetic force H may be defined as the across variable. The driving force H is

$$H = \frac{Ni}{l} \qquad \text{A-turns/m} \tag{1.29}$$

where l is the length of the magnetic path in meters. The *flux density* produced by such a magnetic force, in vacuum, is

$$B = \mu_v H \qquad \text{Wb/m}^2 \tag{1.30}$$

where μ_v is again the space permeability $= 4\pi \times 10^{-7}$.

The flux of a coil, and therefore its inductance, can be greatly increased by providing a flux path made of magnetic steel or of powdered magnetic material compressed to shape, as in Fig. 1.8. Such ferromagnetic materials have tiny internal permanent magnets, called *domains*. Domains have fluxes which are randomly oriented, but as the field of the coil penetrates the material, they tend to align themselves so that their fields are added to the field induced by the current in the coil. The flux density inside the coil is

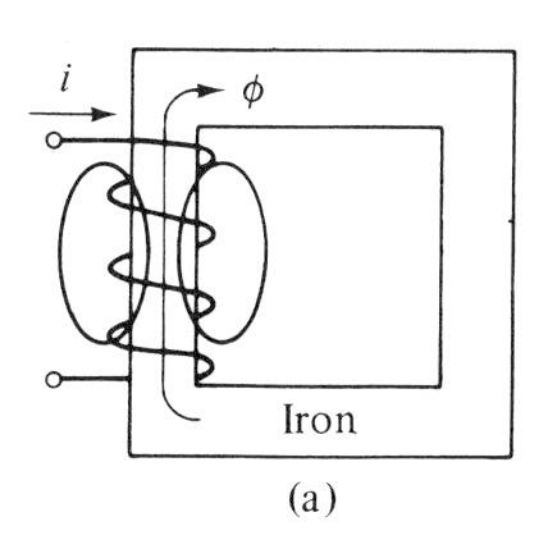

(a)

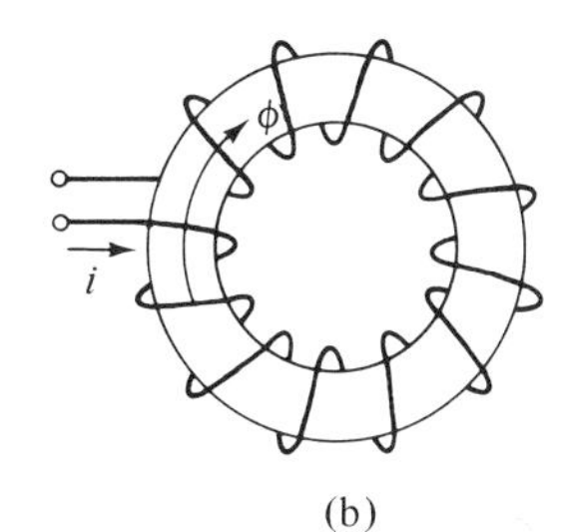

(b)

FIGURE 1.8

(a) Iron-cored inductor; (b) toroid coil.

then greater than if the flux path were only of air. The difference is accounted for by modifying Eq. 1.30 to

$$B = \mu_v \mu_r H \qquad \text{Wb/m}^2 \tag{1.31}$$

where μ_r is the *relative permeability* of the magnetic material used in the flux path. The relative permeability is unity for air or vacuum but reaches values of some thousands for many magnetic materials.

We may combine Eqs. 1.25 and 1.26 to yield

$$L = N\frac{d\Phi}{di} \tag{1.32}$$

In magnetic materials the value of μ_r is a function of B, and the flux density B becomes a nonlinear function of H or of the current. This relation is represented by a *magnetization curve* for each type of iron or steel, as in Fig. 1.9 for "Hipersil" transformer steel. Equation 1.32 shows that the

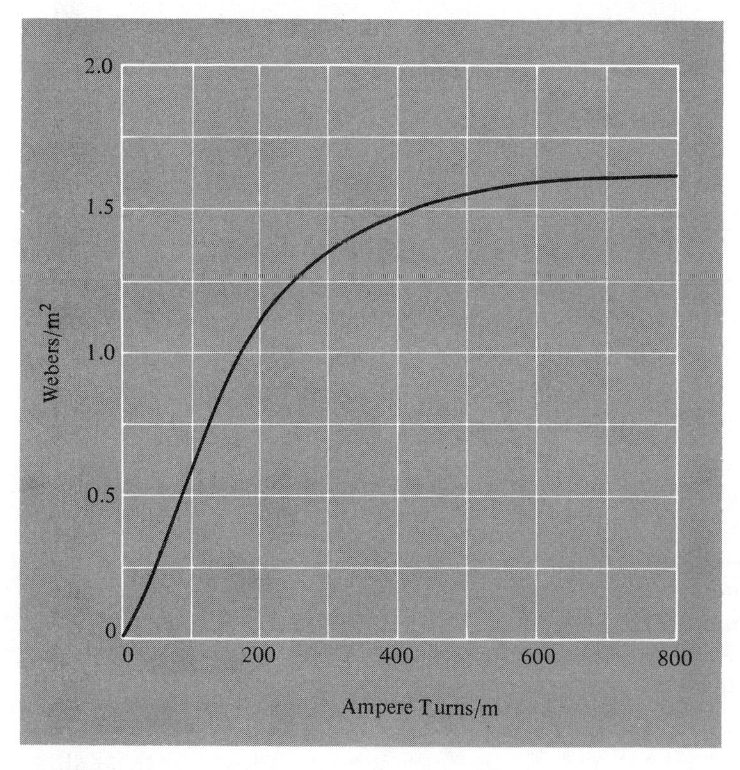

FIGURE 1.9

Magnetization curve for oriented-grain transformer sheet steel.

inductance of a coil is proportional to $d\Phi/di$, or the slope of the magnetization curve. At high H values the slope of a magnetization curve decreases, or the material is said to *saturate*. Physically this means that the domains are all substantially aligned. The inductance of a coil wound on the material of Fig. 1.9 would not be constant over all values of exciting current.

We have said that we shall consider in this text only inductors in which L is independent of current; these are *linear inductors*. The maintaining of this assumption for iron-cored inductors requires that the iron core be operated at values of B or H on the linear, nearly constant-slope portion of the magnetization curve. Air-core coils are inherently linear, or L is constant, since $d\Phi/di$ is constant.

Figure 1.7(a) indicates that part of the flux from N_1 does not link all the coil turns and is not fully effective. If a coil is wound around a circular form or torus so as to bring its lower end around to its upper end, as in Fig. 1.8(b), the symmetry requires that as much flux leave the interior of the coil between each pair of turns as returns by that path. The net outward flux is zero and the flux remains within the coil, where it links all the turns. Such coils are *toroids* and are useful in reducing cross coupling when coils are mounted near each other, since the external flux is zero.

EXAMPLE. A toroidal inductor has 20 turns on a powdered iron core with a 1.0-cm^2 cross section and an average radius of 2 cm. The relative permeability is 5000 and constant. Find the flux in the core and the inductance with a current of 2 A.

$$H = \frac{Ni}{L} = \frac{20 \times 2}{2\pi \times 2 \times 10^{-2}} = 3.18 \times 10^2 \text{ A-turns/m}$$

$$B = \mu_v \mu_r H = 4\pi \times 10^{-7} \times 5 \times 10^3 \times 3.18 \times 10^2$$

$$= 2.0 \text{ Wb/m}^2$$

The total flux in the core is

$$\Phi = BA = 2.0 \times 1 \times 10^{-4} = 2 \times 10^{-4} \text{ Wb}$$

If μ_r is constant, then Eq. 1.32 becomes for constant permeability:

$$L = \frac{N\Phi}{i} = \frac{20 \times 2 \times 10^{-4}}{2} = 2.0 \times 10^{-3} \text{ H} = 2.0 \text{ mH}$$

1.14 Mutual inductance

Magnetic flux established by a changing current in one coil may link with turns of a second coil, inducing a voltage in the second coil, as indicated in Fig. 1.10. The coils are *coupled* by the flux, and the device is known as a *transformer*. The induced voltage is

$$v_2 = \frac{d(M_{12} i_1)}{dt} = M_{12} \frac{di_1}{dt} \tag{1.33}$$

the second form following because of an assumption of the constancy of M_{12}, the *mutual inductance*, measured in henrys. The polarity of v_2 is such as to

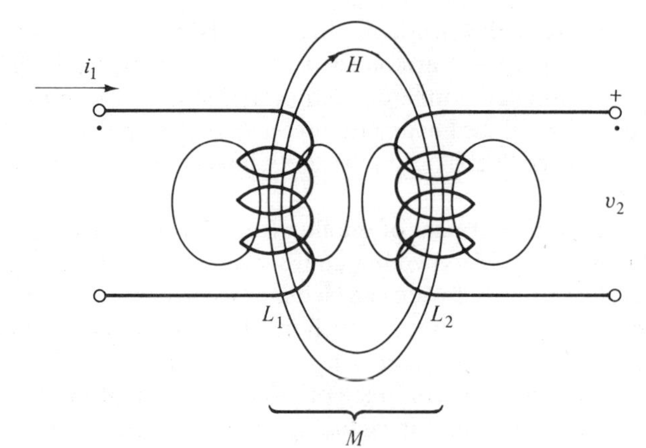

FIGURE 1.10
Mutual inductance and leakage flux between coils.

cause a current i_2 in the connected secondary circuit, which will produce a magnetic flux Φ_2 in coil L_2 that will oppose the original flux Φ_1 established by coil L_1. This is Lenz's law.

In the analysis of circuits employing mutual inductance coupling, the direction of the windings is not always known, yet the polarity of the secondary voltage v_2 must be specified. Dots indicate those terminals assumed to have the same simultaneous positive polarity, as shown in Fig. 1.10. The desired polarity can be realized physically by choice of connections at the secondary terminals.

Except for the toroid it is not possible to confine all flux to the desired magnetic path and thus to link both coils. Some flux will surround one coil but fail to link with the turns of the second coil, and this constitutes the *leakage flux*. It contributes to the self-inductance of L_1 or L_2 but does not form a part of the coupled flux Φ_{12} leading to inductance M_{12}.

Combining Eq. 1.32 for the first circuit with the mutual term similarly written leads to

$$\frac{M_{12}}{L_1} = \frac{N_2\, d\Phi_{12}/di_1}{N_1\, d\Phi_1/di_1} = \frac{N_2\, d\Phi_{12}}{N_1\, d\Phi_1} \tag{1.34}$$

When only a fraction k_1 of the flux Φ_1 links with the turns of N_2, we have $\Phi_{12} = k_1\Phi_1$ and so

$$\frac{M_{12}}{L_1} = k_1 \frac{N_2}{N_1} \tag{1.35}$$

Starting with a current i_2 and writing Eq. 1.33 for v_1, we could carry out similar operations and obtain

$$\frac{M_{21}}{L_2} = k_2 \frac{N_1}{N_2} \tag{1.36}$$

Equations 1.35 and 1.36 can be combined so that

$$k_1 k_2 = \frac{M_{12}M_{21}}{L_1 L_2} \tag{1.37}$$

With constant values of M and L, by symmetry $M_{12} = M_{21} = M$. Defining a *coefficient of coupling* k as

$$k = \sqrt{k_1 k_2} \tag{1.38}$$

then

$$k = \frac{M}{\sqrt{L_1 L_2}} \tag{1.39}$$

The value of k is expressed as a decimal, usually small if the flux path is in air.

TABLE 1.2

ELECTRICAL PROPERTIES OF SOME MATERIALS

RESISTANCE

Material	*Resistivity* (ρ), *Ω-m at 20°C*	*Temp. Coeff., Ω/Ω/°C*	*Conductivity* (σ), *mho/m*
Silver	1.63×10^{-8}	0.0038	61.4×10^6
Copper (hard-drawn)	1.77×10^{-8}	0.0038	56.5×10^6
Aluminum	2.83×10^{-8}	0.004	35.4×10^6
Tungsten	5.5×10^{-8}	0.0045	18.2×10^6
Nickel	7.77×10^{-8}	0.006	12.8×10^6
Manganin	44.0×10^{-8}	0.00001	2.3×10^6
Nichrome	100×10^{-8}	0.00044	1.02×10^6
Germanium	0.60		1.67
Silicon	1.5×10^3		0.67×10^{-3}

DIELECTRICS

	Resistivity (ρ), *Ω-m at 20°C*	ε_r *at* 10^3 *Hz*
Air		1.00058
Glass (Pyrex)		4.97
Mica (ruby)	5×10^{13}	5.4
Paper		2–4
Polyethylene	10^{17}	2.26
Polystyrene	10^{18}	2.56
Quartz	$> 10^{19}$	3.78
Formica		6.0
Titanium dioxide		100
Water	10^6	80

MAGNETICS

	Max μ_r
Hipernik	80,000
Mumetal	110,000
Permalloy (78% nickel)	100,000
Electrical steel sheets	6,150
Oriented transformer steel	40,000

The value of k may be near unity when good magnetic material with high μ_r is used in the flux path; with $k = 1$ the magnetic circuit is perfect and no leakage flux exists. We can combine Eqs. 1.35 and 1.36 to yield

$$\frac{L_1}{L_2} = \frac{N_1^2}{N_2^2} = a^2 \tag{1.40}$$

where a is the ratio of turns of the first coil to the turns of the second coil. This relation will be used later in the study of transformers.

1.15 Current and voltage sources

The *voltage source* and the *current source* are means by which we supply energy to our circuits in electrical form; one-port circuit symbols are shown in Fig. 1.11. That at (a) in Fig. 1.11 applies to a steady source without

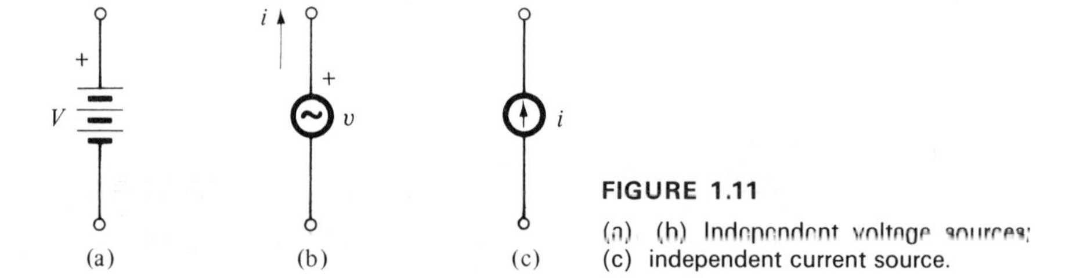

FIGURE 1.11
(a), (b) Independent voltage sources; (c) independent current source.

time variation, while those at (b) and (c) indicate sources with predetermined variations in time. In these sources there is no definable relation between port voltage and current, as has been written for the R, L, and C elements. An ideal voltage source, as in (a) or (b), will maintain the specified port voltage regardless of the external circuit which may be connected at the port, and the ideal current source of (c) will supply its designated current, again independent of the external circuit. These are *independent sources*, with current or voltage magnitudes subject to arbitrary specification.

Practical sources have limitations, and the infinite current of a voltage source under short circuit, or the infinite voltage of a current source with an open circuit at the port, is not possible in a practical sense. However, many of our sources do approximate the ideal sources over usual ranges of operation. Even when our sources are not quite ideal, as with a voltage source whose port voltage falls slightly with increasing current, we can often represent the actual source by an ideal source in conjunction with one or more passive circuit elements. This situation is discussed in Section 6.8.

The *dependent sources* of Fig. 1.12 are two-port sources, in which the current or voltage at the output port is controlled by and linearly dependent on a circuit variable, current or voltage, at the input port. The controlled variable has no effect on the controlling variable, and the source is unidirectional. In (a) in Fig. 1.12 the voltage source has an output μv_1, where

v_1 is the voltage at the input port and μ is a constant generator parameter.

Likewise in (d), the current source has an output $h_f i_1$, where i_1 is a current at the input port and h_f is the constant control parameter. The types shown in (b) and (c) may be derived from the other forms.

The polarities assigned to the dependent sources are dependent on the polarities of the primary controlling voltage or current. Dependent sources are of importance in the modeling of transistors, and the proportionality parameters h_f and g_m are derived from electronic application. The subject is discussed further in Section 6.11.

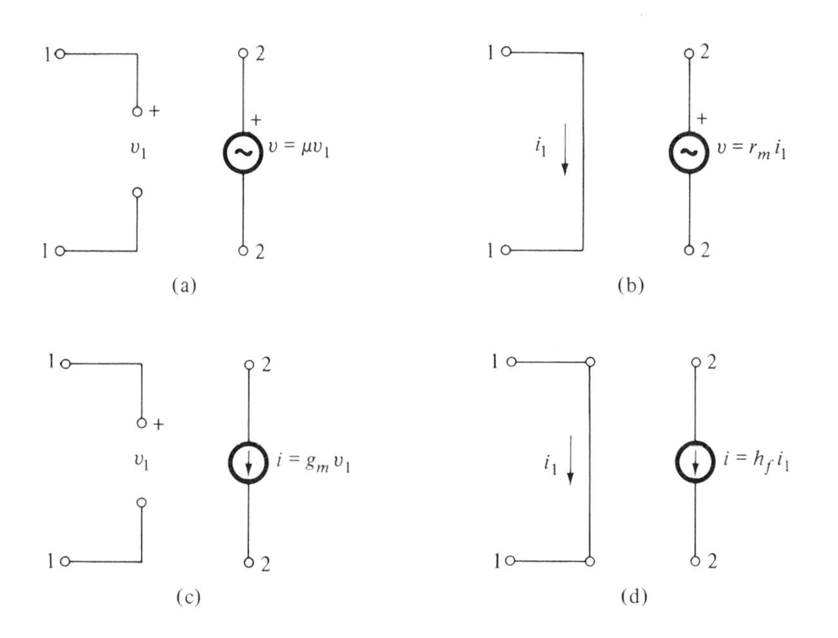

FIGURE 1.12
Dependent voltage and current sources.

1.16 Defined properties of circuits and circuit elements

Practically, the proportionality factors R, L, and C may not remain constant over great ranges of voltage or current, but as ideal elements they will be assumed to be independent of these variables. They will also be assumed to be time-invariant as well. For our circuit elements we have relations between current and voltage or between these variables and their derivatives which contain a constant coefficient:

$$v = Ri; \qquad v = L\frac{di}{dt}; \qquad i = C\frac{dv}{dt}$$

These are linear relations and describe *linear elements.*

A *linear network* is an interconnected set of linear circuit elements whose response to an input variable can be described by first degree integro-differential equations. The equations to be studied here will be those with constant coefficients, by our assumption of ideal elements. The assumption of linearity for the system implies validity for the principle of superposition. That is, if input and response are linearly related, the total response to several inputs acting together is equal to the sum of the responses of the individual inputs acting one at a time. In essence, a complex problem may be broken into a set of simpler and more easily analyzed problems, with the overall result coming from the sum of the solutions.

A complete dynamical system requires energy sources, energy reservoirs, and energy sinks or dissipators. *Active* circuit elements serve as energy sources, converting mechanical, thermal, or chemical energy to electrical form. *Passive* circuit elements either store energy for later return to the circuit or irreversibly dissipate energy. A circuit of passive elements is a *passive circuit*; the R, L, and C elements are passive. *Bilateral circuit elements* have conduction properties in one current direction equal to those in the reverse direction; our linear elements R, L, and C are bilateral.

A *short circuit* is an element of finite current and zero voltage; it is an element of zero resistance. If a circuit element is short-circuited, the voltage across the element is reduced to zero. Conversely, an *open circuit* is an element having finite voltage and zero current; it is created by breaking a connection and represents an infinite resistance.

Actual circuit elements may not be simple in form: A resistor may introduce some inductance and capacitance, an inductor will have some resistance in its windings and some capacitance between wires, and a capacitance may have some inductance in its connecting leads and some resistance in its dielectric. Figure 1.13 shows the ideal element and a more

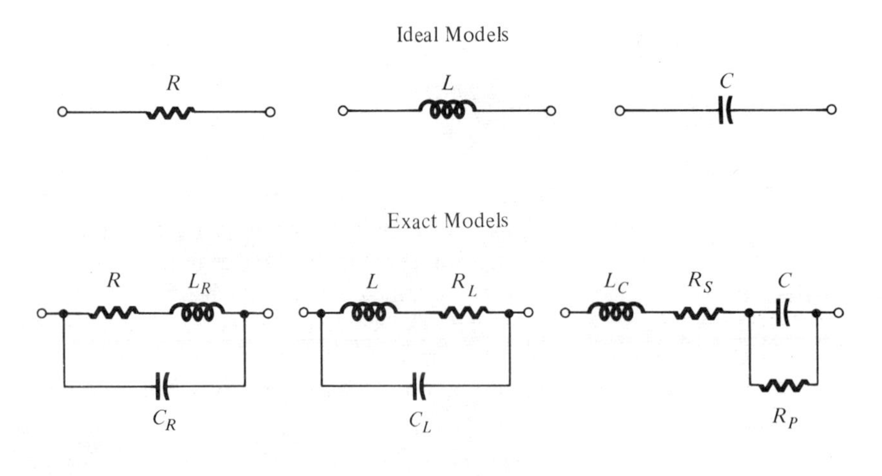

FIGURE 1.13

Models of the passive circuit elements.

exact form of representation. Progression from the ideal to the exact models involves increasing complexity in representation and in analysis. Since designs of the elements are such as to minimize the unwanted parameters, their effects are small at usual frequencies of operation and the ideal forms will be used here.

1.17 Summary

Charge, potential, current, and electrical power are the primary electrical variables. Resistance, capacitance, and inductance are the basic circuit parameters. Their voltage-current relations

$$v = Ri \qquad v = L\frac{di}{dt} \qquad v = \frac{1}{C}\int i\,dt + V_o$$

$$i = \frac{v}{R}; \qquad i = \frac{1}{L}\int v\,dt + I_o; \qquad i = C\frac{dv}{dt}$$

constitute the building blocks in analysis of circuit performance. To these we add current sources and voltage sources, independent and controlled, as converters of energy to electrical form.

Networks to be studied can largely be classed as one-port and two-port in form; n-port study is beyond the scope of this text.

PROBLEMS

1.1 A constant force of $3N$ is applied to a mass M of 20 kg, initially at rest. (a) At $t = 3$ s, determine the acceleration, velocity, and displacement of the mass. (b) What is the energy supplied to the mass at that time?

1.2 What is the weight of the electrons which flow through the tungsten filament of a lamp taking 0.6 A at 115 V during a 24-hour (h) period?

1.3 In a germanium semiconductor there is a positive charge density of 3.68×10^{23} particles/m^3, with the positive charges each equal to that of the electron in magnitude. There is a negative electron density of 1.57×10^{15}/m^3. The electrons move toward a positive terminal at a velocity of 5×10^{-4} m/s, and the positive charges move to the negative terminal at half that velocity. What is the total current per square meter?

1.4 One cubic centimeter of copper is drawn into a wire of 0.01-cm^2 cross-sectional area. If there is a current of 2 A in the wire, how long will it take an individual electron to travel the full length of the wire, not considering the effect of atomic collisions.

1.5 An inductor of 1.2 H is supplied with a current of 2.4 A. (a) What is the stored energy? (b) If the current is increasing at 0.25 A/ms, what is the magnitude of voltage across the inductor?

1.6 In Fig. 1.3(b) the battery supplies 12 V and is connected for 125 ms. The resistors total 15 Ω. (a) How much energy has been dissipated in R_1 and R_2? (b) Also state the energy in kilowatt-hours.

1.7 At $t = 0$ the current through a 4-mH inductor is 30 mA and is decreasing at a rate of 3 mA/μs. (a) What energy is stored in the inductor at $t = 0$? (b) What is the voltage across the inductor at $t = 0$? (c) What is the voltage at $t = 3$ μs, assuming the rate of change to be constant?

1.8 A steady potential of 40 V is applied to the terminals of a 10-H inductance with current $i_o = 0$ at $t = 0$. What will be the current at $t = 3$ s? Plot current against time to $t = 5$ s.

1.9 A current varying as $i = 0.30 \sin 377t$ passes through a 1-H inductance. (a) Find the inductor voltage at $t = 0.003$ s. (b) What will be the positive maximum voltage across the inductor, and at what time will it occur? (c) How is the time of maximum voltage related to the time of maximum current?

1.10 A toroidal coil employs a core steel having the magnetization properties of Fig. 1.9. The core cross section is 1.8 cm^2 and the mean diameter is 4 cm. With 1000 turns and 0.04 A, find (a) the magnetic flux density and the total flux; (b) the coil inductance, assuming that μ_r is constant in the region of the operating current; and (c) the same quantities if the steel is replaced with air.

1.11 The voltage applied across a 1.5-H inductance at $t = 0$ is 200 V; the initial current at $t = 0$ is zero. (a) Determine the current at $t = 2$ s. (b) Write an expression for the stored energy as a function of time.

1.12 A capacitor of 5×10^{-6} F passes a current due to a voltage which is changing at the rate of 100 V/s. What is the current?

1.13 At $t = 1$ s the voltage across a capacitor is varying at 10 V/s and the current is 0.0015 A. What is the value of the capacitance?

1.14 The capacitor of Problem 1.13 employs polyethylene as a dielectric. What would the current have been at $t = 1$ s if the dielectric had been air?

1.15 At $t = 0$ a current $i = 2 \sin 2 \times 10^4 t$ is applied to an uncharged capacitor of 1 μF. (a) Find the voltage across the capacitor at $t = 30$ μs. (b) What is the maximum voltage value, and at what time does it occur? (c) What is the time relation between the maximum voltage and the maximum current?

1.16 A copper bus bar is 12 m long, with a cross section 1.0 × 7 cm. (a) Find its resistance at 20°C. (b) What is the power loss when carrying a current of 3500 A?

1.17 An aluminum conductor 15 cm long is 0.25 × 0.25 cm at one end and varies linearly to 0.25 × 1.5 cm at the other end. Find the resistance of the bar. Hint: the resistance of the bar is not that due to its average cross-section.

1.18 A copper bus bar is 0.4×5.0 cm in cross section and very long. If it can dissipate heat at the rate of 0.01 W/cm^2 from its exterior surfaces without excessive temperature rise, how much current can it carry?

1.19 An electric heating element is to be wound of nichrome wire having a diameter of 1.02 mm. If it is to dissipate 600 W at 117 V, what length of wire is needed?

1.20 A copper motor coil is supplied at 105 V and takes a current of 5.5 A at 20°C. After operating for 2 h, the current is found to have dropped to 4.7 A. What is the apparent temperature throughout the coil?

1.21 A 100-Ω resistor carries a current of $i = 10t$ A. Find the energy supplied to the resistor from $t = 0$ to $t = 2.5$ s.

1.22 A 100-Ω resistor is connected to a source of $v = 20 \sin 377t$. (a) Find the current at $t = 0.003$ s. (b) Find the energy supplied between $t = 0$ and $t = 0.003$ s. (c) What is the maximum value of instantaneous power supplied to the resistor by the source?

1.23 Capacitors C_a and C_b are in series. Show that the equivalent single capacitor must have a value given by $C_aC_b/(C_a + C_b)$.

REFERENCES

1. "IEEE Recommended Practice for Units in Published Scientific and Technical Work." *IEEE Spectrum*, 3, 169 (1966).
2. *IEEE Proposed Standard Definitions of General Electrical and Electronic Terms.* IEEE, New York, 1966.
3. *American National Standard Graphic Symbols for Electrical and Electronics Diagrams*, No. 315, ANSI Y32.2. IEEE, New York, 1971.

circuit responses: natural and transient

two

For useful purposes the circuit elements R, L, and C are connected into networks and are driven by arbitrary voltage or current sources. The response or output of a network will be found dependent on the initially stored energies and on the driving sources applied.

2.1 Kirchhoff's laws

In this text a *branch* is a circuit element with two terminals to which connections can be made, a *node* is a point at which two or more branches join, and a *loop* is a closed path formed by connected branches.

From experiment, Kirchhoff (1824–1887) formulated two basic laws of the electric circuit. The first is

At any instant, the algebraic sum of the voltages around a closed loop in an electric circuit is zero.

Mathematically stated, we have

$$\sum v = 0 \tag{2.1}$$

In general terms, the across measurements around a circuit sum to zero.

In Fig. 2.1(a) we show the three ideal circuit parameters with their assumed currents and resultant voltages marked for reference. In (b) several of these elements and a battery, as a constant voltage source, are combined into a simple circuit. Assumed positive current directions are indicated by arrows, placed so that a current is designated through every element. In solution, a current found to be positive is in agreement with the assumed direction, and a negative result indicates an oppositely directed current.

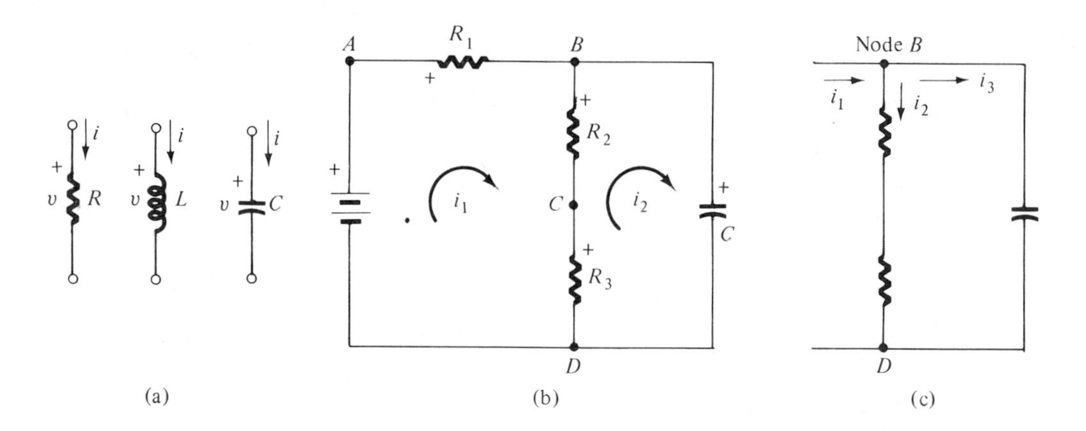

FIGURE 2.1
(a) Voltage-current conventions; (b) voltage summation; (c) current summation.

Kirchhoff's voltage law derives from the requirement for conservation of energy and the definition of voltage as energy per unit charge. Consider the source, a battery, for example, in which the conversion of chemical energy raises positive charges (current i_1) from a reference energy level D to a level of V_s at A. These charges suffer a loss of energy in passing through R_1. The energy per unit charge or the potential falls from $V_s - v_{AD}$ at A to a potential v_{BD} at B.

The currents through R_2 and R_3 result in a net downward current of $i_1 - i_2$. The energy of the charges or the potential falls from B due to energy loss in conduction through R_2 to a level of v_{CD}. The charges lose more energy in passing through R_3 and return to point D. This is the level of reference and they must have the same energy or potential with which they started. Therefore, in measuring the energy of the charges by measurement of the voltages, the voltage rises through the battery to V_s, written as $+V_s$, and the charge energy falls through resistors R_1, R_2, and R_3, with voltages written as $-v_{AB}$, $-v_{BC}$, $-v_{CD}$. For the complete loop,

$$V_s - v_{AB} - v_{BC} - v_{CD} = 0 \tag{2.2}$$

which satisfies *Kirchhoff's voltage law*, as well as the conservation of energy requirement.

The statement may be expanded by use of the voltage-current relations for the elements, giving

$$V_s - R_1 i_1 - (i_1 - i_2)R_2 - (i_1 - i_2)R_3 = 0 \tag{2.3}$$

as the voltage summation for the loop.

In the second loop we again use D as the reference level. The charges in the current $(i_1 - i_2)$ lose energy in moving downward through R_2 and R_3; the charge energy at B is above that at D. Charges in i_2 through the capacitor have higher energy at the plate connected to B than at the negative plate connected to D. These energy losses must be equal, and so we may write

$$(i_1 - i_2)R_2 + (i_1 - i_2)R_3 = \frac{1}{C}\int i_2\,dt + V_o$$

or

$$(i_1 - i_2)(R_2 + R_3) - \frac{1}{C}\int i_2\,dt - V_o = 0 \qquad (2.4)$$

This is the statement of Kirchhoff's voltage law around the second loop.

Instrument connections should correlate the above reasoning with actual measurements. If an ammeter reads upscale or positive, the current is moving from the plus meter terminal to the terminal marked minus; a voltmeter plus terminal and a positive or upscale reading indicates this in the circuit.

Kirchhoff's second circuit law is

At any instant, the algebraic sum of all currents entering (or leaving) a circuit junction is zero.

That is, at a node

$$\sum i = 0 \qquad (2.5)$$

We look at the node indicated at B of the circuit in Fig. 2.1(c). At this point three currents meet and the above law requires that

$$-i_1 + i_2 + i_3 = 0 \qquad (2.6)$$

The positive sign indicates a current from the node, and the negative sign implies a current to the node. Of course, either convention could be used. Proof of the law is furnished by the fact that charge cannot accumulate at B; the current must be continuous around the circuit.

The writing of Kirchhoff's laws assumes that each R, L, or C element is distinct and bounded. It is assumed that magnetic flux is present only in inductances, that all electric field effects appear only in capacitances, and that the connecting wires have zero resistance. We then say that the elements of the circuit are *lumped* into discrete R, L, or C elements.

2.2 Applications of Kirchhoff's laws; current division; potential division

Kirchhoff's equations lead to relationships which usually permit solution for the separate currents and voltages, as will be shown from Fig. 2.2(a). It is helpful if we always assume that the circuital currents are in the same direction; those shown here are clockwise. Small plus signs have been placed at the point of current entry into each element as a reminder of the higher potential there with the designated current; practice will later make this step unnecessary.

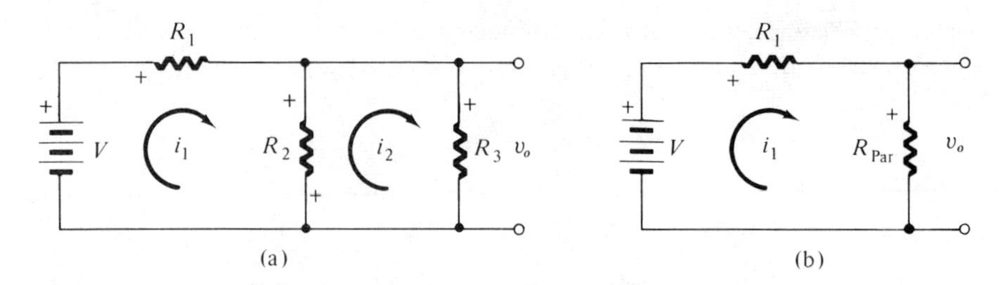

FIGURE 2.2
Examples for Kirchhoff law application.

Applying the voltage law by summing the potentials around the first loop, we obtain

$$V - R_1 i_1 - R_2(i_1 - i_2) = 0$$

Note that i_1 and i_2 have been assumed to be in opposite directions in R_2, the common element between loop 1 and loop 2, and that the current in R_2 is then $i_1 - i_2$, directed downward. Around the second current loop,

$$(i_1 - i_2)R_2 - i_2 R_3 = 0$$

These equations can be systematically assembled as

$$V = (R_1 + R_2)i_1 - R_2 i_2 \tag{2.7}$$

$$0 = -R_2 i_1 + (R_2 + R_3)i_2 \tag{2.8}$$

Written in matrix form, they are

$$\begin{bmatrix} V \\ 0 \end{bmatrix} = \begin{bmatrix} R_1 + R_2 & -R_2 \\ -R_2 & R_2 + R_3 \end{bmatrix} \begin{bmatrix} i_1 \\ i_2 \end{bmatrix}$$

We may solve Eq. 2.8 for i_2 as

$$i_2 = \frac{R_2}{R_2 + R_3} i_1 \tag{2.9}$$

Insertion of this result in Eq. 2.7 leads to

$$i_1 = \frac{R_2 + R_3}{R_1 R_2 + R_1 R_3 + R_2 R_3} V \tag{2.10}$$

Assuming V to be known, the current i_1 has been obtained, and current i_2 follows by use of Eq. 2.9.

Now rewrite Eq. 2.10 as

$$V = \left(R_1 + \frac{R_2R_3}{R_2 + R_3}\right)i_1 \tag{2.11}$$

This result indicates that the original resistive circuit can be represented by two resistors, R_1 and the resultant of R_2 and R_3 in parallel, as

$$R_{\text{par}} = \frac{R_2R_3}{R_2 + R_3} \tag{2.12}$$

Thus two resistors in parallel can be replaced by *a single resistor whose value is equal to the product of the two divided by the sum of the two resistances*; the result appears in Fig. 2.2(b).

A second useful result is apparent by rewriting Eq. 2.9 as

$$\frac{i_2}{i_1} = \frac{R_2}{R_2 + R_3} \tag{2.13}$$

This is the *current division factor*, which shows that when a current, such as i_1, divides into two paths, *the ratio of the desired current, i_2 in R_3, to the total current is given by the resistance of the other path divided by the sum of the two path resistances.*

From Fig. 2.2(b) we may write

$$V = (R_1 + R_{\text{par}})i_1$$

and also

$$v_o = R_{\text{par}}i_1$$

from which

$$\frac{v_o}{V} = \frac{R_{\text{par}}}{R_1 + R_{\text{par}}} \tag{2.14}$$

This is the *voltage division factor*, which states that in a series circuit, *the ratio of a desired voltage to the total applied voltage can be expressed as the ratio of the resistance across which the desired voltage appears to the total series resistance.*

As another example, consider the circuit of Fig. 2.3(a). The currents have been arbitrarily assigned in the positive direction, and Kirchhoff's current law may be applied at a node, giving the nodal equation

$$i - i_1 - i_2 - i_3 = 0$$

Each current can be written by Ohm's law, leading to

$$i = \frac{V}{R_1} + \frac{V}{R_2} + \frac{V}{R_3} = V\left(\frac{R_1R_3 + R_2R_3 + R_1R_2}{R_1R_2R_3}\right) \tag{2.15}$$

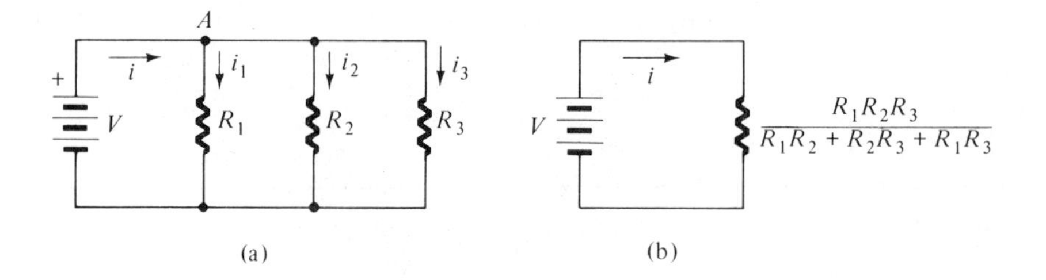

FIGURE 2.3

Parallel form of circuit solution.

In a parallel connection of elements, the current law frequently leads to a simpler solution than if the three loop-current equations had been written by use of the voltage law. This choice of method will be discussed further in Section 4.11, which covers network topology.

Equation 2.15 provides the form to be used if three resistors are to be combined in parallel. The result is given by the triple product, divided by the sum of the double products of the resistors in parallel.

2.3 Natural response; transient response; the steady state

Kirchhoff's laws apply on an instant-by-instant basis; they apply for voltage and current sources that are constant or that vary with time. For study of the response of circuits to variation of the energy sources, we have three classifications.

The *natural response* is the reaction of a circuit to its own internally stored energy; the external excitation is zero. The magnitude of the response will depend on the amount of internal energy, and the manner of variation in time of the response will depend on the circuit configuration.

Connection of an external driving source involves a switching operation, which will create a natural response due to transfer of internal energies to new locations and a second response due to the newly connected driving source. By reason of the principle of superposition these responses are additive; the total response is known as the *transient*.

The internal energies ultimately transfer to new and stable storage locations or are dissipated; the remaining response is that due to the driving source, and this is the *steady state*. The natural response and the steady state correspond to the solutions of the integrodifferential equations of the circuit known as the *complementary function* and the *particular integral*.

In the transition between states, it is necessary to know the circuit condition at $t = 0-$, if the circuit change occurs at $t = 0$, and just after the circuit change at $t = 0+$. The current will not change abruptly through an inductance due to the effect of the integral in Eq. 1.27. More fundamentally, the flux linkages tend to remain continuous in time. Similarly, Eq. 1.21

shows that the voltage across a capacitance is constant over small time increments; fundamentally it takes time to accumulate charge.

These two statements provide criteria for the development of appropriate boundary conditions for response determination.

2.4 Responses in *RC* circuits

We shall now discuss the natural and steady-state responses of several simple circuits. The mathematical relations by which we have modeled our circuit elements provide the starting points, and Kirchhoff's laws provide the method.

Consider the *RC* series circuit of Fig. 2.4. Switch *S* has been closed on point 1 for a long time so that the capacitance is charged to the potential of

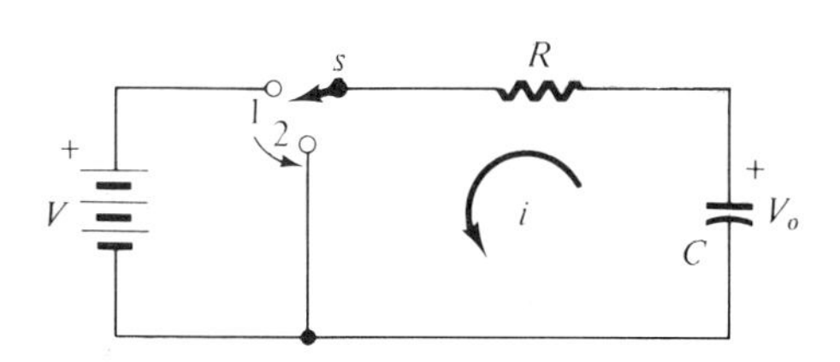

FIGURE 2.4

RC circuit with an initial voltage V_o on *C*.

the battery. The voltage law would give simply $V - V_o = 0$, and the capacitance current would be zero, since the rate of change of voltage is zero. This is an initial steady state. At $t = 0$ the switch is moved to position 2, reducing the external source to zero, and this is the condition for the natural response of the circuit. This will occur while the stored energy in *C* is being dissipated.

Assuming a positive current *i* as shown, the Kirchhoff voltage law and Eq. 1.21 give us

$$V_o - \frac{1}{C}\int_0^t i\,dt - Ri = 0 \tag{2.16}$$

To remove the integral, we differentiate Eq. 2.16 and have

$$\frac{di}{dt} + \frac{i}{RC} = 0 \tag{2.17}$$

Having no external sources, this equation leads to the natural response of the circuit.

Separation of variables gives

$$\frac{di}{i} = -\frac{dt}{RC} \tag{2.18}$$

and integration yields

$$\ln i = -\frac{1}{RC}t + \ln A \tag{2.19}$$

$$i = A\varepsilon^{-t/RC} \tag{2.20}$$

To satisfy Eq. 2.17 we see that the current variable and its derivative must have the same form as functions of time; the exponential at which we have arrived is the only function meeting that requirement.

At $t = 0+$, just after the switch is closed on 2, Eq. 2.16 gives

$$i(0+) = \frac{V_o}{R}$$

and since $i(0+) = A$ at $t = 0$ from Eq. 2.20, our natural response of the circuit is complete:

$$i = \frac{V_o}{R}\varepsilon^{-t/RC} \tag{2.21}$$

The initial steady state had $i = 0$ with $V = V_o$; after a long time, or $t = \infty$, it can be seen from Eq. 2.21 that $i = 0$ again, and this is the second steady state with switch S connected to 2. The variation of current with time is plotted in Fig. 2.5 in terms of time units of RC.

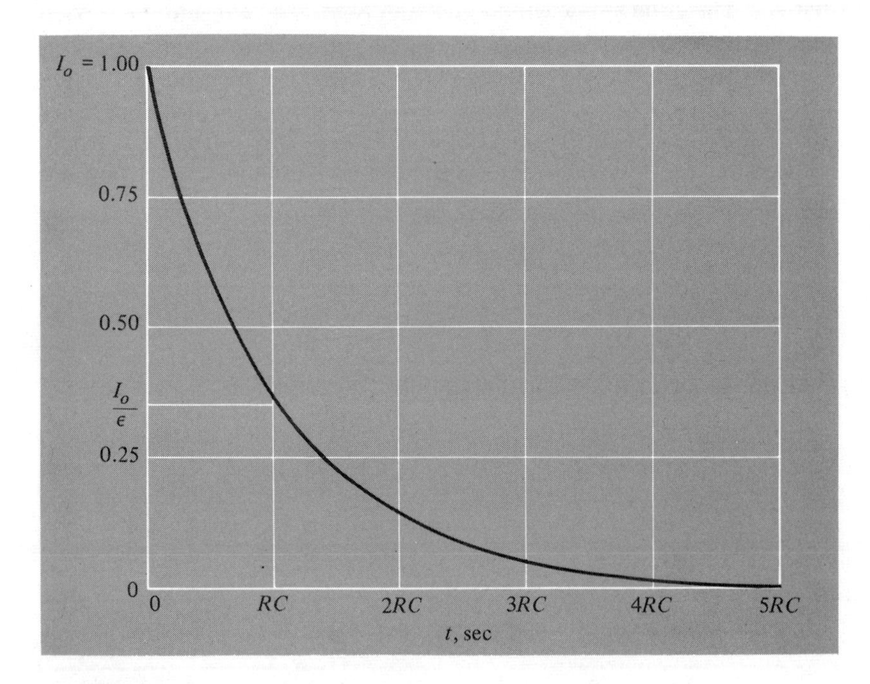

FIGURE 2.5

Natural response of the series RC circuit, $i = I_o\varepsilon^{-t/RC}$.

We have

$$v = \frac{1}{C}\int_0^t i\,dt$$

as our basic capacitance relation from Eq. 1.21. While discharging the energy is outward from C, and the current is negative. Using Eq. 2.21 for i, we can write

$$v = \frac{1}{C}\int_0^t \left(-\frac{V_o}{R}\right)\varepsilon^{-t/RC}\,dt$$

The result is

$$v = V_o\varepsilon^{-t/RC} \tag{2.22}$$

and so the form of Fig. 2.5 can be applied for capacitor voltage as well as current during discharge.

Having reached the second steady state, let us reverse the switch to position 1 at a new value of t, again called $t = 0$ for convenience, and determine how the potential across C reaches V_o again. With the circuit that of Fig. 2.6 and with $V_o = 0$ at $t = 0$, we arbitrarily choose a current summation at point A:

$$-i_1 + i_2 = 0$$

$$-\frac{V - v_{AB}}{R} + C\frac{dv_{AB}}{dt} = 0$$

$$\frac{dv_{AB}}{dt} + \frac{v_{AB}}{RC} = \frac{V}{RC} \tag{2.23}$$

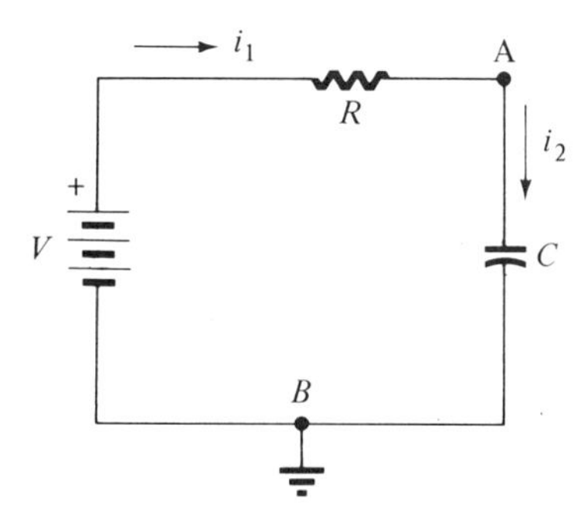

FIGURE 2.6

Charging circuit for C.

The solution to this nonhomogeneous equation will consist of two parts, the complementary function and the particular integral, or, in electrical terminology, the natural response and the steady-state response. That is,

$$v_{ss} + v_n = v$$

The natural response is due to the internal energies, and to obtain it, we set the external driving source, V, equal to zero in Eq. 2.23, yielding

$$\frac{dv_{AB}}{dt} + \frac{v_{AB}}{RC} = 0 \tag{2.24}$$

This is identical in form to Eq. 2.17 and will have an identical form of solution; that is,

$$v_C = v_{AB} = B\varepsilon^{-t/RC}$$

for the potential across the capacitor.

For the steady-state solution at $t = \infty$, it is apparent that the capacitance will be fully charged at a potential V. Writing the complete solution to Eq. 2.23, we have

$$v_C = v_{AB} = v_{ss} + v_n = V + B\varepsilon^{-t/RC}$$

At $t = 0$, $v_{AB} = 0$, and so $B = -V$. Substitution for B then yields

$$v_C = V(1 - \varepsilon^{-t/RC}) \tag{2.25}$$

and this result is plotted as a function of time in Fig. 2.7. The result represents the natural response, varying as $\varepsilon^{-t/RC}$ and disappearing, and the ultimate

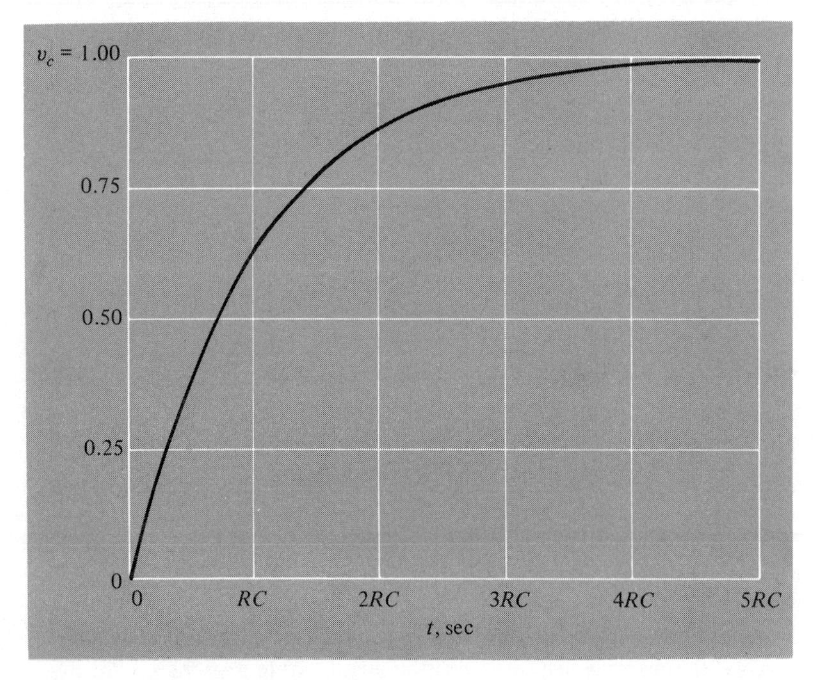

FIGURE 2.7

Charging of a capacitor in an RC circuit, $v = V_o(1 - \varepsilon^{-t/RC})$.

steady-state condition as voltage V across the capacitance, due to the forcing function V.

If the capacitor had a remaining charge and potential V_x at $t = 0$ instead of $V_o = 0$, the result would be modified by $B = V - V_x$, and

$$v_C = V - (V - V_x)\varepsilon^{-t/RC} \tag{2.26}$$

2.5 Response of the *RL* circuit

In the circuit of Fig. 2.8 the switch S has been in position 2 for a long time; all energy in the magnetic field of the inductor has been dissipated in R and the circuit is quiescent with $i = 0$. Switch S is then turned to 1, connecting the source V, and by Kirchhoff's law we can write

$$V - Ri - L\frac{di}{dt} = 0 \tag{2.27}$$

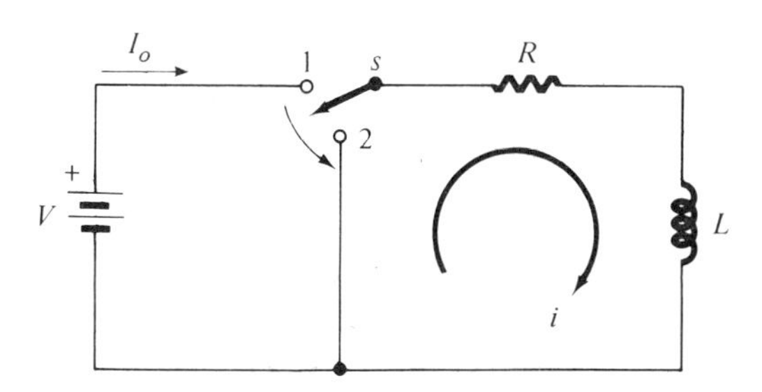

FIGURE 2.8
RL circuit.

The variables may be separated, yielding

$$\frac{di}{i - (V/R)} = -\frac{R}{L}dt$$

Integration gives

$$\ln\left(i - \frac{V}{R}\right) = -\frac{R}{L}t + \ln D$$

from which

$$i = D\varepsilon^{-(R/L)t} + \frac{V}{R} \tag{2.28}$$

Since $i = 0$ before the switch was connected to 1 and since the current in an inductance cannot change suddenly, the current i continues at 0 at $t = 0+$

just after the switch is changed to 1. Using $i = 0, t = 0$ as initial conditions in Eq. 2.28, we find

$$D = -\frac{V}{R}$$

and the solution for the current is

$$i = \frac{V}{R}(1 - \varepsilon^{-(R/L)t}) \tag{2.29}$$

This result has the form of the plot in Fig. 2.7, but with the ordinate changed to represent $i/I_o = i/(V/R)$, and with the abscissa ranged in units of L/R seconds.

After sufficient time the exponential term becomes small with respect to unity and the current reaches a constant limiting value $i = V/R$. Since di/dt is then zero, $L\,di/dt$ is also zero, and with zero voltage across the inductance it follows that the circuit current in this steady state must be $i = V/R$. That is, Eq. 2.29 states

$$i = i_{ss} + i_n$$

a form previously established for the RC circuit voltages.

As before, the natural response could have been obtained from Eq. 2.27 by setting $V = 0$, leading to

$$L\frac{di}{dt} + Ri = 0$$

Solution by the method of the previous section gives

$$i_n = D\varepsilon^{-(R/L)t}$$

Addition of this natural response term to the reasoned steady-state value of $i_{ss} = V/R$ gives us Eq. 2.28 by a second method.

We might then solve the inverse problem of current decay in an inductive circuit by returning switch S to 2, at a time $t = 0$. The circuit equation is

$$L\frac{di}{dt} + Ri = 0$$

with $i = V/R$ at $t = 0-$. It follows that $i = V/R$ at $t = 0+$ after S is connected to 2, because of the inductance L. Solving the above equation as before, and inserting the initial conditions, gives

$$i_n = \frac{V}{R}\varepsilon^{-(R/L)t}$$

Ultimately the energy in the magnetic field of L will have been dissipated in resistance R; the current will be zero and this is again the steady state with $V = 0$. The solution for decaying current in an RL circuit is

$$i = i_n + i_{ss} = i_n + 0 = \frac{V}{R}\varepsilon^{-(R/L)t} \tag{2.30}$$

Thus Fig. 2.5 will serve as a plot of i against time, with the ordinate as $i/I_o = i/(V/R)$, and the abscissa in units of L/R.

The results for RC and RL circuits may be generalized. Examination of the various expressions for voltage and current shows that the transient period is always exponential and occurs between certain initial and final circuit states. In an RL circuit these conditions are determined by the fact that current does not instantaneously change in an inductance, that the rate of rise of current is limited by L, and that the final current value is limited by the circuit resistance. In an RC circuit the voltage does not instantaneously change across a capacitance, the initial value of current is limited by the resistance, and the final value of current must be zero. If we let $g(t)$ be a response which starts at $g(0)$ and ends at $g(\infty)$, there will be an exponential transition between these limits. A general expression can then be written from

$$g(t) = g(\infty) - [g(\infty) - g(0)]\varepsilon^{-t/\tau} \tag{2.31}$$

where τ is the time constant, defined in the next section.

With the conditions set up in Section 2.4 we have $g(\infty) = 0$ and $g(0) = V_o/R$, $\tau = RC$, and we can write Eq. 2.21 directly. Equation 2.22 also follows from this general relation with $g(\infty) = 0$, $g(0) = V_o$.

2.6 The time constant

Figures 2.5 and 2.7 illustrate an exponential form of variation with time which is typical of the response of systems governed by first-order differential equations. The *time constant* τ of a circuit with such a response is the time at which the response has completed $(1 - 1/\varepsilon) = 1 - 0.368 = 0.632$ of its total change, as in Fig. 2.9. It is the value of time t at which the exponent of ε equals -1. For the resistance-capacitance circuit the time constant is

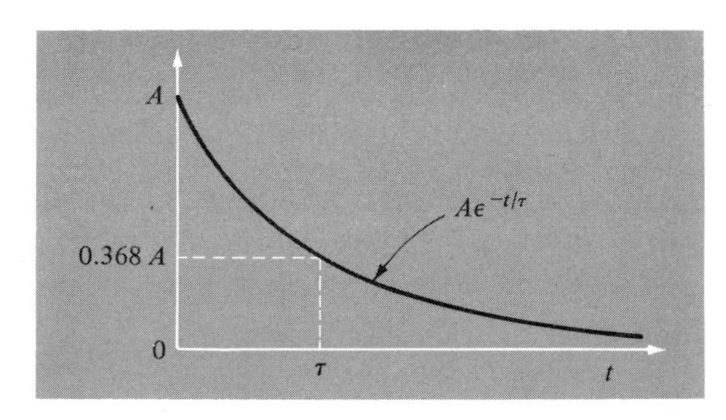

FIGURE 2.9

Time-constant relationships.

expressed by $RC = \tau$; for the resistance-inductance circuit the time constant is $L/R = \tau$. The units of RC or L/R are seconds.

The time constant provides a simple measure of the relative time scale for the natural response of the circuit. While the actual response approaches the limit asymptotically, the transient may be considered to have been completed after $t = 5\tau$ s; 99 percent of the change will then have taken place.

2.7

Second-order system response

A somewhat different situation arises when we have a network with two energy-storing elements, such as L and C in Fig. 2.10(a). A practical problem results if we suddenly apply a constant voltage V_o to the network

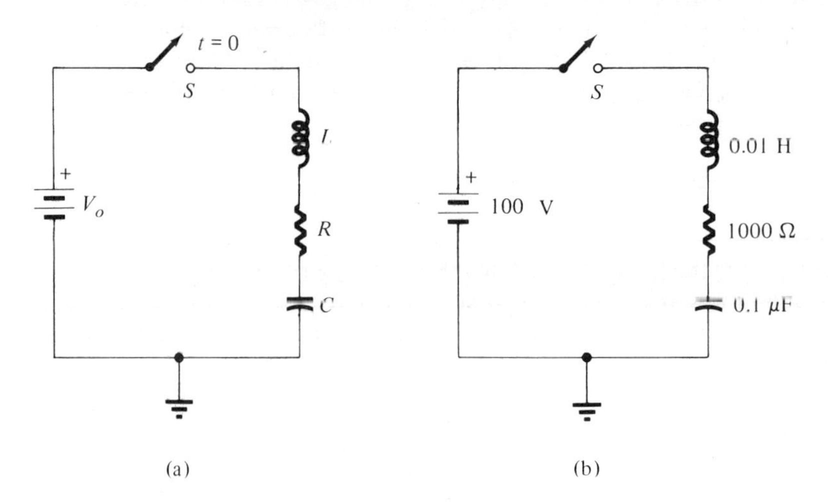

FIGURE 2.10
(a) *RLC* series circuit; (b) example.

at time $t = 0$ by closure of switch S. At that time we assume C to be uncharged. We can sum the voltages around the circuit, due to the assumed current i, as

$$L\frac{di}{dt} + Ri + \frac{1}{C}\int_0^t i\,dt = V_o \tag{2.32}$$

The equation can be simplified in mathematical form if we differentiate with respect to time, giving

$$L\frac{d^2i}{dt^2} + R\frac{di}{dt} + \frac{i}{C} = 0 \tag{2.33}$$

The highest-order derivative is the second, and the circuit represents a *second-order system*.

Again, we expect the complete solution to consist of a natural response, due to rearrangement of the internal stored energies, and a steady-state response, due to the source V_o. That is,

$$i = i_n + i_{ss}$$

We can reason that in ultimate time sufficient charge will have been driven around the circuit to raise the potential of C to V_o (positive terminal uppermost). Under that condition the current i will drop to zero.

Equation 2.33, being equated to zero, is already in appropriate form for study of the natural response. Again, a solution in exponential form is called for, because i and its first and second derivatives with respect to time must have similar time variations. We may assume that

$$i = M\varepsilon^{st} \tag{2.34}$$

where M and s are, respectively, the *amplitude* and the *complex frequency*. Using this current in Eq. 2.33, we can write

$$\left(s^2L + sR + \frac{1}{C}\right)i = 0$$

Since i is obviously not zero,

$$s^2L + sR + \frac{1}{C} = 0 \tag{2.35}$$

This is the *characteristic equation*.

Solution of this quadratic for s yields

$$s_1, s_2 = -\frac{R}{2L} \pm \sqrt{\frac{R^2}{4L^2} - \frac{1}{LC}} \tag{2.36}$$

and the assumption of Eq. 2.34 is shown to be incomplete, since there are two values of s which satisfy Eq. 2.33. The complete current expression may then be written as the sum of two components,

$$i = M_1\varepsilon^{s_1t} + M_2\varepsilon^{s_2t} \tag{2.37}$$

where

$$s_1 = -\frac{R}{2L} + \sqrt{\frac{R^2}{4L^2} - \frac{1}{LC}} = -\sigma + \sqrt{\omega^2} \tag{2.38}$$

$$s_2 = -\frac{R}{2L} - \sqrt{\frac{R^2}{4L^2} - \frac{1}{LC}} = -\sigma - \sqrt{\omega^2} \tag{2.39}$$

The two coefficients M_1 and M_2 are arbitrary and required by the second-order differential equation; they are to be found by application of known circuit conditions.

In Eqs. 2.38 and 2.39 the terms

$$\sigma = \frac{R}{2L}$$

$$\omega^2 = \frac{R^2}{4L^2} - \frac{1}{LC} \qquad (2.40)$$

are employed. Interesting possibilities appear, since it is possible for ω^2 to be positive, zero, or negative. The three cases are

$$\text{I.} \quad \omega^2 > 0; \qquad \frac{R^2}{4L^2} > \frac{1}{LC}$$

and s_1 and s_2 are negative real and unequal.

$$\text{II.} \quad \omega^2 = 0; \qquad \frac{R^2}{4L^2} = \frac{1}{LC}$$

with $s_1 = s_2$ and negative real.

$$\text{III.} \quad \omega^2 < 0; \qquad \frac{R^2}{4L^2} < \frac{1}{LC}$$

and s_1 and s_2 are complex conjugate roots.

2.8 Solutions for the natural response

Case I. With the resistance sufficiently large that $R^2/(4L^2) > 1/(LC)$, the circuit is *over-damped*; roots s_1 and s_2 are real and negative and the solution of Eq. 2.37 represents the sum of two decaying exponential functions. The constants M_1 and M_2 can be evaluated if the initial conditions are known.

The current in L was zero at $t = 0-$, and with the switch closing at $t = 0$, we know that

$$i]_{t=0+} = 0 \qquad (2.41)$$

because of the circuit inductance. With zero current there will be zero voltage across R and zero voltage across C, since the integral of $i(0+)$ is also zero. Consequently, at $t = 0+$ the circuit voltages reduce to

$$L\frac{di}{dt} = V_o$$

or

$$\left.\frac{di}{dt}\right]_{t=0+} = \frac{V_o}{L} \tag{2.42}$$

Using the condition of Eq. 2.41 in Eq. 2.37 we can write, for $t = 0+$,

$$M_1 + M_2 = 0$$

Taking the derivative of Eq. 2.37 and introducing the condition of Eq. 2.42, we have, at $t = 0+$,

$$\left.\frac{di}{dt}\right]_{t=0+} = s_1 M_1 + s_2 M_2 = \frac{V_o}{L}$$

Simultaneous solution leads to

$$M_1 = -M_2 \tag{2.43}$$

$$M_1 = \frac{V_o}{L}\left(\frac{1}{s_1 - s_2}\right) \tag{2.44}$$

and the arbitrary coefficients are determined in terms of circuit parameters.

EXAMPLE. In Fig. 2.10(b) $V_o = 100$ V, $C = 0.1\ \mu$F, $L = 10$ mH, and $R = 1000\ \Omega$. We seek the current after the switch closes at $t = 0$.

We must first determine that the circuit is actually overdamped, or that $R^2/(4L^2) > 1/(LC)$. That is,

$$\frac{1}{4}\left(\frac{10^3}{0.01}\right)^2 > \frac{1}{0.01 \times 10^{-7}}$$

$$2.5 \times 10^9 > 10^9$$

and the requirement is satisfied. The values of s_1 and s_2 can be found as

$$s_1 = -\frac{R}{2L} + \sqrt{\frac{R^2}{4L^2} - \frac{1}{LC}}$$

$$= -\frac{1000}{0.02} + \sqrt{\frac{10^6}{4 \times 10^{-4}} - \frac{1}{0.01 \times 10^{-7}}}$$

$$= -50{,}000 + 38{,}700 = -11{,}300$$

$$s_2 = -50{,}000 - 38{,}700 = -88{,}700$$

We may determine M_1 and M_2 from Eqs. 2.43 and 2.44:

$$M_1 = \frac{V_o}{L}\left(\frac{1}{s_1 - s_2}\right) = \frac{100}{0.01}\,\frac{1}{[-11.3 - (-88.7)]10^3}$$

$$= \frac{10}{77.4} = 0.129$$

$$M_2 = -M_1 = -0.129$$

The natural response of the circuit to the suddenly applied voltage V_o is then

$$i = 0.129\varepsilon^{-1.13\times10^4 t} - 0.129\varepsilon^{-8.87\times10^4 t}$$

The above equation represents the addition of a positive exponential current of a relatively large time constant, $\tau_1 = 10^{-4}/1.13$, and a negative exponential current of a small time constant, $\tau_2 = 10^{-4}/8.87$. These currents are plotted as a function of time in Fig. 2.11. The

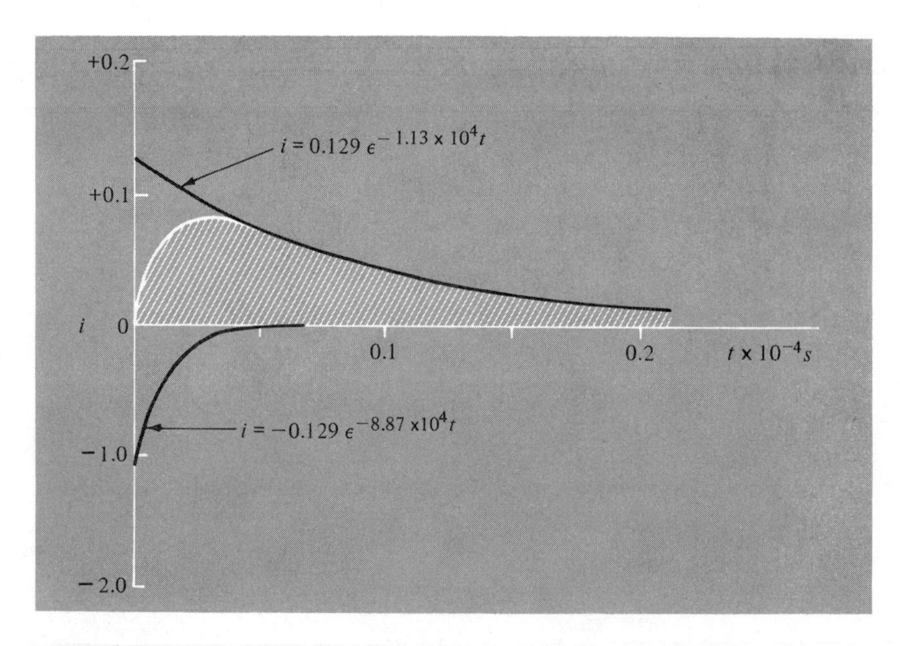

FIGURE 2.11

Case I, solution for the circuit of Fig. 2.10(b).

resultant sum is cross-hatched and shows a current whose integral would be the charge moved to the capacitor by voltage V_o. After C is charged to V_o, the current remains at zero in the steady state.

Case II. The problem with $s_1 = s_2$ is of mathematical interest, but in a physical sense an *exact* equality of s_1 and s_2 is not to be expected. The current can be obtained by the usual methods* for equal roots and is

$$i = M_1\varepsilon^{st} + M_2 t\varepsilon^{st} \tag{2.45}$$

and M_1 and M_2 can be evaluated from knowledge of the initial circuit conditions.

This is the condition of *critical damping* separating Case I from Case III. Since the radical in s must go to zero, we have

$$\frac{R^2}{4L^2} = \frac{1}{LC}$$

which is $\alpha^2 = \omega^2$, and the value of R at critical damping will be

$$R_c = 2\sqrt{\frac{L}{C}} \tag{2.46}$$

If R is greater than R_c, the circuit is overdamped (case I), and if R is less than R_c, the circuit is underdamped (case III).

Case III. With $R^2/(4L^2) < 1/(LC)$ the radicand of Eq. 2.36 is a negative quantity. The roots s_1 and s_2 become complex conjugates because

$$s_1, s_2 = -\frac{R}{2L} \pm \sqrt{-1}\sqrt{\frac{1}{LC} - \frac{R^2}{4L^2}}$$

$$= -\frac{R}{2L} \pm j\sqrt{\frac{1}{LC} - \frac{R^2}{4L^2}} \tag{2.47}$$

The symbol $j = \sqrt{-1}$, since i (often used by mathematicians) symbolizes current in electrical practice. The above can be written

$$s_1, s_2 = -\sigma \pm j\omega$$

and the solution for the underdamped natural response is

$$i = M_1\varepsilon^{(-\sigma + j\omega)t} + M_2\varepsilon^{(-\sigma - j\omega)t}$$

$$= \varepsilon^{-\sigma t}(M_1\varepsilon^{j\omega t} + M_2\varepsilon^{-j\omega t}) \tag{2.48}$$

As before, the steady-state response is zero.

*See H. B. Phillips, *Differential Equations*. John Wiley & Sons, Inc., New York, 1951.

To provide a better understanding of the situation, we employ Euler's equation,*

$$\varepsilon^{\pm j\omega t} = \cos \omega t \pm j \sin \omega t$$

and write Eq. 2.48 as

$$i = \varepsilon^{-\sigma t}[(M_1 + M_2)\cos \omega t + j(M_1 - M_2)\sin \omega t] \tag{2.49}$$

Using Eq. 2.44 and the values for s_1 and s_2, we have

$$M_1 = \frac{V_o}{L}\left[\frac{1}{-\sigma + j\omega - (-\sigma - j\omega)}\right] = \frac{V_o}{j2\omega L} \tag{2.50}$$

From Eq. 2.43, $M_1 + M_2 = 0$, and the cosine term of Eq. 2.49 disappears. The coefficient for the sine term is

$$M_1 - M_2 = 2M_1 = \frac{V}{j\omega L} \tag{2.51}$$

The final solution for the current is

$$i = \frac{V_o}{\omega L}\varepsilon^{-\sigma t} \sin \omega t \tag{2.52}$$

This equation represents a sinusoid of amplitude $V_o/\omega L$ decaying with time due to the exponential term; the result is plotted in Fig. 2.12(c).

*The student may readily prove Euler's equation by use of the series expansion for the cosine and sine as

$$\cos \omega t = 1 - \frac{(\omega t)^2}{2!} + \frac{(\omega t)^4}{4!} - \frac{(\omega t)^6}{6!} + \cdots$$

$$j \sin \omega t = j\omega t - \frac{j(\omega t)^3}{3!} + \frac{j(\omega t)^5}{5!} - \frac{j(\omega t)^7}{7!} + \cdots$$

and then

$$\cos \omega t + j \sin \omega t = 1 + j\omega t - \frac{(\omega t)^2}{2!} - \frac{j(\omega t)^3}{3!} + \frac{(\omega t)^4}{4} + \frac{j(\omega t)^5}{5!} - \cdots$$

$$= 1 + j\omega t + \frac{(j\omega t)^2}{2!} + \frac{(j\omega t)^3}{3!} + \frac{(j\omega t)^4}{4!} + \frac{(j\omega t)^5}{5!} + \cdots$$

This is just the series expansion for the exponential with $j\omega t$ as the argument, and so

$$\varepsilon^{j\omega t} = \cos \omega t + j \sin \omega t$$

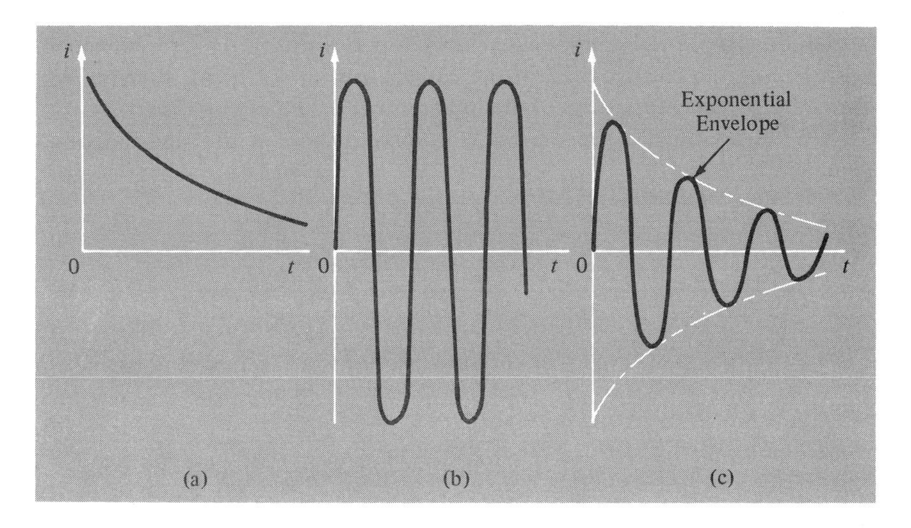

FIGURE 2.12

(a) Exponential term; (b) sinusoidal term; (c) total response.

2.9

Natural frequencies; the root locus

The natural response of the series RLC circuit has been obtained as the sum of two exponential functions of time. With real coefficients in the differential equation, these exponents must be real or must occur in complex conjugate pairs. Depending on the magnitude relation between $R^2/4L^2 = \sigma^2$ and $1/LC = \omega_0^2$, the exponents are real or complex. In the overdamped case, with negative real exponents, the response is the sum of two exponential quantities decaying with time. At the critical value $\sigma = \omega_0$, the response changes; in the underdamped case the exponents become complex conjugates and the response is oscillatory, but with an exponential rate of decay.

The variable ω can be recognized as an angular velocity, called the *angular frequency* in electrical usage. For the underdamped case we have the *damped angular frequency*, written from Eq. 2.47 as

$$\omega_d = \sqrt{\frac{1}{LC} - \frac{R^2}{4L^2}} = \sqrt{\frac{1}{LC}}\sqrt{1 - \frac{R^2C}{4L}} \quad \text{r/s} \tag{2.53}$$

As the resistance of the circuit is reduced, this frequency increases and at $R = 0$ reaches a limit called the *undamped natural frequency* of the circuit:

$$\omega_0 = \sqrt{\frac{1}{LC}} \tag{2.54}$$

In this limiting case the circuit is lossless. The sinusoidal oscillations start as a result of initial energy present in L or C and continue indefinitely at full amplitude. The initial energy is exchanged periodically: At zero current it is present in the electric field of the capacitance and at peak current it is present in the magnetic field of the inductance.

With T equal to the time of a complete cycle of the sine function, we have

$$\omega T = 2\pi$$

$$T = \frac{2\pi}{\omega}$$

Frequency is defined as the number of oscillations per second, and so

$$f = \frac{1}{T} = \frac{\omega}{2\pi} \qquad \text{Hz (Hertz)} \tag{2.55}$$

is stated in cycles per second or *hertz*. We then have

$$\omega = 2\pi f \qquad \text{r/s} \tag{2.56}$$

as the definition of *angular frequency*.

The several forms of the exponential response are dependent on the values of the roots of the characteristic equation:

$$s_1, s_2 = -\sigma \pm j\omega \tag{2.57}$$

With the identification of ω as an angular frequency, to be dimensionally consistent we have chosen to call s a *complex frequency*. The characteristic equation of a circuit is written in terms of this variable.

Our concept of complex frequency may be broadened by considering the complex plane plots of Fig. 2.13. The abscissa of the plane is the axis of

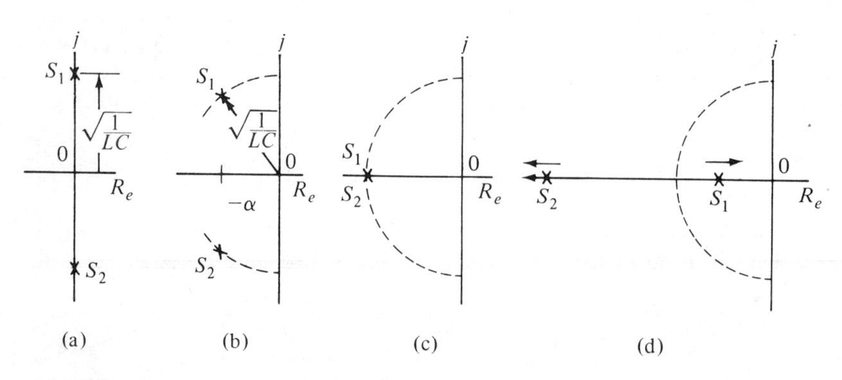

FIGURE 2.13

(a) $R = 0$, oscillatory response; (b) $R^2/4L^2 < 1/LC$, response underdamped; (c) $R^2/4L^2 = 1/LC$, critically damped; (d) $R^2/4L^2 > 1/LC$, response overdamped.

real numbers, and the ordinate is the axis of j quantities. The values for s_1 and s_2 were

$$s_1, s_2 = -\frac{R}{2L} \pm \sqrt{\frac{R^2}{4L^2} - \frac{1}{LC}}$$

and with $R = 0$, these become

$$s_1, s_2 = \pm j\sqrt{\frac{1}{LC}} \tag{2.58}$$

which appear as conjugate points on the j axis, equidistant from the origin, in Fig. 2.13(a). There is no real component, there is no resistance and no damping, and the response is a continuing sinusoidal oscillation at frequency ω_0.

Now let R increase, but with $R^2/4L^2 < 1/LC$, so that the complex frequencies are

$$s_1, s_2 = -\frac{R}{2L} \pm j\sqrt{\frac{1}{LC} - \frac{R^2}{4L^2}} \tag{2.59}$$

This expression represents an x, y plot in which

$$x = -\sigma \qquad \text{and} \qquad y^2 = \omega_0^2 - \sigma^2$$

Substituting, we have

$$x^2 + y^2 = \omega_0^2$$

and as R varies, the locus of s appears as a circle with its center at the origin and a radius $= \omega_0 = \sqrt{1/LC}$. The locus of the roots passes through the root locations on the j axis in (a) in Fig. 2.13, and as R increases, the complex frequencies move around the dashed circle, as in (b). The response is a sinusoid, with damping increasing as the real part of the complex frequency increases.

At critical damping the complex frequencies become equal and real as in (c) in Fig. 2.13; oscillation just ceases. With further increase in R the roots are negative real with s_1 moving toward the origin and s_2 moving toward $-\infty$. This is the overdamped condition of (d) with the complex frequencies s_1 and s_2 as negative reals.

The units of ω are radians per second, but a radian is a dimensionless angle so that ω has dimensions of reciprocal time. The same must be required of α, since αt and ωt are the necessarily dimensionless exponents of ε. Therefore, s must have dimensions of $1/t$, and our usage of the name *complex frequency* is supported. In effect, the function ε^{st} combines the constant sinusoid, the increasing or decreasing sinusoid, and the real exponential time function within one expression. The general utility of s in circuit analysis will appear in the next chapter.

For circuits with one energy-storing parameter we developed a first-order differential equation with a single complex frequency s having its j component equal to zero. With two energy-storing elements we found the circuit response described by a second-order differential equation with two complex frequencies s_1 and s_2. In writing equations for more complicated circuits we expect each L or C to contribute a derivative; when simplified, an equation for one of the variables will be of nth order, where n is the sum of the inductors and capacitors present. If we assume ε^{st} for each natural response, we shall obtain a polynomial of degree n as the characteristic equation, and there will be n roots or n complex natural frequencies for the network. Similar results would have been obtained if we had started with a parallel RLC circuit instead of a series form.

The real parts of the complex frequencies have fallen in the negative or left half of the complex plane. Since a negative exponential has been the result of a decaying energy supply, a positive exponential resulting from a root with a positive real part must imply that energy is increasing with time. The response then increases indefinitely, and nature will not be able to supply the infinite energies called for. Therefore a positive real root creates instability and ultimate disaster.

2.10 Summary

Kirchhoff's laws tell us that we can develop circuit relations between v and i by writing $\sum v = 0$ around a circuit loop or $\sum i = 0$ at a circuit node. The use of these methods with RC, RL, and RLC circuits led us to the natural response of the circuit, due to exchange and/or dissipation of the stored circuit energies; the form of the natural response with time is dependent on the circuit configuration.

The driving source also introduces its own form of response, and this continues in the steady state after decay of the natural response.

In second-order systems we find a natural response due to a decaying exponential as a result of overdamping in the circuit or an oscillatory response form as a result of underdamping of the circuit. Separation of these responses occurs at critical damping for which the circuit resistance has the value $R_c = 2\sqrt{L/C}$.

With exponential forcing functions we introduce the complex frequency s. Using the complex frequency, we predict the actions of the circuit as we vary R by use of the locus followed by the s roots of the circuit equation. Other parameters could have been varied and other loci obtained.

PROBLEMS

2.1 Write two loop and three node equations for the circuit of (a) in Fig. 2.14.

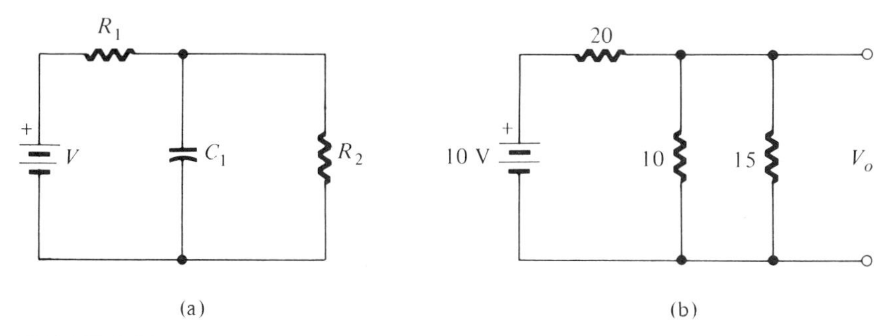

FIGURE 2.14

2.2 Find V_o for the circuit of (b) in Fig. 2.14. Also find the current in the 15-Ω resistor.

2.3 Write all possible loop and node equations for the circuit of (a) in Fig. 2.15.

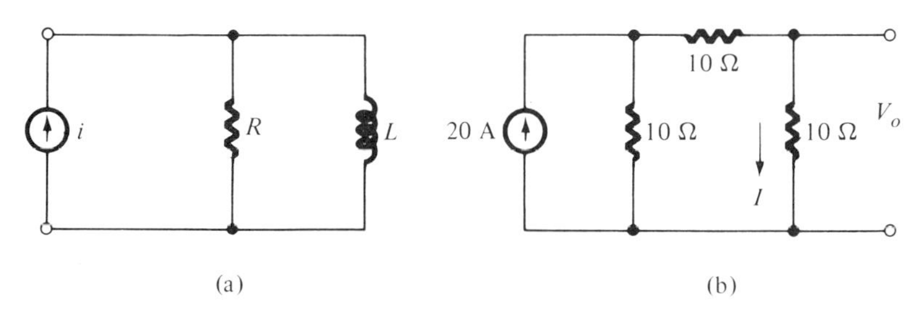

FIGURE 2.15

2.4 Find the current I and the voltage V_o of (b) in Fig. 2.15.

2.5 Find i_1, i_2, and V_o of (a) in Fig. 2.16.

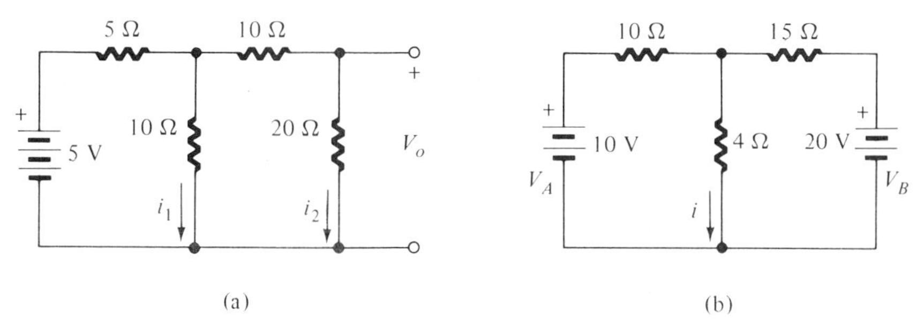

FIGURE 2.16

2.6 Using the circuit of Fig. 2.16(b), find the current in the 4-Ω resistor. Also find the current supplied by each source.

2.7 Find the value of R needed to make $I = 1$ A in Fig. 2.17(a); then determine the value of V_o with that R inserted.

2.8 Find the value of i of (b) in Fig. 2.17 if V_1 is 50 V.

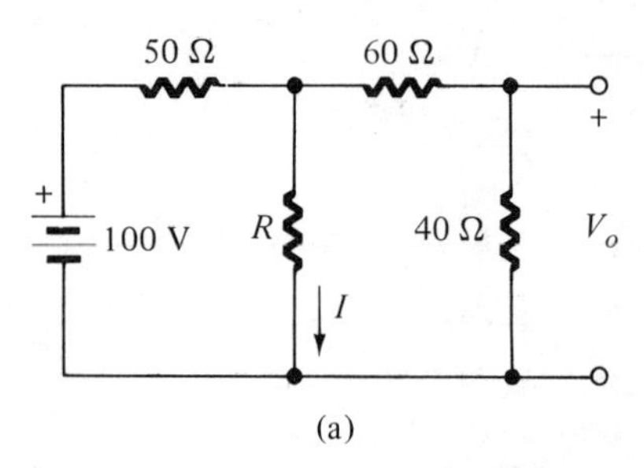

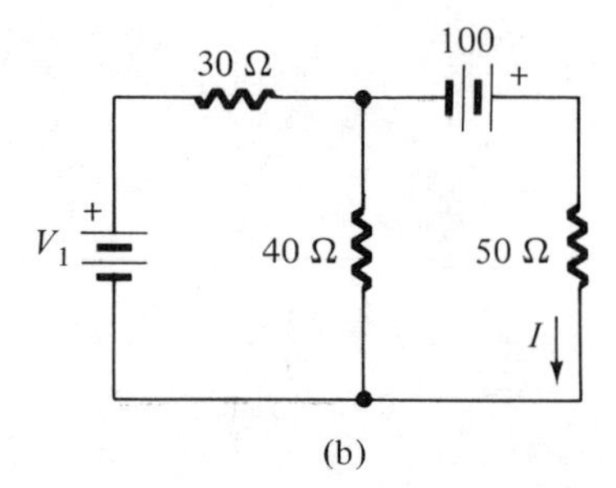

FIGURE 2.17

2.9 In (a) in Fig. 2.18, switch S has been at position 1 for a long time. (a) Write the initial conditions which describe the situation when S is moved to 2 at $t = 0$. (b) Determine the equation which describes the current after S is moved to 2.

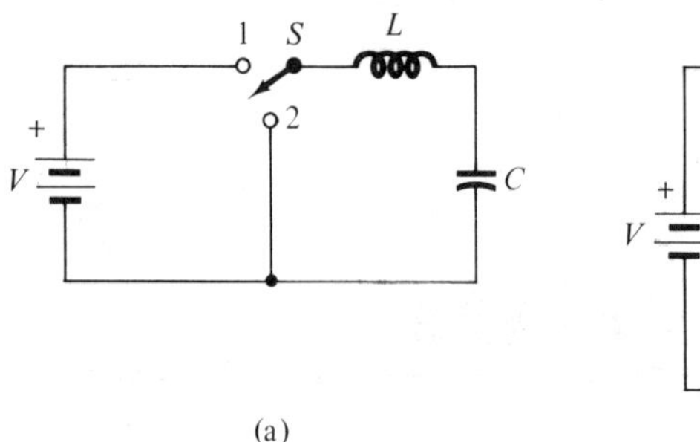

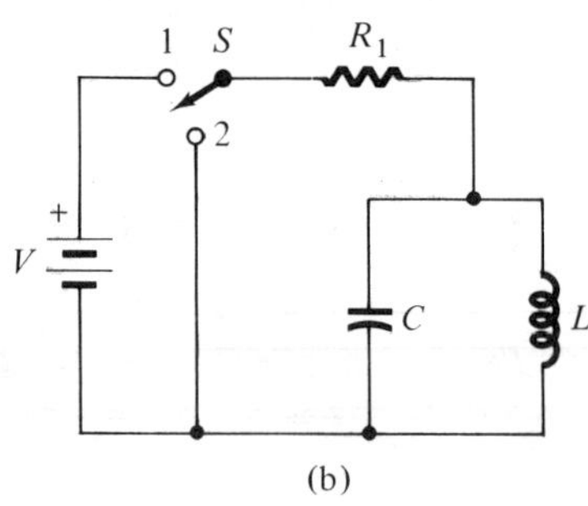

FIGURE 2.18

2.10 Repeat Problem 2.9 for the circuit of (b) in Fig. 2.18.

2.11 The switch S is closed at $t = 0$. Plot v_1 and v_2 to the end of the transient period for (a) in Fig. 2.19, with the capacitor initially uncharged.

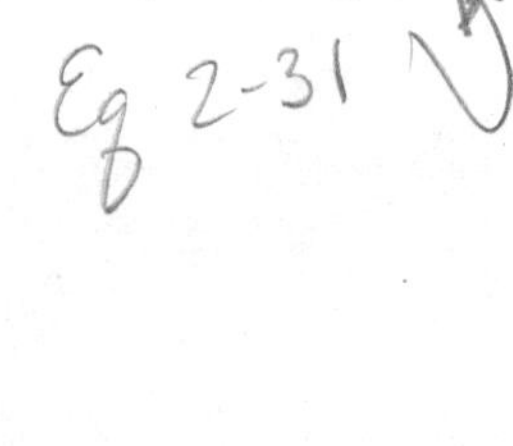

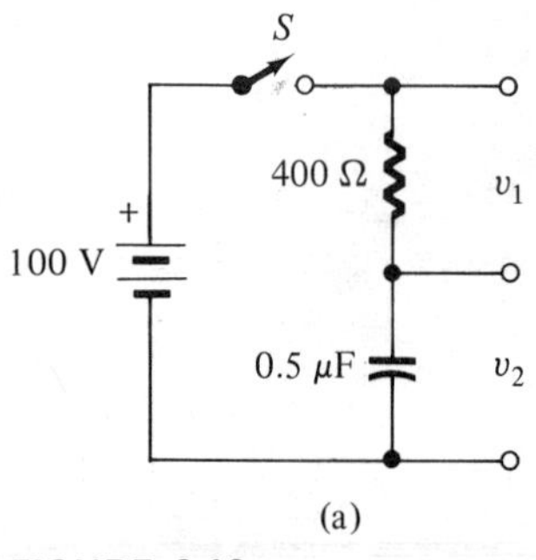

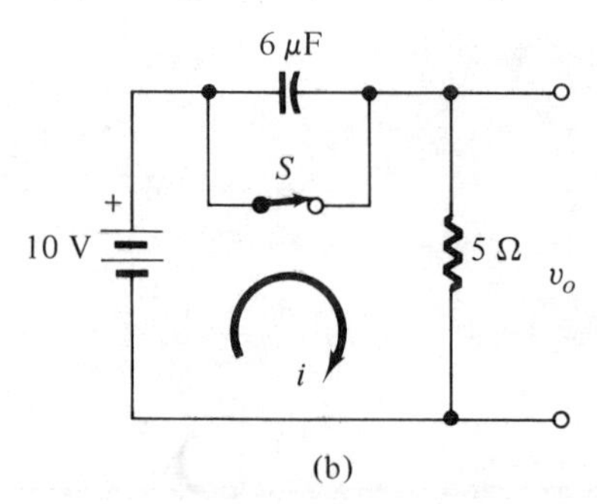

FIGURE 2.19

2.12 In Fig. 2.19(b), the switch S has been closed and is opened at $t = 0$. Find the voltage v_O at $t = 50\ \mu s$.

2.13 An electric motor has a rotary moment of inertia $J = 10$ kg-m^2 and a friction damping constant $B = 2$ N-m/r/s. It is running at 1030 r/min but is then disconnected from the supply line and coasts to a stop.

(a) Write the differential equation during the coasting period and solve for ω. (b) Find the time to an approximate stop (five time constants).

2.14 Use Euler's equation to verify that

$$\sin(\alpha + \beta) = \sin \alpha \cos \beta + \cos \alpha \sin \beta$$
$$\sin \alpha \cos \beta = \tfrac{1}{2} \sin(\alpha + \beta) + \tfrac{1}{2} \sin(\alpha - \beta)$$

2.15 Show that

$$\cos \omega t = \frac{\varepsilon^{j\omega t} + \varepsilon^{-j\omega t}}{2}; \qquad \sin \omega t = \frac{\varepsilon^{j\omega t} - \varepsilon^{j\omega t}}{2j}$$

2.16 The decay of a radioactive element is exponential in time, and the half-life is the time required for the process to decay to one-half amplitude. Express the half-life in terms of the time constant of the process.

2.17 In Fig. 2.20(a), the switch S has been connected to 1 for a long time. At $t = 0$ it is switched to 2, connecting the uncharged capacitor C. Determine i as a function of time for the overdamped case.

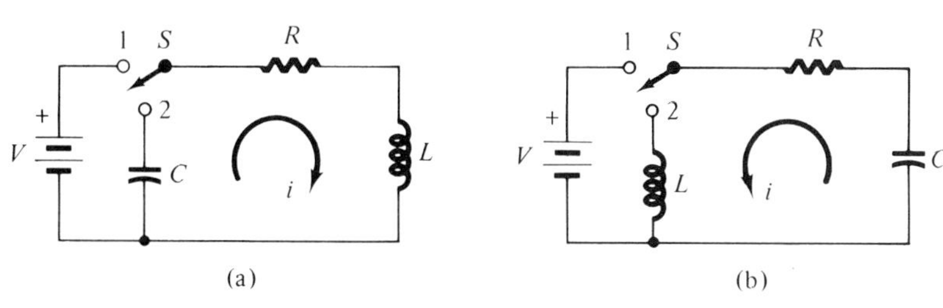

FIGURE 2.20

2.18 In Fig. 2.20(b), the switch has been connected to 1 for a long time. At $t = 0$ it is turned to 2, connecting an inductor L. (a) Determine i as a time function if $R = 1000\,\Omega$, $C = 0.1\,\mu$F, and $L = 1$ H. Find (b) the natural frequency of response, (c) the maximum value of current i for $V = 50$ V, and (d) the maximum voltage across L.

2.19 A constant-current source of 8 A is in parallel with an inductor of 5 H, having a series resistance of 50 Ω. At $t = 0$ a switch is closed, short-circuiting the source. (a) Determine the inductor current at $t = 0.08$ s. (b) What is the maximum voltage across the inductor?

2.20 An inductance L with a series resistance R has 100 V suddenly applied at $t = 0$. At $t = 40$ ms the current is 60 mA, and after a long time the current reaches 120 mA. What are the resistance and inductance values present?

2.21 Solve

$$\frac{d^2y}{dt^2} + 4\frac{dy}{dt} + 3y = 0$$

if $y(0) = 0$ and $dy/dt(0) = 2$.

circuit solution by the Laplace transform

three

Transient phenomena occur during the transition of an electrical circuit from one stable or steady state in time to a second steady state. The previous chapter undertook the analysis of circuit performance during the transient period by classical solution of the circuit differential equations.

The Laplace transform method provides a means of unifying and systematizing these solutions. Essentially, we change the original differential equations in time by writing them in terms of a second variable s, the complex frequency already introduced. The equations involving s are algebraic and can be given algebraic solutions. By transformation of the solution back to the original variable, we obtain the solution to the differential equations with which we started.

In the use of the Laplace transform we find that the initial conditions are included, the work of solution is algebraic, the steady-state and transient components are simultaneously obtained, and the labor is much reduced through reference to a table of transforms.

3.1 The Laplace transformation

We form a Laplace transform of a time function, $f(t)$, by multiplying $f(t)$ by ε^{-st} and integrating the product from zero to infinity. Therein s is again the complex frequency given as $s = \sigma + j\omega$. The Laplace transform is given by

$$\mathscr{L}[f(t)] = F(s) = \int_0^\infty f(t)\varepsilon^{-st}\,dt \tag{3.1}$$

with the script letter $\mathscr{L}$ indicating "the Laplace transform of."

The integral in Eq. 3.1 must converge; this implies that $f(t)$ is piecewise continuous over every finite interval $0 \leqq t_1 \leqq t \leqq t_2$. Other properties which are required for a function to be transformable can be found listed in suitable mathematics textbooks. Fortunately the time functions encountered in electric circuit theory are transformable; functions such as $f(t) = \varepsilon^{t^2}$ are not encountered in our work.

The use of ε^{-st} and integration over all positive time transforms a differential equation into an algebraic function of the complex frequency s. The original equation exists in the *time domain*, and the Laplace transformation changes the equation into one in the *s domain*. Since the limits of integration are from zero to infinity on time, the value of $f(t)$ in negative time is zero.

The integration involved in Eq. 3.1 is usually not difficult, but we shall learn to write the s-domain circuit equations directly, avoiding the differential equation for most circuit variables. Manipulation by algebraic means can lead to solution for the desired circuit variable, as a function of s. We then complete the solution by taking the *inverse transform*,

$$\mathscr{L}^{-1}[F(s)] = f(t) \tag{3.2}$$

and implying that

$$\mathscr{L}^{-1}\{\mathscr{L}[f(t)]\} = f(t) \tag{3.3}$$

Usually this step can be carried out by reference to a table of transform pairs, where values of $F(s)$ and the related time functions are tabulated. A small table of transforms appears as Table 3.1. We shall discuss the development of the pairs for such a table, as well as the mathematical processes by which the transform of a complicated function may be altered to a form which is recognizable in terms of the standard transforms of the tables.

An example will illustrate the transformation process.

EXAMPLE. The RL series circuit with an exponential driving source V^{-at} applied at $t = 0$ has the circuit equation

$$L\frac{di}{dt} + Ri = V^{-at} \tag{3.4}$$

Writing the Laplace transform for the above,

$$L\int_0^\infty \frac{di}{dt}\varepsilon^{-st}\,dt + R\int_0^\infty i\varepsilon^{-st}\,dt = V\int_0^\infty \varepsilon^{-(a+s)t}\,dt \tag{3.5}$$

Solution of the first integral by parts gives

$$Li\varepsilon^{-st}\Big]_0^\infty + sL\int_0^\infty i\varepsilon^{-st}\,dt$$

At $t = \infty$ the first term is zero, and at $t = 0+$ it becomes $-Li(0+)$, where $i(0+)$ means the current in the inductance just after the driving function is applied at $t = 0$. If $i = 0$ before the source was connected, then $i(0+) = 0$ just after the connection. The initial value of the current in the inductance is introduced directly into the solution instead of being added at the end, as in the classic method of solution. Then the complete form of Eq. 3.5 is

$$sL \int_0^\infty i\varepsilon^{-st}\,dt - Li(0+) + R\int_0^\infty i\varepsilon^{-st}\,dt = \frac{V}{s+a} \tag{3.6}$$

after applying the limits to the term on the right in Eq. 3.5. We can write the current as

$$I(s) = \int_0^\infty i\varepsilon^{-st}\,dt \tag{3.7}$$

and after rearrangement we have

$$I(s) = \frac{1}{R + sL}\left[\frac{V}{s+a} + Li(0+)\right] \tag{3.8}$$

The use of the Laplace transform has changed the differential equation in time, Eq. 3.4, into an algebraic equation in s, Eq. 3.8. Inversion to a function in time could follow by reference to a table of transform pairs. This will be demonstrated later.

3.2 Theorems

A few theorems are needed in the manipulation of the Laplace transform.

LINEARITY

If a is independent of s and t, then

$$\mathscr{L}[af(t)] = aF(s) \tag{3.9}$$

SUPERPOSITION

The principle is illustrated by

$$\mathscr{L}[f_1(t) \pm f_2(t)] = \mathscr{L}[f_1(t)] \pm \mathscr{L}[f_2(t)] = F_1(s) \pm F_2(s) \tag{3.10}$$

TRANSLATION IN TIME

If $f(t)$ has a transform $F(s)$ and $f(t) = 0$ when $t < 0$, then

$$\mathscr{L}[f(t-b)] = \varepsilon^{-bs}F(s) \tag{3.11}$$

provided that b is positive and real. Translation in time b is equivalent to multiplication of the transform by ε^{-bs}.

TRANSLATION IN THE s DOMAIN

If $f(t)$ has a transform $F(s)$ and a is a real or complex number, then

$$\mathscr{L}[\varepsilon^{at}f(t)] = F(s-a) \tag{3.12}$$

which states that multiplication by ε^{at} in the t domain is equivalent to translation by $-a$ in the s domain.

REAL DIFFERENTIATION

The transform of the derivative of $f(t)$ as $d[f(t)]/dt$ can be shown by integration of Eq. 3.1 by parts, with $u = \varepsilon^{-st}$ and $dv = f(t)\,dt$, giving

$$\int_0^\infty f(t)\varepsilon^{-st}\,dt = \frac{1}{s}f(0+) + \frac{1}{s}\int_0^\infty \frac{d[f(t)]}{dt}\varepsilon^{-st}\,dt$$

By Eq. 3.1 the left side is $F(s)$, so that the result of the theorem is

$$\mathscr{L}\left[\frac{df(t)}{dt}\right] = \int_0^\infty \frac{d[f(t)]}{dt}\varepsilon^{-st}\,dt = sF(s) - f(0+) \tag{3.13}$$

In the above, $f(0+)$ has the value of $f(t)$ as t approaches zero from the positive side. Therefore the result includes the initial condition.

REAL INTEGRATION

The transform of the integral of a transformable function $f(t)$ is possible, giving

$$\mathscr{L}\left[\int f(t)\,dt\right] = \frac{1}{s}\int f(t)_{(t=0+)}\,dt + \frac{F(s)}{s} \tag{3.14}$$

where the first term on the right is to be integrated and evaluated at $t = 0+$, the time immediately after switching action occurs. We start with Eq. 3.1 and integrate by parts, using

$$u = \int f(t)\,dt, \qquad dv = \varepsilon^{-st}\,dt$$

$$du = f(t)\,dt, \qquad v = -\frac{1}{s}\varepsilon^{-st}$$

so that

$$\mathscr{L}\left[\int f(t)\,dt\right] = \int_0^\infty f(t)\,\varepsilon^{-st}\,dt$$

$$= -\frac{1}{s}\varepsilon^{-st}\int f(t)\,dt\Big]_0^\infty + \frac{1}{s}\int_0^\infty \varepsilon^{-st}f(t)\,dt \tag{3.15}$$

The first term on the right is zero at the upper limit and at the lower limit has the value of the integral evaluated at the lower limit, $t = 0+$. The second term on the right can be recognized as $1/s$ times $F(s)$, and the theorem is proved; that is,

$$\mathscr{L}\left[\int f(t)\,dt\right] = \frac{1}{s}\int f(t)_{(t=0+)}\,dt + \frac{F(s)}{s} \tag{3.16}$$

3.3 Development of a table of transforms

The development of the transforms of a few frequently occurring functions of t will illustrate the manner in which a table of transform pairs is prepared.

UNIT STEP FUNCTION Definition of the transform of the unit step function, $U(t)$ of Fig. 3.1(a), requires care in handling the discontinuity due to the switching action at the origin in time.

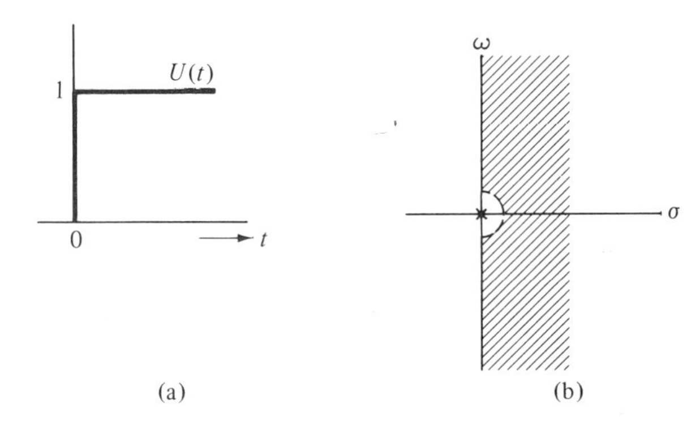

FIGURE 3.1

Unit step function; (b) region of convergence, shaded.

We can avoid this by choosing a very small time δ at which the function has the value 1 and writing

$$U(s) = \mathscr{L}[U(t)] = \int_0^\infty U(t)\varepsilon^{-st}\,dt = \lim_{\delta\to 0}\left[\frac{\varepsilon^{-st}}{-s}\right]_\delta^\infty \tag{3.17}$$

The discontinuity can be approached as closely as we please by letting δ become very small. The upper limit can be considered by writing

$$\varepsilon^{-st}\Big]^\infty = \varepsilon^{-\sigma t}\varepsilon^{-j\omega t}\Big]^\infty$$

and this quantity is zero at the limit if the real part of s is positive; i.e., $\sigma > 0$. Then from the lower limit with $\delta \to 0$,

$$U(s) = \mathscr{L}[U(t)] = \frac{1}{s} \tag{3.18}$$

and we have the transform for the unit step function in time. This function is that which arises when a steady unit voltage or a steady unit current is suddenly switched onto a circuit.

EXPONENTIAL FUNCTION $\varepsilon^{\pm at}$

The transform of this function can be written as

$$\mathscr{L}[\varepsilon^{\pm at}] = \int_0^\infty \varepsilon^{-(s \mp a)t}\, dt = \frac{1}{s \mp a} \tag{3.19}$$

for which $\sigma > -a$, to make the real part of the exponent negative. The result was previously derived in a more specialized form in Eq. 3.6.

COSINE AND SINE OF ωt

We require that ω be positive. Since

$$\cos \omega t = \frac{\varepsilon^{j\omega t} + \varepsilon^{-j\omega t}}{2} \tag{3.20}$$

Then by superposition

$$\mathscr{L}[\cos \omega t] = \mathscr{L}\left[\frac{\varepsilon^{j\omega t} + \varepsilon^{-j\omega t}}{2}\right] = \mathscr{L}\left[\frac{\varepsilon^{j\omega t}}{2}\right] + \mathscr{L}\left[\frac{\varepsilon^{-j\omega t}}{2}\right] \tag{3.21}$$

and by use of Eq. 3.19 we have

$$\mathscr{L}[\cos \omega t] = \frac{1}{2}\left(\frac{1}{s - j\omega} + \frac{1}{s + j\omega}\right) = \frac{s}{s^2 + \omega^2} \tag{3.22}$$

since $j(-j) = \sqrt{-1} \times -\sqrt{-1} = 1$. In a similar manner we can obtain

$$\mathscr{L}[\sin \omega t] = \frac{\omega}{s^2 + \omega^2} \tag{3.23}$$

The above requires that $\sigma > 0$.

By similar use of the exponential definitions of the sine and cosine, we can derive the transforms

$$\mathscr{L}[\varepsilon^{-bt} \sin \omega t] = \frac{\omega}{(s + b)^2 + \omega^2} \tag{3.24}$$

$$\mathscr{L}[\varepsilon^{-bt} \cos \omega t] = \frac{s + b}{(s + b)^2 + \omega^2} \tag{3.25}$$

The hyperbolic function transforms also follow by use of Eq. 3.19.

RAMP FUNCTION $f(t) = At$

This function is zero at $t = 0$ and has a constant slope A for $t > 0$, as in Fig. 3.2. From Eq. 3.1,

$$\mathscr{L}[At] = A \int_0^\infty t\varepsilon^{-st}\, dt \tag{3.26}$$

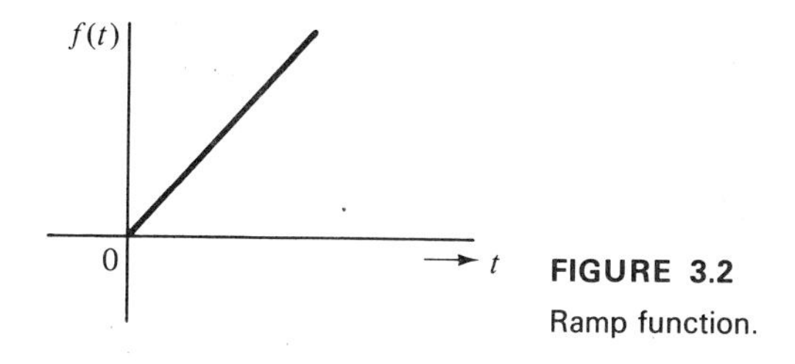

FIGURE 3.2
Ramp function.

Integration by parts with $u = t$ and $dv = \varepsilon^{-st}\,dt$ leads to

$$A\int_0^{\infty} t\varepsilon^{-st}\,dt = A\left[\varepsilon^{-st}\left(-\frac{t}{s} - \frac{1}{s^2}\right)\right]_0^{\infty}$$

and so

$$\mathscr{L}[At] = \frac{A}{s^2} \tag{3.27}$$

for σ positive.

Carrying out such extensive integrations for every problem would be tedious. However, in each case we have an $f(t)$ and its corresponding transform $F(s)$ as a transform pair. Recognizing this one-to-one correspondence, we prepare tables of transform pairs, of which Table 3.1 is a brief collection of commonly used forms. By reduction of the transform function in the s domain to a standard algebraic form, its related time function can usually be found in a table, and a solution in the time domain can be written directly.

EXAMPLE. Suppose that

$$I(s) = \frac{K_1}{s+2} + \frac{K_2}{s+3}$$

This is pair 3 in Table 3.1, and the function in the time domain is

$$i(t) = K_1\varepsilon^{-2t} + K_2\varepsilon^{-3t}$$

EXAMPLE. Let

$$F(s) = \frac{3}{(s+2)^2 + 9}$$

From pair 13 of Table 3.1, the time-domain function can be written directly as

$$f(t) = \varepsilon^{-2t}\sin 3t$$

TABLE 3.1 LAPLACE TRANSFORM PAIRS

	$F(s)$	$f(t)$
	OPERATIONS	
	$\int_0^\infty f(t)\varepsilon^{-st}\,dt$	$f(t)$
	$KF(s)$	$Kf(t)$
	$F_1(s) + F_2(s)$	$f_1(t) + f_2(t)$
	$sF(s) - f(0+)$	$\frac{d}{dt}f(t)$
	$s^2F(s) - sf(0+) - \frac{d}{dt}f(0+)$	$\frac{d^2}{dt^2}f(t)$
	$\frac{1}{s}\int_0^\infty f(t)_{(t=0+)}\,dt + \frac{F(s)}{s}$	$\int_0 f(t)\,dt$
	$\varepsilon^{-as}F(s)$	$f(t-a)$
	$\frac{F(s/a)}{a}$	$f(at)$
	$F(s-a)$	$\varepsilon^{at}f(t)$
	$\lim_{s\to\infty} sF(s)$	$\lim_{t\to 0+} f(t)$
	$\lim_{s\to 0} sF(s)$	$\lim_{t\to\infty} f(t)$
	TRANSFORMS	
1.	$\frac{1}{s}$	$U(t)$
2.	$\frac{1}{s^n}$	$\frac{t^{(n-1)}}{(n-1)!}$ (integer n)
3.	$\frac{1}{s \pm a}$	$\varepsilon^{\mp at}$
4.	$\frac{1}{(s \pm a)^n}$	$\frac{t^{(n-1)}\varepsilon^{\mp at}}{(n-1)!}$
5.	$\frac{1}{s(s+a)}$	$\frac{1}{a}(1 - \varepsilon^{-at})$
6.	$\frac{1}{s(s+a)^2}$	$\frac{1}{a^2}[1 - (1 + at)\varepsilon^{-at}]$
7.	$\frac{1}{s^2(s+a)}$	$\frac{1}{a^2}(at - 1 + \varepsilon^{-at})$
8.	$\frac{1}{(s+a)(s+b)}$	$\frac{1}{(b-a)}(\varepsilon^{-at} - \varepsilon^{-bt})$
9.	$\frac{s}{(s+a)(s+b)}$	$\frac{1}{(b-a)}(b\varepsilon^{-bt} - a\varepsilon^{-at})$
10.	$\frac{s}{s^2 + \omega^2}$	$\cos \omega t$

Table 3.1 continued

	$F(s)$	$f(t)$
11.	$\dfrac{\omega}{s^2+\omega^2}$	$\sin \omega t$
12.	$\dfrac{s+a}{(s+a)^2+\omega^2}$	$\varepsilon^{-at} \cos \omega t$
13.	$\dfrac{\omega}{(s+a)^2+\omega^2}$	$\varepsilon^{-at} \sin \omega t$
14.	$\dfrac{s^2-\omega^2}{(s^2+\omega^2)^2}$	$t \cos \omega t$
15.	$\dfrac{s}{(s^2+\omega^2)^2}$	$\dfrac{t}{2\omega} \sin \omega t$
16.	$\dfrac{s}{s^2-\omega^2}$	$\cosh \omega t$
17.	$\dfrac{\omega}{s^2-\omega^2}$	$\sinh \omega t$
18.	$\dfrac{1}{(s+b)(s^2+\omega^2)}$	$\dfrac{\varepsilon^{-bt}}{b^2+\omega^2} + \dfrac{1}{\omega\sqrt{\omega^2+b^2}} \sin(\omega t - \Phi)$ $\Phi = \tan^{-1}\dfrac{\omega}{b}$
19.	$\dfrac{s}{(s+b)(s^2+\omega^2)}$	$\dfrac{-b}{\omega^2+b^2}\varepsilon^{-bt} + \dfrac{1}{\sqrt{\omega^2+b^2}} \cos(\omega t - \Phi)$ $\Phi = \tan^{-1}\dfrac{\omega}{b}$

3.4 Application to circuit equations

The simple RC series circuit of Fig. 3.3(a) can be solved for the current using a transform from the table. Consider voltage V applied to the circuit by switch S at $t = 0$. The Kirchhoff voltage law may be used, giving the differential equation

$$Ri + \frac{1}{C}\int i\,dt = VU(t) \tag{3.28}$$

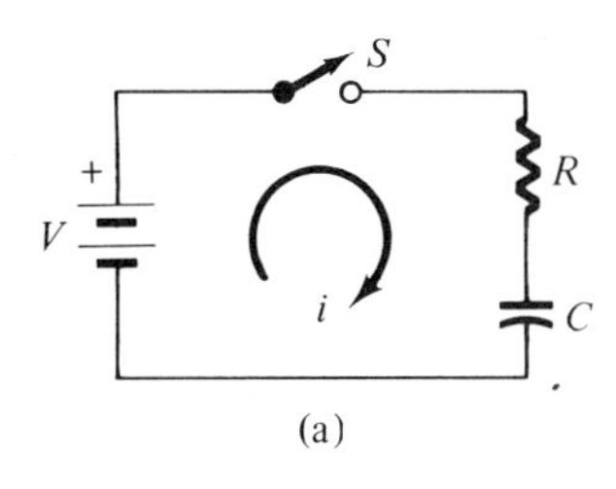

(a)

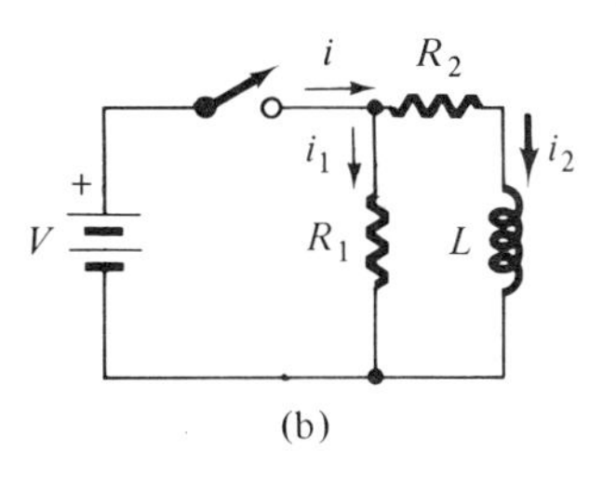

(b)

FIGURE 3.3
Examples.

noting that the closure of the switch applies V as a step function in time, $U(t)$.

Taking the transform of the first term,

$$\mathscr{L}[Ri] = R\int_0^{\infty} i\varepsilon^{-st}\,dt \tag{3.29}$$

Calling the current integral $I(s)$, as before,

$$I(s) = \int_0^{\infty} i\varepsilon^{-st}\,dt$$

we have

$$\mathscr{L}[Ri] = RI(s) \tag{3.30}$$

Using the integration theorem of Eq. 3.14 for the second term of the differential equation, we can write

$$\begin{aligned}\mathscr{L}\left[\frac{1}{C}\int i\,dt\right] &= \frac{1}{sC}q_C(0+) + \frac{1}{sC}\int_0^{\infty} i\varepsilon^{-st}\,dt \\ &= \frac{V_C(0+)}{s} + \frac{I(s)}{sC}\end{aligned} \tag{3.31}$$

where $V_C(0+) = q_C(0+)/C$ is the voltage across the capacitance just after the switch closes. For finite voltages this is equal to the voltage across C before the switch closes, at $t = 0-$.

Using the transform of $U(t) = 1/s$, we can write the transformed equation with Eqs. 3.30 and 3.31 as

$$RI(s) + \frac{V_C(0+)}{s} + \frac{1}{sC}I(s) = \frac{V}{s} \tag{3.32}$$

This equation is algebraic in nature and automatically includes the initial condition for the capacitor.

Now assume that $V_C(0+) = 0$ for simplicity or that the capacitance is initially uncharged; we then have

$$I(s)\left(R + \frac{1}{sC}\right) = \frac{V}{s}$$

Solving for $I(s)$ as the desired variable and rearranging the terms, we have

$$I(s) = \frac{V/R}{s + 1/RC} \tag{3.33}$$

The return to the time domain can be accomplished by taking the inverse transform:

$$\mathscr{L}^{-1}[I(s)] = \mathscr{L}^{-1}\left[\frac{V/R}{s + 1/RC}\right] = i(t) \tag{3.34}$$

The $F(s)$ is of the form

$$\frac{V}{R}\frac{1}{s + a}$$

and can be identified as transform pair 3 of Table 3.1. The solution in the time domain is then

$$i = \frac{V}{R}\varepsilon^{-t/RC} \tag{3.35}$$

This is the complete solution, showing the steady-state current to be zero, with the arbitrary constant already evaluated. The equation corresponds to the solution of Eq. 2.21.

Now consider the circuit at (b) in Fig. 3.3, with S closing at $t = 0$. Current i_1 is obtained from

$$Ri_1 = VU(t)$$

and the resultant transform is

$$\mathscr{L}[R_2 i_1] = \mathscr{L}[VU(t)]$$

$$R_2 I_1(s) = \frac{V}{s} \tag{3.36}$$

The second current can be obtained from

$$R_2 i_2 + L\frac{di_2}{dt} = VU(t) \tag{3.37}$$

By use of the derivative theorem, Eq. 3.13, we can write the transformed equation as

$$R_2 I_2(s) + sL\int_0^\infty i_2\varepsilon^{-st}\,dt - Li_2(0+) = \frac{V}{s} \tag{3.38}$$

the third term being $f(0+)$, or $f(t)$ evaluated at $t = 0+$. That is $i_2(0+)$ is the value of i_2 at $t = 0+$, just after the switch is connected at $t = 0$. We then have

$$I_2(s)(R_2 + sL) = \frac{V}{s} + Li_2(0+) \tag{3.39}$$

Let the initial inductor current be zero, $i_2(0+) = 0$, and by use of Eqs. 3.36 and 3.39 the current summation can be formed as

$$I(s) = I_1(s) + I_2(s)$$

$$I(s) = \frac{V}{R_1}\frac{1}{s} + \frac{V}{s(R_2 + sL)} = \frac{V}{R_1}\frac{1}{s} + \frac{V/L}{s(s + R_2/L)} \tag{3.40}$$

The first term on the right is readily identified as pair 1 of Table 3.1, and we can use it to write

$$i_1 = \frac{V}{R_1} \tag{3.41}$$

However, the second term is more complicated. The inverse transform appears as

$$\mathscr{L}^{-1}\left[\frac{V}{L}\frac{1}{s}\frac{1}{s + R_2/L}\right]$$

Let us assume that we do not yet have the complete table available and need to evaluate the product of the s terms. By a well-used algebraic procedure we can require that

$$\frac{1}{s}\frac{1}{s + R_2/L} = \frac{K_1}{s} + \frac{K_2}{s + R_2/L} \tag{3.42}$$

where K_1 and K_2 are to be determined. Placing the terms on the right over a common denominator, we have

$$1 = K_1(s + R_2/L) + K_2 s \tag{3.43}$$

By equating the coefficients of like functions,

$$\frac{K_1 R_2}{L} = 1$$

and from the s terms

$$K_1 + K_2 = 0$$

We then find the needed coefficient values:

$$K_1 = \frac{L}{R_2} \tag{3.44}$$

$$K_2 = -\frac{L}{R_2} \tag{3.45}$$

The inverse transform can be written as

$$\mathscr{L}^{-1}\left[\frac{V}{L}\left(\frac{L}{R_2}\frac{1}{s} - \frac{L/R_2}{s + R_2/L}\right)\right]$$

and the terms have the form of pairs 1 and 3 of Table 3.1. Then the complete solution, including Eq. 3.41, is

$$i = \frac{V}{R_1} + \frac{V}{R_2} - \frac{V}{R_2}\varepsilon^{-R_2t/L}$$

$$= \frac{V}{R_1} + \frac{V}{R_2}(1 - \varepsilon^{-R_2t/L}) \tag{3.46}$$

This includes the steady-state currents. In the solution we have derived transform pair 5.

The method of Eq. 3.42 will be considered further in Section 3.5.

3.5 Introduction to poles and zeros

Equation 3.40 applies to the RL circuit of Fig. 3.3(b). With $i_2(0+) = 0$, we can write the equation in the form

$$I(s) = \frac{V}{R_1}\left[\frac{s + (R_1 + R_2)/L}{s(s + R_2/L)}\right] \tag{3.47}$$

$$= H\left(\frac{s + s_1}{s(s + s_2)}\right) \tag{3.48}$$

where s_1 and s_2 are dependent on the circuit parameters. The coefficient H is a real *scale factor*.

Since s is dependent on the source and is variable, the equation demonstrates some interesting actions as s varies. When $s = -s_1 = -(R_1 + R_2)/L$, the numerator vanishes; accordingly, $s = -s_1$ is called a *zero of the function.* The condition $s + s_1 = 0$ represents a complex frequency at which the current $I(s)$ is zero. At a complex frequency $s = 0$, the current $I(s)$ becomes infinite, and the function is said to have a *pole* at $s = 0$. Another pole arises when the complex frequency is $s = -s_2 = -R_2/L$.

The function also has a zero at $s = \infty$; it takes the form s/s^2 as $s \to \infty$. When the poles at zero and infinity are included, the total number of zeros equals the total number of poles.

Poles and zeros are critical frequencies at which something happens to the function; as a result of poles and zeros the function changes as s varies. The importance of poles and zeros in network functions will increase as we progress further.

3.6

Partial fraction expansion

The circuit function in s may not always be in a simple form from which its inverse, $f(t)$, can be found through reference to a table of transforms. However, the transformed function, $F(s)$, may be expanded as a sum of partial fractions, and each term may then be more suited to inversion by finding a transform pair.

It has just been shown that by a Laplace transformation a circuit differential equation can be put into the form of a ratio of two polynomials in s. In general form the relation could be written

$$F(s) = \frac{A(s)}{B(s)} = H\frac{s^n + a_1 s^{n-1} + a_2 s^{n-2} + \cdots + a_n}{s^k + b_1 s^{k-1} + b_2 s^{k-2} + \cdots + b_n} \tag{3.49}$$

where H is a real *scale factor*.

The denominator polynomial will be factored so that

$$B(s) = (s + s_1)(s + s_2)(s + s_3)\cdots(s + s_n) \tag{3.50}$$

where $s_1, s_2, \ldots, s_n$ are the n roots of the nth order polynomial. Each factor, being in the denominator, locates a pole of $F(s)$.

If there are no multiple roots, $F(s)$ can then be expanded into additive terms

$$\frac{A(s)}{(s + s_1)(s + s_2)(s + s_3)\cdots(s + s_n)} = \frac{K_1}{s + s_1} + \frac{K_2}{s + s_2} + \cdots + \frac{K_n}{s + s_n} \tag{3.51}$$

The roots may be simple and real or in complex conjugate pairs. The right side is cross-multiplied so as to appear over the same denominator as on the left. The numerators must then be equal for all values of s. By evaluating for s at each pole, a set of simultaneous equations involving the Ks is written, and the K coefficients can be obtained.

Insertion of the K values in Eq. 3.51 gives terms of a form suited to inversion by use of a table of transforms, and the resulting function of time can be written as the sum of the several time functions.

Usually the order of the numerator will be less than that of the denominator. If this condition is not met, the denominator must be divided into the numerator to obtain

$$\frac{A(s)}{B(s)} = C_o + C_1 s + C_2 s^2 + \cdots + C_{n-d} s^{n-d} + \frac{A_1(s)}{B(s)} \tag{3.52}$$

where n is the order of the numerator and d is the order of the denominator. The last term is a new polynomial with the order of the numerator less than the order of the denominator; the usual method then applies.

EXAMPLE. Find the inverse transform of

$$F(s) = \frac{s^2 + 3s + 1}{(s + 1)(s + 2)(s + 3)} = \frac{K_1}{s + 1} + \frac{K_2}{s + 2} + \frac{K_3}{s + 3}$$

Placing the right side over a common denominator gives

$$s^2 + 3s + 1 = K_1(s + 2)(s + 3) + K_2(s + 1)(s + 3) + K_3(s + 1)(s + 2)$$

If we make $s = -1$, the above reduces to

$$K_1 = \left.\frac{s^2 + 3s + 1}{(s + 2)(s + 3)}\right]_{s=-1} = -\frac{1}{2}$$

Setting $s = -2$, we have

$$K_2 = \left.\frac{s^2 + 3s + 1}{(s + 1)(s + 3)}\right]_{s=-2} = 1$$

With s at the remaining pole at -3, we have

$$K_3 = \left.\frac{s^2 + 3s + 1}{(s + 1)(s + 2)}\right]_{s=-3} = \frac{1}{2}$$

Using these values for the coefficients, we can write

$$F(s) = -\frac{1/2}{s + 1} + \frac{1}{s + 2} + \frac{1/2}{s + 3}$$

The time-domain solution follows by reference to the table of transforms, where suitable pairs are found to give

$$f(t) = -0.5\varepsilon^{-t} + \varepsilon^{-2t} + 0.5\varepsilon^{-3t}$$

The accuracy of the calculations for the K coefficients can be checked by cross multiplication of $F(s)$, giving

$$s^2 + 3s + 1 = -0.5(s + 2)(s + 3) + (s + 1)(s + 3) + 0.5(s + 1)(s + 2)$$

$$= s^2 + 3s + 1$$

EXAMPLE. Find the time-domain expression for

$$F(s) = \frac{3s^2 + 2s + 8}{(s^2 + 4)(s + 1)}$$

The term $s^2 + 4$ could be factored to yield $(s + j2)(s - j2)$, after which the procedure would be as in the example above. However, another procedure is to write

$$F(s) = \frac{3s^2 + 2s + 8}{(s^2 + 4)(s + 1)} = \frac{K_1 s + K_2}{s^2 + 4} + \frac{K_3}{s + 1}$$

Cross-multiplying and sorting by powers of s gives

$$3s^2 + 2s + 8 = (K_1 + K_3)s^2 + (K_1 + K_2)s + (K_2 + 4K_3)$$

Equating the coefficients of like powers of s yields

$$K_1 + K_3 = 3$$

$$K_1 + K_2 = 2$$

$$K_2 + 4K_3 = 8$$

and simultaneous solution of these relations gives

$$K_1 = 1.2$$

$$K_2 = 0.8$$

$$K_3 = 1.8$$

Then we can write

$$F(s) = \frac{1.2s}{s^2 + 4} + \frac{0.8}{s^2 + 4} + \frac{1.8}{s + 1}$$

Consultation of the table of transform pairs yields the time-domain solution:

$$f(t) = 1.2 \cos 2\omega t + 0.4 \sin 2\omega t + 1.8\varepsilon^{-t}$$

3.7 Circuit parameters and impedance in the s domain

We define a *driving-point impedance* at a port of an element or a circuit as *the ratio of the exponential voltage across the port to the exponential current through the port.* The impedance is the ratio of the across and through variables; here it is designated $Z(s)$, since we are using exponential forcing functions containing the complex frequency s.

Starting with the Laplace transform of a time-varying current i, we write

$$\mathscr{L}[i] = \int_0^\infty i\varepsilon^{-st}\,dt = I(s) \tag{3.53}$$

and with use of Eq. 3.9 we obtain

$$\mathscr{L}[Ri] = RI(s) = V(s) \tag{3.54}$$

for the transformed voltage across a resistance R. Then the *impedance* of the resistor is

$$Z_R(s) = \frac{V(s)}{I(s)} = R \qquad \Omega \tag{3.55}$$

Impedance, being expressed as volts per ampere, has units of ohms. The schematic for the resistor in the s domain is presented in Fig. 3.4(a).

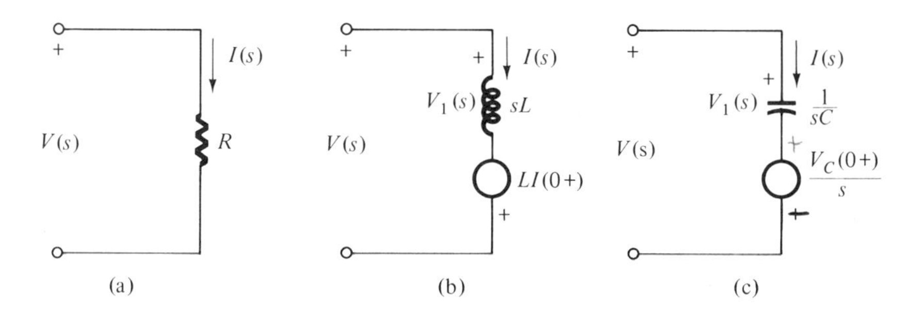

FIGURE 3.4
Transformed circuit impedances, including initial values.

The time-domain voltage relation for an inductance is

$$v = L\frac{di}{dt}$$

and by use of Eq. 3.13 the transform of the voltage can be written as

$$\mathscr{L}\left[L\frac{di}{dt}\right] = L[sI(s) - i(0+)] = V(s) \tag{3.56}$$

Then

$$sLI(s) = V(s) + Li(0+) \tag{3.57}$$

The terms on the right represent the transform of the driving voltage and a transformed voltage source due to any initial current through the inductor. Let $V_1(s)$ be the transform voltage across the inductance element; then

$$sLI(s) = V_1(s) = V(s) + Li(0+)$$

and the impedance of the inductance parameter in the s domain is

$$Z_L(s) = \frac{V_1(s)}{I(s)} = sL \qquad \Omega \tag{3.58}$$

again being expressed in ohms.

Using the time-domain relation for a capacitance

$$v = \frac{1}{C}\int i\,dt$$

and the theorem for the transform of an integral as in Eq. 3.16, we obtain

$$\mathscr{L}\left[\frac{1}{C}\int i\,dt\right] = \frac{q_C(0+)}{sC} + \frac{I(s)}{sC} = V(s) \tag{3.59}$$

Since $v_C = q/C$, we have

$$\frac{I(s)}{sC} = V(s) - \frac{v_C(0+)}{s} \tag{3.60}$$

Calling $V_1(s) = V(s) - [v_C(0+)/s]$, then

$$Z_C(s) = \frac{V_1(s)}{I(s)} = \frac{1}{sC} \qquad \Omega \tag{3.61}$$

In the above, $V(s)$ is the transform of the voltage applied to the port, and $v_C(0+)$ is a transform voltage source resulting from the step application of any potential on the capacitance at $t = 0$ when the switching action occurs; note the $1/s$ form of this function as being the transform of a step input.

Summarizing,

$$Z_R(s) = R \tag{3.62}$$

$$Z_L(s) = sL \tag{3.63}$$

$$Z_C(s) = \frac{1}{sC} \tag{3.64}$$

These are impedance definitions for the circuit parameters in the transform domain.

We can also define the *admittance* of a circuit parameter as the reciprocal of its impedance. For the resistance,

$$Y_R(s) = \frac{I(s)}{V(s)} = \frac{1}{R} = G \qquad \text{mhos} \tag{3.65}$$

in units of reciprocal ohms, or mhos.

For the inductance, we can write Eq. 3.57 as

$$\frac{V(s)}{sL} = I(s) - \frac{i(0+)}{s}$$

which is a current summation, with $i(0+)/s$ as a transform current source due to the initial inductor current. Referring to Fig. 3.5(a),

$$I_1(s) = I(s) - \frac{i(0+)}{s}$$

is the current in the inductance, or

$$\frac{V(s)}{sL} = I_1(s) \tag{3.66}$$

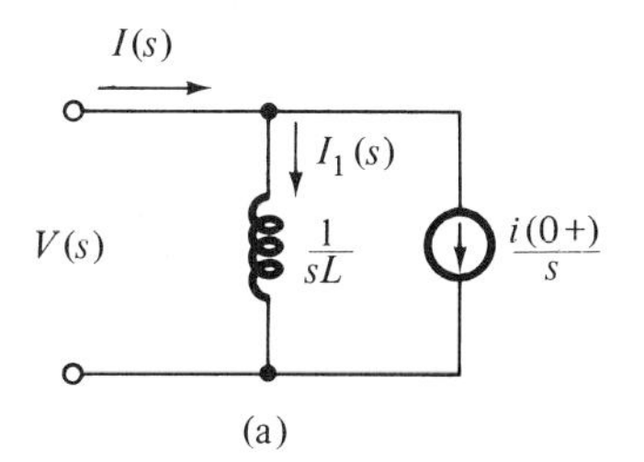

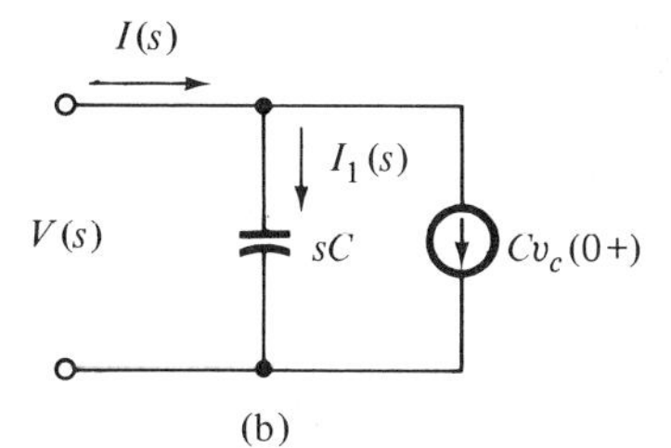

FIGURE 3.5
(a) Admittance of an inductance; (b) admittance of a capacitance.

Therefore the admittance of the inductance is

$$Y_L(s) = \frac{I_1(s)}{V(s)} = \frac{1}{sL} \qquad \text{mhos} \tag{3.67}$$

In a similar manner, Eq. 3.60 for the capacitance can be regrouped as

$$sCV(s) = I(s) - Cv_C(0+) \tag{3.68}$$

with $Cv_C(0+)$ being a transformed current source due to any initial potential across the capacitance. In Fig. 3.5(b),

$$I_1(s) = I(s) - Cv_C(0+)$$

and so

$$Y_C(s) = \frac{I_1(s)}{V(s)} = sC \qquad \text{mhos} \tag{3.69}$$

is the transform admittance of a capacitance C.

Summarizing, we have

$$Y_R(s) = G \tag{3.70}$$

$$Y_L(s) = \frac{1}{sL} \tag{3.71}$$

$$Y_C(s) = sC \tag{3.72}$$

for the admittances of the three circuit parameters.

It is then possible to use these impedances and admittances and the respective currents to write current or voltage equations directly in the s domain.

3.8 Application to *RC* and *RL* circuits

The series RC circuit of Fig. 3.6(a) has a capacitance with an initial charge or potential V_c when the switch S is closed at $t = 0$. Using the transform impedances developed in the previous section, the circuit may be redrawn in the s domain as in (b) in Fig. 3.6. The step function $VU(t)$ introduced by the battery V at the closure of switch S

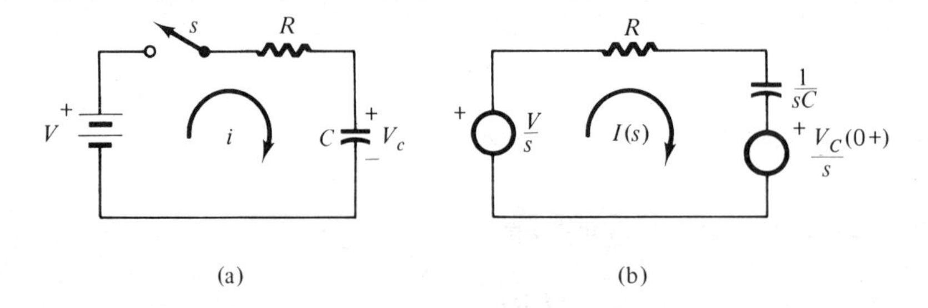

FIGURE 3.6
(a) Series ***RC*** circuit, time domain; (b) parameters transformed to the ***s*** domain.

at $t = 0$ is shown as V/s. A circuit equation may then be written in the s domain, avoiding the differential equation step. That is,

$$RI(s) + \frac{1}{sC}I(s) + \frac{V_c(0+)}{s} = \frac{V}{s} \tag{3.73}$$

The solution proceeds as

$$I(s) = \left[\frac{1}{s + (1/RC)}\right]\left(\frac{V - V_c(0+)}{R}\right) \tag{3.74}$$

Letting $V_c(0+) = V_c$ for simplicity, the time-domain solution follows from transform 3 of Table 3.1. Then

$$i = \left(\frac{V - V_c}{R}\right)\varepsilon^{-t/RC} \tag{3.75}$$

This result checks Eq. 2.21, modified by the presence of the initial voltage on the capacitance.

EXAMPLE. Figure 3.7(a) represents a circuit which is transformed to the s domain in (b) by use of the impedance relations of Section 3.7. The

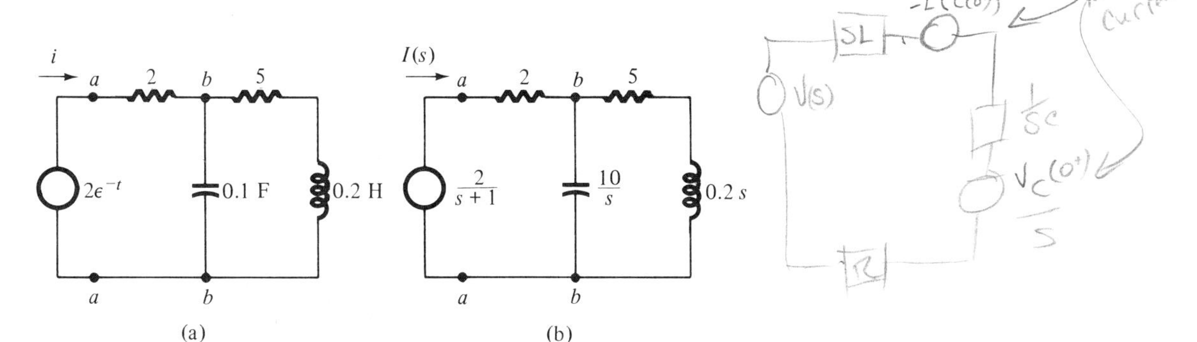

FIGURE 3.7
(a) Circuit in t domain; (b) transformed circuit.

circuit is assumed to be quiescent, that is, with no initial currents or charges present at $t = 0$. The driving source is transformed by reference to pair 3 of Table 3.1; this example illustrates a valuable reciprocal use of such tables.

The impedance of the parallel elements to the right of terminals bb is

$$Z_{bb}(s) = \frac{(5 + 0.2s)(10/s)}{5 + 0.2s + (10/s)} = \frac{10(s + 25)}{s^2 + 25s + 50}$$

The impedance at port aa includes the 2-Ω resistance and so

$$Z_{aa}(s) = 2 + \frac{10(s + 25)}{s^2 + 25s + 50}$$

$$= \frac{2(s^2 + 30s + 175)}{s^2 + 25s + 50}$$

The current supplied to the circuit is $I(s)$, as shown, given by

$$I(s) = \frac{V_{aa}(s)}{Z_{aa}(s)} = \frac{2}{s + 1}\,\frac{s^2 + 25s + 50}{2(s^2 + 30s + 175)}$$

Factoring the quadratic in the denominator leads to $s_1 = -7.93$, $s_2 = -22.07$, and so

$$I(s) = \frac{s^2 + 25s + 50}{(s+1)(s+7.93)(s+22.07)}$$

By partial fractions

$$\frac{s^2 + 25s + 50}{(s+1)(s+7.93)(s+22.07)} = \frac{K_1}{s+1} + \frac{K_2}{s+7.93} + \frac{K_3}{s+22.07}$$

$$s^2 + 25s + 50 = K_1(s+7.93)(s+22.07) + K_2(s+1)(s+22.07) + K_3(s+1)(s+7.93)$$

Letting $s = -1$,

$$1 - 25 + 50 = 26 = K_1(6.93)(21.07)$$

$$K_1 = \frac{26}{146} = 0.178$$

With $s = -7.93$,

$$-85.1 = K_2(-6.93)(14.14)$$

$$K_2 = \frac{-85.1}{-98} = 0.87$$

Similarly, with $s = -22.07$,

$$-15 = K_3(-21.07)(-14.14)$$

$$K_3 = \frac{-15}{298} = -0.050$$

Then

$$I(s) = \frac{0.178}{s+1} + \frac{0.87}{s+7.93} - \frac{0.050}{s+22.07}$$

Reference to transform pair 3 of Table 3.1 leads to the solution in time:

$$i = 0.178\varepsilon^{-t} + 0.87\varepsilon^{-7.93t} - 0.050\varepsilon^{-22.07t}$$

3.9 Circuit solution with sinusoidal excitation

The response of a network when a sinusoidal source is suddenly connected is a matter of importance in power switching. An *RL* circuit appears in Fig. 3.8, with switch *S* closed at $t = 0$. With $i = 0$ at $t = 0-$, $i = 0$ at $t = 0+$, since the current in the inductance cannot change suddenly.

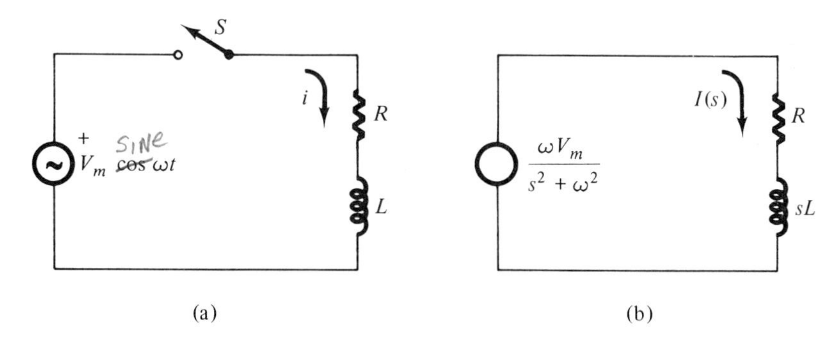

FIGURE 3.8
Driving function and transformed circuit.

The transformed circuit is shown at (b) in Fig. 3.8, with the sinusoidal driving source transformed by reference to pair 11 of Table 3.1. The transformed circuit equation is

$$RI(s) + sLI(s) = \frac{\omega V_m}{s^2 + \omega^2} \tag{3.76}$$

so that

$$I(s) = \frac{\omega V_m}{(s^2 + \omega^2)(R + sL)} = \frac{V(s)}{Z(s)} \tag{3.77}$$

The term $R + sL$ is the series impedance of the circuit in the s domain. Further manipulation leads to

$$I(s) = \frac{\omega V_m/L}{(s^2 + \omega^2)(s + R/L)} \tag{3.78}$$

Noting that $(s^2 + \omega^2) = (s + j\omega)(s - j\omega)$, we can write

$$I(s) = \frac{\omega V_m/L}{(s + j\omega)(s - j\omega)(s + R/L)}$$

and solve by the method of partial fractions. However, reference to form 18 of the transform table will save labor, and we then have

$$i = \mathscr{L}^{-1}[I(s)] = \frac{\omega V_m}{L}\left[\frac{\varepsilon^{-Rt/L}}{\omega^2 + R^2/L^2} + \frac{1}{\omega\sqrt{\omega^2 + R^2/L^2}}\sin(\omega t - \Phi)\right]$$

where

$$\Phi = \tan^{-1}\frac{\omega L}{R}$$

This may be simplified to

$$i = \frac{V_m}{\sqrt{R^2 + \omega^2 L^2}}\left[\sin(\omega t - \Phi) + \frac{\omega L}{\sqrt{R^2 + \omega^2 L^2}}\varepsilon^{-Rt/L}\right] \tag{3.79}$$

$$\Phi = \tan^{-1}\frac{\omega L}{R} \tag{3.80}$$

The angle Φ appears as that of a triangle having sides ωL and R and a hypotenuse $\sqrt{R^2 + \omega^2 L^2}$. Therefore the coefficient $\omega L/\sqrt{R^2 + \omega^2 L^2}$ represents $\sin \Phi$ and

$$i = \frac{V_m}{\sqrt{R^2 + \omega^2 L^2}}[\sin(\omega t - \Phi) + \sin \Phi \varepsilon^{-Rt/L}] \tag{3.81}$$

with Φ as defined above.

The result, indicated in Fig. 3.9, includes the natural response varying as $\varepsilon^{-Rt/L}$ and the steady-state term due to the sinusoidal forcing function. Excessive currents may occur in the circuit due to the sum of the natural response and the sine function at $t = 0$.

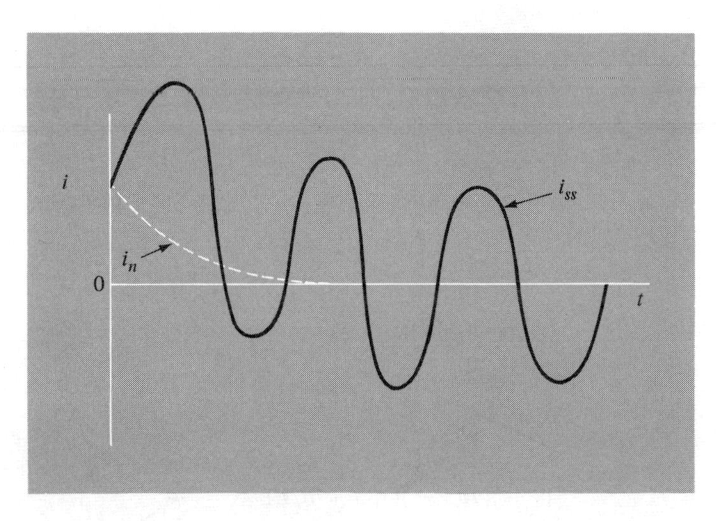

FIGURE 3.9
Sinusoidal transient at switching.

3.10 Initial and final conditions

The Laplace method requires a knowledge of initial currents in inductors and initial voltages on capacitances. These quantities can be found through the *initial value theorem*, stated as

$$f(0+) = \lim_{s \to \infty} [sF(s)] \tag{3.82}$$

The proof follows from the theorem

$$\mathscr{L}\left[\frac{d}{dt}f(t)\right] = sF(s) - f(0+)$$

Taking the limit,

$$\lim_{s\to\infty} \mathscr{L}\left[\frac{d}{dt}f(t)\right] = \lim_{s\to\infty} [sF(s)] - f(0+)$$

where $f(0+)$ is not a function of s. But

$$\lim_{s\to\infty} \mathscr{L}\left[\frac{d}{dt}f(t)\right] = \lim_{s\to\infty} \int_0^\infty \frac{d}{dt}f(t)\varepsilon^{-st}\,dt = 0 \tag{3.83}$$

from which

$$0 = \lim_{s\to\infty} [sF(s)] - f(0+)$$

We have arrived at our proof, since

$$f(0+) = \lim_{s\to\infty} [sF(s)] \tag{3.84}$$

Suppose that

$$F(s) = \frac{9bs}{(s+1)(s+2)}$$

Then by the initial value theorem,

$$f(0+) = \lim_{s\to\infty} \left[\frac{9bs^2}{(s+1)(s+2)}\right] = 9b$$

In the circuit the initial conditions can be found by remembering that an inductance appears as a constant current over the interval $t = 0-$ to $t = 0+$, since magnetic flux cannot change instantaneously. A capacitor represents a constant voltage across the same interval, since charge cannot be moved instantaneously.

The *final value theorem* follows in a similar manner:

$$\lim_{t\to\infty} [f(t)] = \lim_{s\to 0} [sF(s)] \tag{3.85}$$

3.11

The unit impulse

The rectangular function in time of Fig. 3.10(a) has unit area $a \times 1/a$ and is called the *unit pulse*. If the dimension a is allowed to approach zero, then $1/a$ approaches infinity; the pulse still has unit area but now approaches the situation in (b), when it is known as the *unit impulse* or the *delta function*.

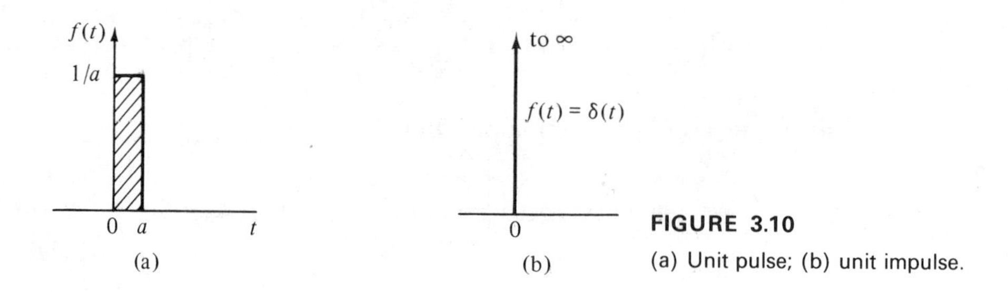

FIGURE 3.10
(a) Unit pulse; (b) unit impulse.

Its value is zero except in an arbitrarily small region near $t = 0$, where it becomes infinite such that

$$\int_{0-}^{b} \delta(t)\, dt = 1 \tag{3.86}$$

where b is any positive quantity. Such an impulse in current is theoretically obtained if we suddenly connect a voltage source at $t = 0$ directly to an uncharged capacitance, since with $R = 0$ the time constant is zero. In theory, at least, we may drive a circuit with an impulse, although in a practical sense we never have a circuit of zero resistance.

It is possible to determine the Laplace transform of $\delta(t)$ and this allows direct circuit application. Taking the transform of Eq. 3.86,

$$\mathscr{L}[\delta(t)] = \lim_{\beta \to a} \int_{0-}^{\beta} \frac{1}{a} \varepsilon^{-st}\, dt$$

$$= \lim_{\beta \to a} \left[\frac{\varepsilon^{-st}}{-as} \right]_{0-}^{\beta} = \frac{1 - \varepsilon^{-as}}{as} \tag{3.87}$$

If we now force a to zero, we obtain an indeterminate 0/0; this can be evaluated by differentiation, leading to

$$\mathscr{L}[\delta(t)] = \lim_{a \to 0} \left[\frac{s\varepsilon^{-as}}{s} \right] = 1.0 \tag{3.88}$$

The Laplace transform of the unit impulse is unity.

3.12

Summary

The Laplace method transforms a differential circuit equation in time (or other variable) into an algebraic equation in the variable s, a complex frequency, with the initial conditions of the system included. The transform equations can be written directly through use of the appropriate s-domain impedance functions and usual circuit techniques. The driving function is transformed to the s domain by reference to a table of transform pairs. The equation is then ready for algebraic solution for the desired variable. Use of an inverse transform, ordinarily found in a table of transforms, leads back to the solution as a function of time.

When a given function of s is not found in the table, the s equation may be expanded into a sum of partial fractions, obtaining separate components which may then be recognized as simpler transforms in the table.

The poles of $F(s)$ determine the general type of response: a decaying exponential for the overdamped situation and an exponential plus a sinusoidal term for the underdamped case.

PROBLEMS

3.1 Transform the following to the s domain:

$$i(t) = \frac{V}{R} + \left[i(0+) - \frac{V}{R} \right] \varepsilon^{-Rt/L}$$

3.2 Show that

$$L\frac{di}{dt} + Ri = V\varepsilon^{at}$$

will yield

$$I(s) = \frac{1}{R + sL}\left[\frac{V}{s - a} + Li(0+)\right]$$

3.3 Determine $f(t)$ from

(a) $\dfrac{1}{s^2 + 12s + 9}$

(b) $F(s) = \dfrac{s^2 + 9}{s(s + 1)(s + 2)}$

(c) $F(s) = \dfrac{s^2 - 3s + 4}{(s + 3)(s^2 - 4)}$

(d) $F(s) = \dfrac{s^2 - s + 10}{(s + 1)(s^2 + 100)}$

(e) $F(s) = \dfrac{s^2 + 2}{s(s^2 + 1)}$

(f) $F(s) = \dfrac{s^2 - 1}{(s + 2)(s^2 + 2s + 10)}$

(g) $F(s) = \dfrac{3s + 4}{(s + 3)(s + 2)}$ (h) $F(s) = \dfrac{s + 3}{s^3 + 4s^2 - s - 22}$

(i) $F(s) = \dfrac{11s + 50}{s^2 + 5s}$ (j) $F(s) = \dfrac{1 + 5s}{10s + 5s^2}$

3.4 Develop $F(s)$ from

(a) $f(t) = 1 - \varepsilon^{at}$ (b) $f(t) = \dfrac{1 - \cos bt}{b^2}$

(c) $f(t) = 1 + \cosh t$

3.5 Develop

$$F(s) = \frac{\omega^2}{s(s^2 + 2a\omega s + \omega^2)}$$

as a function of time.

3.6 A series RL circuit has $R = 40\ \Omega$, $L = 0.1$ H, and $V(t) = 100 \cos 377t$. After being in the steady state, the resistance is changed to $10\ \Omega$ at $t = 0$. Find an expression for the current as a function of time for $t > 0$.

3.7 Given that $F(s) = (s - 2)/(s + 1)$, determine $f(0+)$.

3.8 A voltage $V\varepsilon^{-at} \sin \omega t$ is applied to an RL series circuit at $t = 0$; $i = i(0)$ when $t = 0$. Find the value of $i(t)$ for $t > 0$.

3.9 A 10-V battery is connected to an RLC series circuit with $R = 50\ \Omega$, $L = 1$ H, and $C = 100\ \mu$F. The circuit is quiescent at $t = 0$ [$i(0) = 0$, $q_o = 0$]. Find an expression for $i(t)$ for $t > 0$.

3.10 A steady voltage of 100 V has been applied to an RLC series circuit for a long time; $R = 10\ \Omega$, $L = 0.1$ H, and $C = 50\ \mu$F. At $t = 0$ the voltage starts to fall as $v(t) = 100\varepsilon^{-0.01t}$. Find the current and the capacitor voltage at $t = 0.3$ s.

3.11 You have $F(s) = (s - 2)/s(s + 1)$; determine $f(t)$.

3.12 Determine $f(t)$ for

(a) $F(s) = \dfrac{10s - 30}{(s^2 + 4)(s + 1)}$ (b) $F(s) = \dfrac{s + 3}{s^3 + 3s^2 - s - 3}$

3.13 For the circuit of Fig. 3.11(a), find the value of $i_2(t)$, assuming that $V_{c1}(0) = V_{c2}(0) = 0$ at $t = 0$.

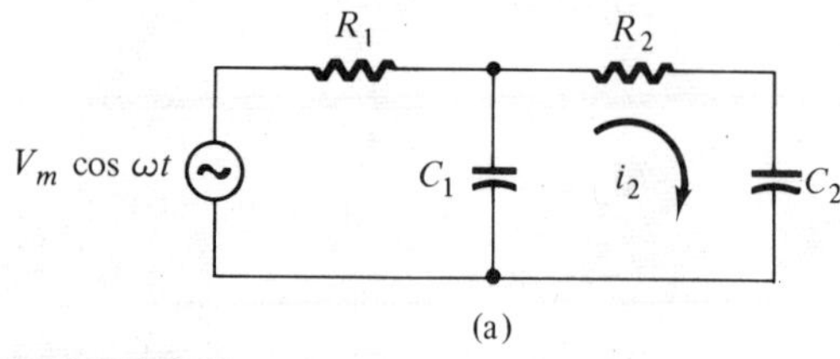

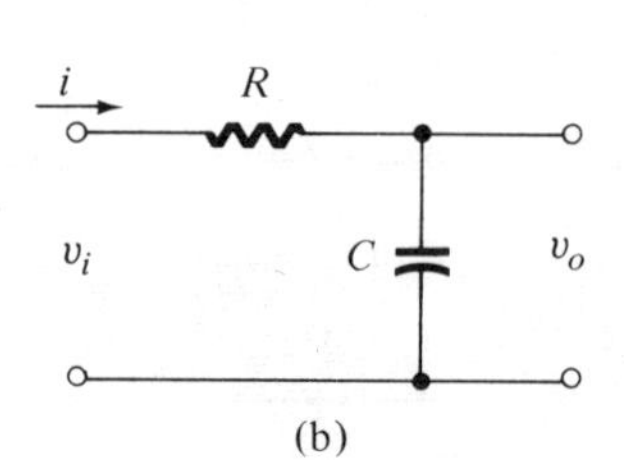

FIGURE 3.11

3.14 A voltage v_i rises linearly from zero to $+100$ V in 1000 μs; it is applied to the circuit of Fig. 3.11(b), where $R = 100{,}000\ \Omega$ and $C = 0.005\ \mu$F. (a) Find the current i at the end of the period of v_i. (b) Determine v_o at that time.

3.15 The relay coil of Fig. 3.12(a) has negligible inductance and 300 Ω of resistance. It closes its armature contact at a coil current of 0.1 A; resistor $R = 600\ \Omega$. Find C so that the relay contact closes 0.55 s after switch S closes.

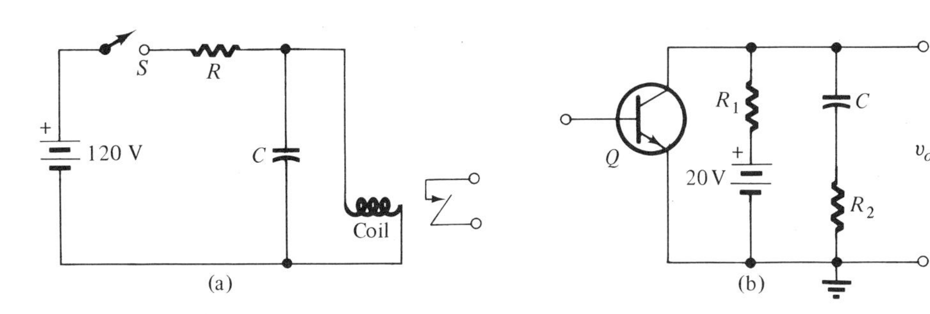

FIGURE 3.12

3.16 In Problem 3.15 the coil inductance is found to be 150 mH. Redetermine the value of C needed.

3.17 The transistor Q of the circuit in Fig. 3.12(b) serves as a switch. With $R_1 = 10{,}000\ \Omega$, $R_2 = 2000\ \Omega$, and $C = 0.1\ \mu$F, plot the output wave form v_o for 400 μs after the transistor switches from closed to open.

3.18 In the circuit of Fig. 3.13(a), the switch is ordinarily closed on 1. Close the switch on 2 at $t = 0$ and find v as a function of time.

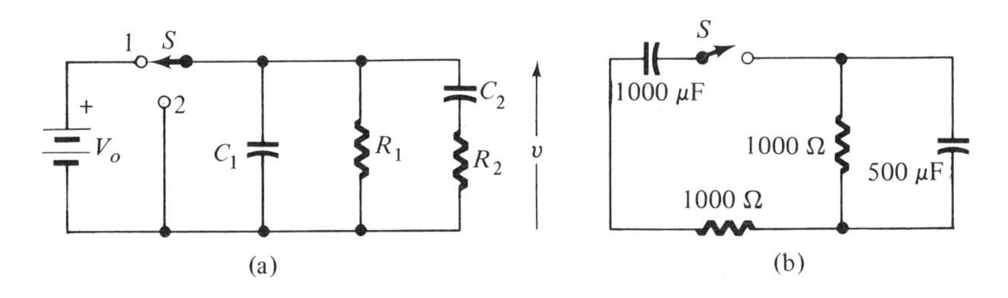

FIGURE 3.13

3.19 In Fig. 3.13(b), the 1000-μF capacitor is charged to 500 V. At $t = 0.1$ s after S is closed, find the voltage across the 500-μF capacitor.

3.20 The circuit of Fig. 3.14 represents the equivalent of a transistor amplifier. If v_1 as the input is a unit step function $U(t)$, calculate v_o as a time function and plot this against time. The value of g_m is 20,000 μmhos.

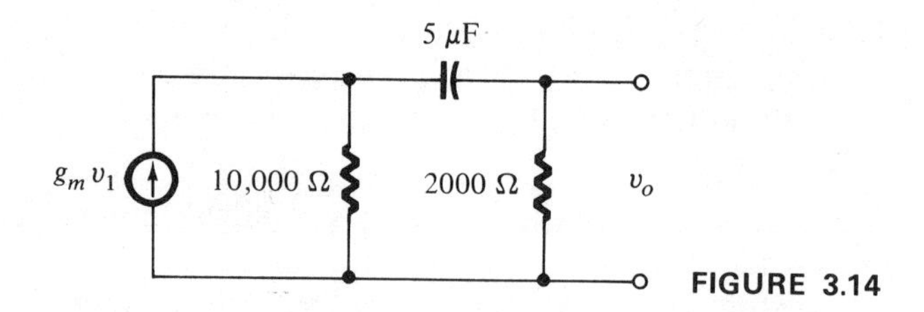

FIGURE 3.14

3.21 With $i(t)$ as a unit step function, $U(t)$, in Fig. 3.15(a) and $R = 4000\ \Omega$, $L = 100\ \mu$H, and $C = 45$ pF, plot $v(t)$ as a function of time for 1 μs after the step. Initial conditions are assumed to be quiescent.

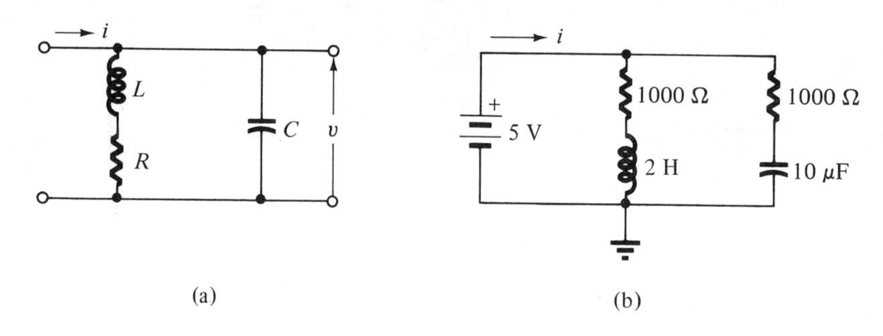

FIGURE 3.15

3.22 In Fig. 3.15(b) at $t = 0$, the inductor has a current of 1 A upward, and the capacitor has a voltage of -3 V to ground. Find the initial value of i.

3.23 A voltage of 10 V has been applied to a series RL circuit with $R = 10\ \Omega$, $L = 1$ H for a long time. At $t = 0$ the applied voltage becomes zero by shorting the voltage source. Write the current as a function of time.

3.24 Find the current i as a time function for the circuit of Fig. 3.16(a) if switch S closes at $t = 0$.

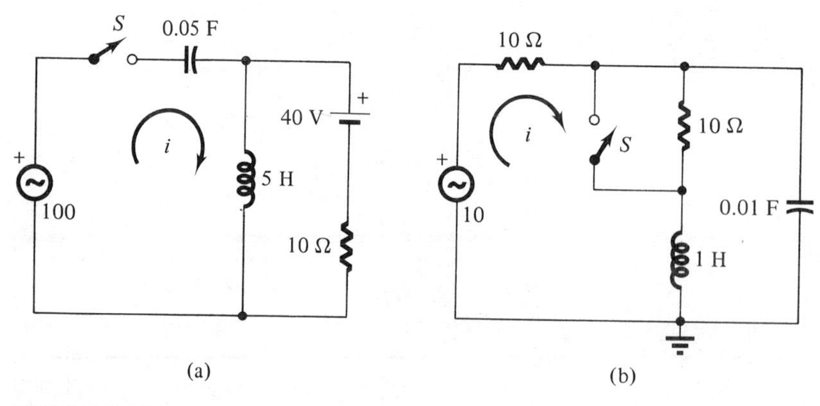

FIGURE 3.16

3.25 In Fig. 3.16(b), switch S has been open for a long time. It closes at $t = 0$. Find i in the time domain for $t > 0$, if the 0.01-F capacitor has a potential of 5 V at $t = 0-$, negative to ground.

3.26 There are no initial charges or currents in the circuit of Fig. 3.17. State v_o as a function of time after S closes at $t = 0$.

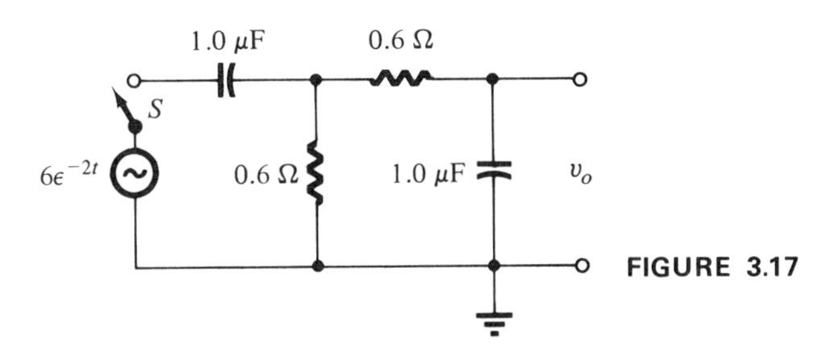

FIGURE 3.17

REFERENCES

1. M. F. Gardner and J. L. Barnes, *Transients in Linear Systems*. John Wiley & Sons, Inc., New York, 1945.
2. F. E. Nixon, *Handbook of Laplace Transformation*. Prentice-Hall, Inc., Englewood Cliffs, N.J., 1960.
3. W. T. Thomson, *Laplace Transformation*. Prentice-Hall, Inc., Englewood Cliffs, N.J., 1960.

steady-state response; phasors; power

four

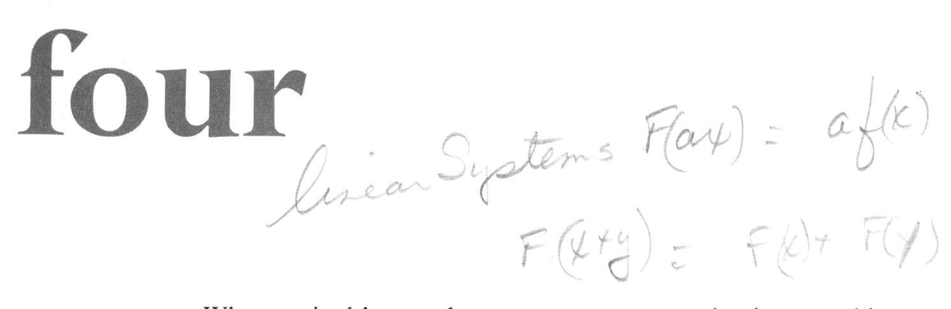

When excited by a voltage or current source having an arbitrary variation with time, an electric circuit has a natural response with a time variation dependent on the circuit configuration and a steady-state response in which the time variation is that of the driving source. While we employ the natural response in some electronic and control circuits, our electrical systems largely operate in and are analyzed in the steady state, after the natural response has decayed to zero.

4.1 The steady state

We recognize several basic forms of time variation as employed in electrical circuits:

1. A *direct current or voltage* (dc) in which the magnitude of the forcing function is constant with time. An average value exists.
2. An *alternating current or voltage* (ac) in which the magnitude of the driving function varies in a sinusoidal manner with time. Its values recur in each period or *cycle*, and the average value is zero.
3. A *periodic current or voltage* having an arbitrary variation in a finite time interval and in which the variation recurs in each interval.

In electronic circuits and electronic processes we usually employ the dc form of supply. In electric power distribution and utilization we employ ac variation, sinusoidal in time. In communication we consider the signals as the sum of many sinusoids, and the rectangular wave is used in control functions, computers, and pulse communications.

4.2

The choice of a wave form

Many different cyclic wave forms could have been employed in establishing the alternating-current electrical system. The fundamental importance of the decision and the advantages of a sinusoidal form were realized at an early date, with the engineering decision based on arguments such as

1. Nature recognizes the sinusoid; the natural response of an undamped *RLC* circuit is sinusoidal.
2. When differentiated or integrated, as happens in circuits with inductance or capacitance, the sinusoidal wave form is unchanged.
3. Two sinusoids of identical period add to form a sinusoid of the same period.
4. By use of the Fourier series, it is possible to carry out an analysis for any periodic wave form as a sum of sinusoidal functions. The analysis for each discrete frequency may be superimposed under the linear network assumption, to give a complete solution.

The sinusoid was chosen and standardized, as a fortunate result of the above arguments. We have already worked with this form in the Laplace transform method and will now learn how to specialize this treatment into sinusoidal steady-state analyses of even simpler form.

4.3

The rotating exponential; phasors

We must now introduce sinusoids into our circuit equations. However, some mathematical complexity results even with the Laplace transform, and it is mathematically simpler to employ the exponential $\varepsilon^{j\omega t}$ as the form of time variation and to transform the results into either sine or cosine form at a later point in problem solution.

That this can be done seems indicated by the current

$$i = I_m\varepsilon^{j\omega t} = I_m(\cos \omega t + j \sin \omega t) \tag{4.1}$$

Applied to a resistance, we have the voltage

$$v = Ri = RI_m\varepsilon^{j\omega t} = RI_m \cos \omega t + jRI_m \sin \omega t \tag{4.2}$$

The first term on the right-hand side is the real part of the voltage, $Re(v)$, and this is the result of the real part of the current, $Re(i)$. The imaginary term, $Im(v)$, of Eq. 4.2 is a result of $Im(i)$.

For an inductance carrying the current of Eq. 4.1, we can write

$$v = L\frac{di}{dt} = LI_m\frac{d\varepsilon^{j\omega t}}{dt} = -\omega LI_m \sin \omega t + j\omega LI_m \cos \omega t \tag{4.3}$$

For a capacitance C with the current of Eq. 4.1, we have

$$v = \frac{1}{C}\int i\,dt = \frac{1}{C}\int I_m \varepsilon^{j\omega t}\,dt$$

$$= \frac{I_m}{\omega C}\sin\omega t - j\frac{I_m}{\omega C}\cos\omega t \tag{4.4}$$

In each of these cases the real part of the voltage results from the real part of the current, and the j term in the voltage is due to the j term of the current. Since this is true for the individual circuit parameters, it is true for combinations of the parameters in circuits. That is,

$$Re(V_m \varepsilon^{j\omega t}) = V_m \cos\omega t$$

$$I_m(V_m \varepsilon^{j\omega t}) = V_m \sin\omega t$$

We can use the simple exponential forcing function of time and upon completion of a computation can choose the real part or the j portion to give the result of a cosine or sine forcing function.

Thus $\varepsilon^{j\omega t}$ seems a desirably simple mathematical form, but we need a better conceptual understanding of $\varepsilon^{j\omega t}$ as well. Since $\varepsilon^{j\omega t}$ is a complex quantity, we have

$$|\varepsilon^{j\omega t}| = \sqrt{\cos^2\omega t + \sin^2\omega t} = 1 \tag{4.5}$$

and we see that the *magnitude of the function is always unity*. In the complex plane plot of Fig. 4.1 we define the positive abscissa as the zero value for the

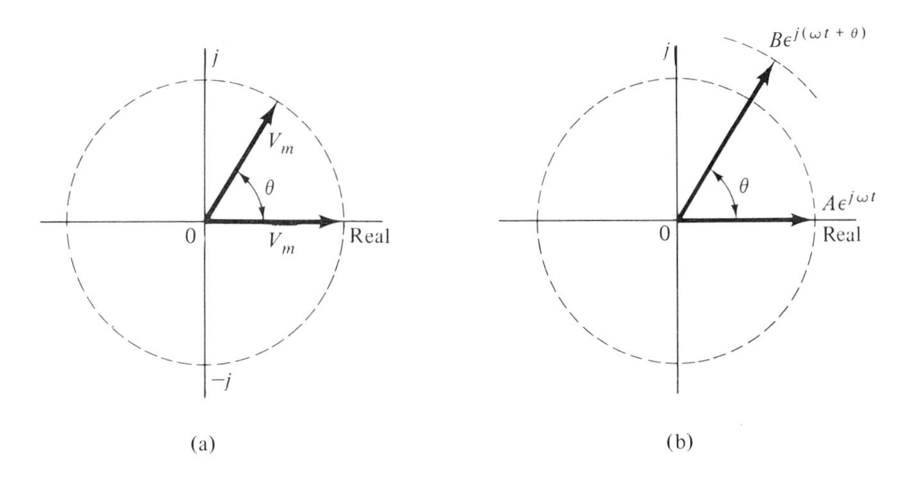

FIGURE 4.1

(a) $\varepsilon^{j\omega t}$ as a rotating operator; (b) rotating phasors.

angle ωt and define the positive direction of measurement of ωt as counter-clockwise. A voltage function may be written as

$$v = V_m \varepsilon^{j\omega t}$$

and at $t = 0$, $v = V_m$, which is represented by a line V_m units long at $\omega t = 0$ or along the positive real axis. Later, at a time when $\omega t_1 = \theta$, v will have the value

$$V_m \varepsilon^{j\omega t_1} = V_m \varepsilon^{j\theta} = V_m \cos\theta + jV_m \sin\theta$$

and the line representing v will have moved to the position at an angle θ. At $\omega t = \pi/2$

$$\varepsilon^{j\pi/2} = 0 + j1$$

and the line for v will have moved to the j axis, through an angle of $\pi/2$ radians or 90°. At $\omega t = \pi$,

$$\varepsilon^{j\pi} = -1 + j0$$

and v is represented by V_m units along the negative real axis. The rotation of v with time can be similarly followed around the remainder of the circle. One rotation is completed in $2\pi/\omega$ s, or one cycle of the variable. The angle ωt grows uniformly with time and *the function $\varepsilon^{j\omega t}$ represents an operator of unit magnitude which rotates around the origin, uniformly with time.*

In circuit analysis we may encounter several sinusoidal currents (or voltages) of the same frequency but which differ in time relation or *phase*. The sum or the difference of these several waves would depend on that phase difference. Let us assume that

$$i_a = A\varepsilon^{j\omega t} = (A\varepsilon^{j0})\varepsilon^{j\omega t} \tag{4.6}$$

$$i_b = B\varepsilon^{j(\omega t + \theta)} = (B\varepsilon^{j\theta})\varepsilon^{j\omega t} \tag{4.7}$$

These expressions for two currents tell us that i_a can be represented on the complex plane by a line A units long being rotated by $\varepsilon^{j\omega t}$ and that i_b can be represented by a line B units long also rotated by $\varepsilon^{j\omega t}$. However, the angular position of i_b is always θ rad ahead of i_a; this is the *phase difference*. We eliminate time as a variable or stop the rotation by taking a photograph of the spinning lines. Figure 4.1(b) represents such a picture, taken at $t = 0$. The θ relation would be unchanged for any other time, and the figure allows study of the magnitude and phase relations between the several currents (or voltages) at an arbitrary time.

The quantities $B\varepsilon^{j\theta}$ and $A\varepsilon^{j0}$, *having magnitude and phase angle* on the complex plane, are called *phasors*. The representation of sinusoidal voltages and currents as phasors eliminates time as a variable in our circuit equations, yet retains the time differences or phase angles between the respective circuit variables. *Phasors will be indicated by boldface uppercase letter symbols.*

The angle θ appears as the angle by which phasor **B** precedes or *leads* phasor **A**; phasor **A** may also be said to *lag* phasor **B** by θ radians.

4.4

Impedance in the exponential steady state

Now consider the *RLC* series circuit of Fig. 4.2, having a driving function $v = V_m \cos(\omega t + \theta)$, where V_m is the peak or maximum voltage. Instead of employing this function, we may use $V_m \varepsilon^{j\theta} \varepsilon^{j\omega t}$ and at a later time take the real part if a function of time is needed.

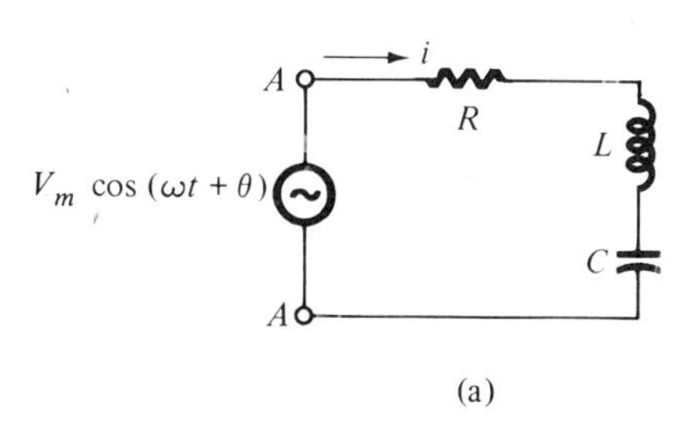

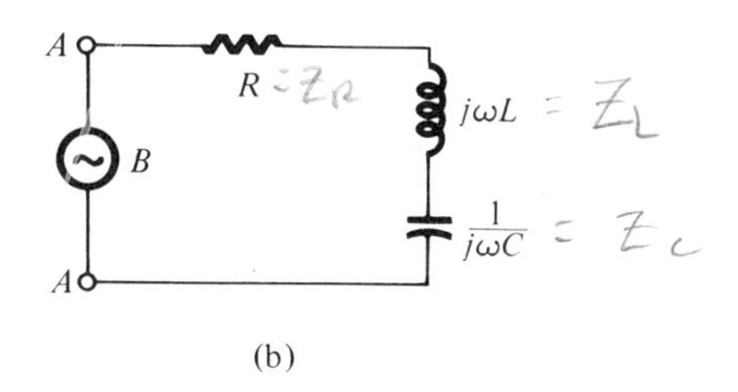

FIGURE 4.2
(a) ***RLC*** circuit; (b) sinusoidal steady-state circuit.

The circuit differential equation in the steady state is

$$Ri + L\frac{di}{dt} + \frac{1}{C}\int i\,dt = V_m \varepsilon^{j\theta} \varepsilon^{j\omega t} \tag{4.8}$$

The steady-state current will have the same time form as the driving function, so that

$$i = I_m \varepsilon^{j\phi} \varepsilon^{j\omega t}$$

By substitution and collection of terms we have

$$\left(R + j\omega L + \frac{1}{j\omega C}\right) I_m \varepsilon^{j\phi} \varepsilon^{j\omega t} = V_m \varepsilon^{j\theta} \varepsilon^{j\omega t}$$

We have defined the input impedance at A,A in the circuit as the ratio of the exponential voltage across the port to the exponential current into the port. Therefore the impedance at A,A is

$$\mathbf{Z}(\omega) = \frac{V_m \varepsilon^{j\theta} \varepsilon^{j\omega t}}{I_m \varepsilon^{j\phi} \varepsilon^{j\omega t}} = R + j\omega L + \frac{1}{j\omega C} \tag{4.10}$$

But $V_m\varepsilon^{j\theta} = \mathbf{B}$, a phasor, and $I_m\varepsilon^{j\phi} = \mathbf{D}$, another phasor; therefore impedance can be written as

$$\mathbf{Z}(\omega) = \frac{\mathbf{B}}{\mathbf{D}} \tag{4.11}$$

the ratio of phasors, *independent of the time function.*

Referring to the Laplace transform definitions, we note that $\varepsilon^{j\omega t}$ is a special case of ε^{st} with $\sigma = 0$; therefore $s = j\omega$. This is confirmed by comparison of the impedance terms in the above equations with those of Section 3.7:

$$\mathbf{Z}_R(\omega) = Z_R(s)]_{s=j\omega} = R \tag{4.12}$$

$$\mathbf{Z}_L(\omega) = Z_L(s)]_{s=j\omega} = j\omega L \tag{4.13}$$

$$\mathbf{Z}_C(\omega) = Z_C(s)]_{s=j\omega} = \frac{1}{j\omega C} \tag{4.14}$$

Similarly, admittance terms could be written for comparison:

$$\mathbf{Y}_R(\omega) = Y_R(s)]_{s=j\omega} = G \tag{4.15}$$

$$\mathbf{Y}_L(\omega) = Y_L(s)]_{s=j\omega} = \frac{1}{j\omega L} \tag{4.16}$$

$$\mathbf{Y}_C(\omega) = Y_C(s)]_{s=j\omega} = j\omega C \tag{4.17}$$

We can transform equations written for the s domain to circuit equations for the sinusoidal steady state by writing $j\omega$ for s wherever it occurs. The circuit, transformed to the sinusoidal steady state using phasors, appears in Fig. 4.2(b).

The steady-state impedance is written $\mathbf{Z}(\omega)$ above, but as we progress this will often be referred to simply as $\mathbf{Z}$. Measured in volts per ampere, the units of $\mathbf{Z}(\omega)$ are ohms. The angular frequency $\omega = 2\pi f$, and the frequency f is expressed in hertz, as cycles per second.

The development of methods of analysis for the ac system began in the last years of the nineteenth century, and the impedance concept was devised to parallel the simple methods then used in resistive networks with dc inputs. Because impedance appears with the application of time-varying inputs to L and C elements and fulfills the requirements of Ohm's law, it was once called *apparent resistance.*

4.5 The algebra of complex numbers

We have been employing $j = \sqrt{-1}$, and any real number multiplied by j is an *imaginary number.** Numbers consisting of real and imaginary parts are called *complex numbers.* The

*The mathematical symbol i is assigned to current in electrical nomenclature.

impedance $\mathbf{Z}(\omega)$ appears as a complex number in Eq. 4.10, and the proper interpretation and manipulation of equations involving impedance quantities requires that we study the algebra of complex numbers.

The complex plane is used to plot complex numbers, and it appears in Fig. 4.3 with an axis of real numbers and a j axis or axis of imaginaries. For

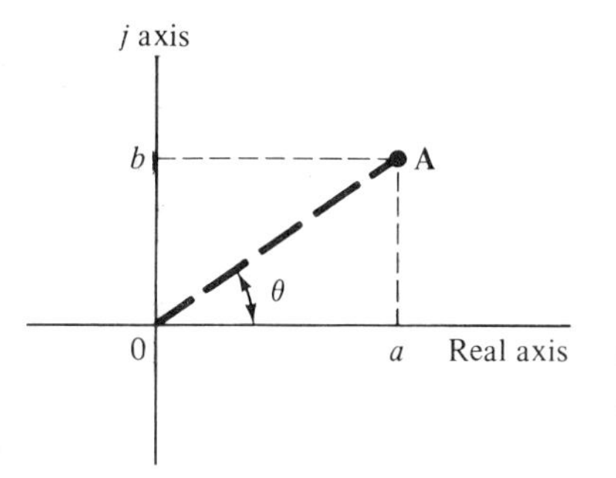

FIGURE 4.3

Complex number $\mathbf{A}\angle\theta$ on the complex plane.

our usage, the imaginary implication of j is unfortunate, since we shall show that j serves as an *operator* performing a counterclockwise rotation of 90° on the complex plane. At the moment we justify this statement by pointing out that $j \times j = \sqrt{-1} \times \sqrt{-1} = -1$. A double operation by j rotates by 180° on the complex plane, and a single operation should rotate by 90°. The symbol j therefore indicates a number measured along the j axis of the complex plane, and we shall refer to this as the *quadrature axis*.

In *rectangular form* a complex number is written as

$$\mathbf{A} = a + jb \tag{4.18}$$

where a and b are real numbers. On the complex plane, the complex number $\mathbf{A}$ is represented as having a distance a on the real axis and a distance of b units measured along the j or quadrature axis. Because they involve a distance and an angle on the complex plane, as does a phasor, *complex numbers are represented by boldface uppercase letters.*

The location of $\mathbf{A}$ on the complex plane may also be specified by a radial distance from the origin and an angle θ measured between the positive real axis and the radial line to $\mathbf{A}$; positive θ is taken as counterclockwise. Thus the *polar form* of the complex number in Eq. 4.18 is

$$\mathbf{A} = |A| \angle\theta \tag{4.19}$$

where

$$|\mathbf{A}| = \sqrt{a^2 + b^2} \tag{4.20}$$

The symbol $|\mathbf{A}|$ is known as the *magnitude* or the *absolute value* of the complex quantity $\mathbf{A}$. Also,

$$\theta = \tan^{-1} \frac{b}{a} \tag{4.21}$$

For θ, special attention should be given to determination of the correct quadrant.

Equations 4.19, 4.20, and 4.21 are sufficient to transform a complex number in rectangular form into its polar form. The process of transformation from the polar form to the rectangular form may be accomplished by noting from the geometry of Fig. 4.3 that

$$a = |\mathbf{A}| \cos \theta \tag{4.22}$$

$$b = |\mathbf{A}| \sin \theta \tag{4.23}$$

and so

$$\mathbf{A} = |\mathbf{A}| (\cos \theta + j \sin \theta) = a + jb \tag{4.24}$$

Thus we have returned to the rectangular form of expression.

Furthermore, by use of Euler's equation,

$$\mathbf{A} = |\mathbf{A}| (\cos \theta + j \sin \theta) = |\mathbf{A}|\varepsilon^{j\theta} \tag{4.25}$$

This is the *exponential form* for a complex number $\mathbf{A}$. The angle θ should properly be written in radians but is often written in degrees.

We shall occasionally write

$$Re\mathbf{A} = |\mathbf{A}| \cos \theta \tag{4.26}$$

as the real component measured along the real axis, and

$$Im\mathbf{A} = |\mathbf{A}| \sin \theta \tag{4.27}$$

as the component of $\mathbf{A}$ lying along the j or quadrature axis.

The rectangular form, the polar form, and the exponential form are equivalent forms of expression for a complex number. That is,

$$\mathbf{A} = a + jb = |\mathbf{A}| (\cos \theta + j \sin \theta) = |\mathbf{A}|\varepsilon^{j\theta} = |\mathbf{A}| \angle \theta$$

We use each form as appropriate to the problem to be solved, and several transformations may be required within a given problem.

Addition of complex numbers is demonstrated in Fig. 4.4 as the addition of the real components and of the j components separately. Let

$$\mathbf{A}_1 = a_1 + jb_1$$

$$\mathbf{A}_2 = a_2 + jb_2$$

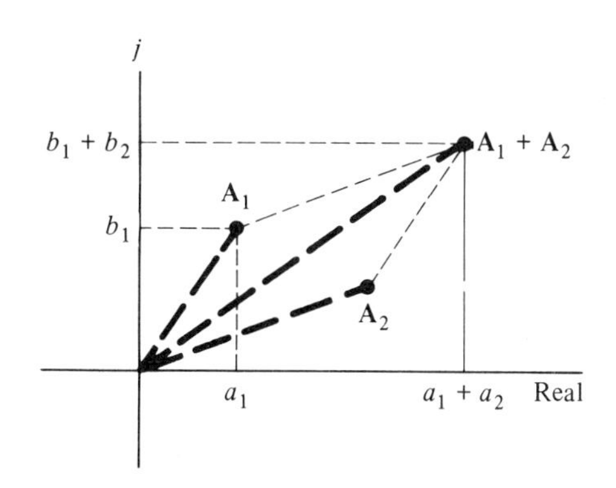

FIGURE 4.4
Sum of two complex numbers.

Then

$$\mathbf{B} = \mathbf{A}_1 + \mathbf{A}_2 = (a_1 + a_2) + j(b_1 + b_2) \tag{4.28}$$

Figure 4.4 shows the graphical addition of the components on each axis. *Subtraction* follows as the difference of the reals and the difference of the j components, taken separately. Thus

$$\mathbf{C} = \mathbf{A}_1 - \mathbf{A}_2 = (a_1 - a_2) + j(b_1 - b_2) \tag{4.29}$$

The processes of addition and subtraction must be performed with the rectangular forms of the complex numbers.

Multiplication may be carried out with either the rectangular form, the polar form, or the exponential form of expression. With the rectangular form we employ the usual method of algebraic multiplication as

$$\begin{aligned}\mathbf{D} = \mathbf{A}_1\mathbf{A}_2 &= (a_1 + jb_1)(a_2 + jb_2) = a_1a_2 + j(a_2b_1 + a_1b_2) + j^2b_1b_2 \\ &= (a_1a_2 - b_1b_2) + j(a_2b_1 + a_1b_2) = c + jd\end{aligned} \tag{4.30}$$

since $j^2 = -1$, as we have pointed out. Our result remains in rectangular form.

With the terms in exponential form we can multiply directly:

$$\mathbf{A}_1\mathbf{A}_2 = |\mathbf{A}_1|\varepsilon^{j\theta_1}|\mathbf{A}_2|\varepsilon^{j\theta_2} = |\mathbf{A}_1||\mathbf{A}_2|\varepsilon^{j(\theta_1 + \theta_2)}$$

The product will appear in a similar manner, with use of the polar form as

$$\mathbf{A}_1\mathbf{A}_2 = |\mathbf{A}_1|\underline{/\theta_1}|\mathbf{A}_2|\underline{/\theta_2} = |\mathbf{A}_1||\mathbf{A}_2|\underline{/\theta_1 + \theta_2} \tag{4.31}$$

The product of two complex numbers in polar form is given by the product of the magnitudes of the numbers, and the angle of the product is equal to the sum of the angles of the individual complex numbers.

The *division operation* is carried out with the exponential or polar forms of the complex numbers. In exponential form,

$$\frac{\mathbf{A}_1}{\mathbf{A}_2} = \frac{|\mathbf{A}_1|\varepsilon^{j\theta_1}}{|\mathbf{A}_2|\varepsilon^{j\theta_2}} = \frac{|\mathbf{A}_1|}{|\mathbf{A}_2|}\varepsilon^{j\theta_1}\varepsilon^{-j\theta_2} = \frac{|\mathbf{A}_1|}{|\mathbf{A}_2|}\varepsilon^{j(\theta_1-\theta_2)} \tag{4.32}$$

The result has a magnitude given by the quotient of the respective magnitudes and an angle equal to the numerator angle minus the denominator angle. The polar form follows similarly.

When the complex numbers are given in rectangular form, it is often easiest to transform each to polar form and proceed as above, especially if the result can be used in polar form.

The *conjugate* of a complex number is defined as a number with the same real part but an imaginary part which is opposite in sign. Designated by an asterisk, the conjugate of $\mathbf{A} = a_1 + jb_1$ is

$$\mathbf{A}^* = a_1 - jb_1$$

By use of the conjugate we can develop a method of division for complex numbers which retains the rectangular form. We multiply both numerator and denominator by the conjugate of the denominator, giving

$$\left(\frac{a_1 + jb_1}{a_2 + jb_2}\right)\left(\frac{a_2 - jb_2}{a_2 - jb_2}\right) = \frac{(a_1a_2 + b_1b_2) + j(a_2b_1 - a_1b_2)}{a_2^2 + b_2^2} \tag{4.33}$$

since

$$(a_2 + jb_2)(a_2 - jb_2) = a_2^2 + ja_2b_2 - ja_2b_2 + b_2^2 = a_2^2 + b_2^2$$

The process of multiplying the numerator and denominator by the conjugate of the denominator is called *rationalization.*

A problem solution is not in proper form when j terms remain in the denominator. Quadrature terms should appear only in the numerator as in that position we can readily give physical meaning to the angle of the complex quantity. Thus we either divide polar forms or rationalize in the rectangular form, to place a final answer in the form of $a + jb$.

Raising a complex number to a power follows as a corollary to the multiplication process:

$$\mathbf{A}^3 = (\mathbf{A}\varepsilon^{j\theta})^3 = |\mathbf{A}|^3\varepsilon^{j3\theta} \tag{4.34}$$

Extracting a root is done in similar manner:

$$\mathbf{A}^{1/3} = (\mathbf{A}\varepsilon^{j\theta})^{1/3} = |\mathbf{A}|^{1/3}\varepsilon^{j\theta/3} \tag{4.35}$$

In polar form,

$$\mathbf{A}^{1/3} = |\mathbf{A}|^{1/3}\underline{/\theta/3}$$

Finally, *equality* of two complex numbers occurs only if the real parts are equal and the imaginary parts are equal. That is,

$$a + jb + c + jd = 3 + j4$$

so that

$$a + c = 3$$
$$b + d = 4$$

4.6 Reactance and the impedance triangle

The input impedance of the circuit of Fig. 4.2(b) may be written from Eq. 4.10 as

$$\mathbf{Z}(\omega) = R + j\left(\omega L - \frac{1}{\omega C}\right) \tag{4.36}$$

The negative sign in the parentheses enters because in rationalization

$$\left(\frac{1}{j\omega C}\right)\left(\frac{-j}{-j}\right) = \frac{-j}{\omega C}$$

Impedance is a time-independent quantity, but its quadrature terms are present because of variation of current with time in L or C elements. We call the quadrature terms *reactances*, with the symbol X, where the *inductive reactance* is

$$X_L = \omega L \qquad \Omega \tag{4.37}$$

and the *capacitive reactance* is

$$X_C = -\frac{1}{\omega C} \qquad \Omega \tag{4.38}$$

Note that reactance possesses an algebraic sign, positive for inductive reactance and negative for capacitive reactance. The impedance of these elements is then

$$\mathbf{Z}_L = jX_L = j\omega L, \qquad \mathbf{Z}_C = -jX_C = \frac{-j}{\omega C}$$

and includes the operator j.

The impedance may be expressed in rectangular form as

$$\mathbf{Z} = R + jX \qquad \Omega \tag{4.39}$$

and when presented on the complex plane, the quantities R, X, and $\mathbf{Z}$ always complete a right triangle. This triangle is shown in Fig. 4.5, with

$$|\mathbf{Z}| = \sqrt{R^2 + X^2} \tag{4.40}$$

$$\theta = \tan^{-1}\frac{X}{R} \tag{4.41}$$

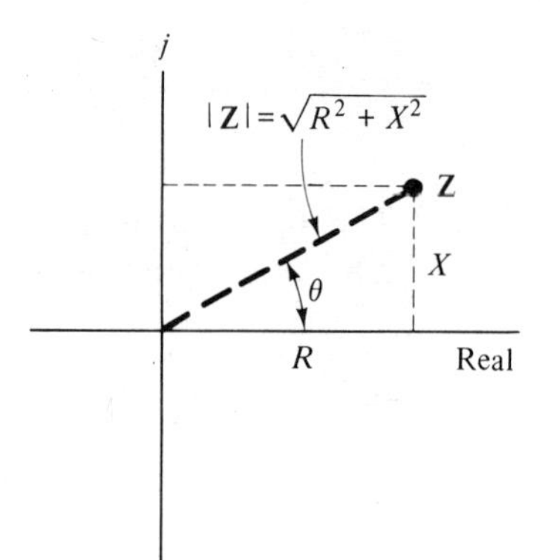

FIGURE 4.5
Impedance triangle.

from which

$$\mathbf{Z} = |\mathbf{Z}|\underline{/\theta} \tag{4.42}$$

The angle of the impedance and of the triangle will be positive for inductive reactance and negative for capacitive reactance. The impedance triangle is a useful geometric concept.

An alternative form of Ohm's law is obtained through use of the admittance $\mathbf{Y} = 1/\mathbf{Z}$:

$$\mathbf{I} = \mathbf{YV} \tag{4.43}$$

We can also express **Y** as a complex number,

$$\mathbf{Y} = G + jB \qquad \text{mhos} \tag{4.44}$$

where G and B are real numbers; G is the *conductance* and B is the *susceptance*, both measured in mhos. Then

$$R + jX = \frac{1}{G + jB} = \frac{G - jB}{G^2 + B^2} \tag{4.45}$$

It can be seen that

$$R = \frac{G}{G^2 + B^2} \qquad \text{and} \qquad X = \frac{-B}{G^2 + B^2} \tag{4.46}$$

The use of admittances simplifies the parallel element relation, written from Eq. 2.12 as

$$\mathbf{Z}_{\text{par}} = \frac{\mathbf{Z}_1\mathbf{Z}_2}{\mathbf{Z}_1 + \mathbf{Z}_2}$$

$$\mathbf{Y}_{\text{par}} = \frac{1}{\mathbf{Z}_{\text{par}}} = \frac{(1/\mathbf{Y}_1) + (1/\mathbf{Y}_2)}{1/\mathbf{Y}_1\mathbf{Y}_2} = \mathbf{Y}_1 + \mathbf{Y}_2 \qquad (4.47)$$

This usage is advantageous when the circuit under analysis has several parallel branches.

With Eqs. 4.46 and 4.47, the admittance of a resistance in parallel with an inductance is

$$\mathbf{Y} = G - \frac{j}{\omega L} \qquad \text{mhos} \qquad (4.48)$$

and that of a resistance in parallel with a capacitance is

$$\mathbf{Y} = G + j\omega C \qquad \text{mhos} \qquad (4.49)$$

The use of complex algebra in solving problems involving impedances is shown by examples.

EXAMPLE. The two impedances

$$\mathbf{Z}_A = 34.4 - j6.07\ \Omega$$

$$\mathbf{Z}_B = 35.0 + j60.6\ \Omega$$

are in parallel, and it is desired to find the resultant impedance value. We can utilize the parallel element relation of Eq. 2.12, writing it as

$$\mathbf{Z}_{\text{par}} = \frac{\mathbf{Z}_A\mathbf{Z}_B}{\mathbf{Z}_A + \mathbf{Z}_B} = \frac{(34.4 - j6.07)(35.0 + j60.6)}{34.4 - j6.07 + 35.0 + j60.6}$$

The process proceeds most easily if the numerator is placed in polar form:

$$\mathbf{Z}_{\text{par}} = \frac{35\underline{/-10^\circ} \times 70\underline{/60^\circ}}{69.4 + j54.5}$$

It is easy to place the denominator in polar form and

$$\mathbf{Z}_{\text{par}} = \frac{2450\underline{/50^\circ}}{\sqrt{69.4^2 + 54.5^2}\underline{/\tan^{-1}(54.5/69.4)}}$$

$$= \frac{2450\underline{/50^\circ}}{88.2\underline{/38^\circ}} = 27.8\underline{/12^\circ}\ \Omega$$

In rectangular form this is

$$\mathbf{Z}_{\text{par}} = 27.8(\cos 12° + j\sin 12°) = 27.2 + j5.8\ \Omega$$

This is the series equivalent to the original parallel circuit.

Such repeated transformation from rectangular form to polar form and back is frequently required in circuit solution.

EXAMPLE. The previous example may be solved by use of admittances. We find

$$\mathbf{Y}_A = \frac{1}{34.4 - j6.07} = \frac{1}{35\angle -10°} = 0.0286\angle 10° \text{ mhos}$$

$$\mathbf{Y}_B = \frac{1}{35.0 + j60.6} = \frac{1}{70\angle 60°} = 0.0143\angle -60° \text{ mhos}$$

The parallel admittance $\mathbf{Y}_{\text{par}}$ is

$$\mathbf{Y}_{\text{par}} = \mathbf{Y}_A + \mathbf{Y}_B = 0.0281 + j0.00495 + 0.0072 - j0.0124$$

after placing $\mathbf{Y}_A$ and $\mathbf{Y}_B$ in rectangular form. Then

$$\mathbf{Y}_{\text{par}} = 0.0353 - j0.00745 = 0.0361\angle -12° \text{ mhos}$$

from which

$$\mathbf{Z}_{\text{par}} = \frac{1}{\mathbf{Y}_{\text{par}}} = \frac{1}{0.0361\angle -12°} = 27.8\angle 12°\ \Omega$$

This checks the result obtained using impedance methods.

4.7 Periodic currents and voltages; average power

In the introduction to this chapter, we recognized several steady-state forms of current or voltage variation with time. Figure 4.6(a) illustrates a steady direct current by a solid line, and this might be the output current of a battery into a resistor. Much of our electronic equipment is supplied with such steady dc. The dashed line indicates a variation about the steady value I, and the result is dc with a superimposed varying component such as occurs in telephone and electronic circuits.

The wave at (b) in Fig. 4.6 is *periodic*, repeating after the period T. The frequency is $1/T$, expressed in hertz. This periodic wave has an *average value* obtained by finding the net area under the curve and dividing by the period.

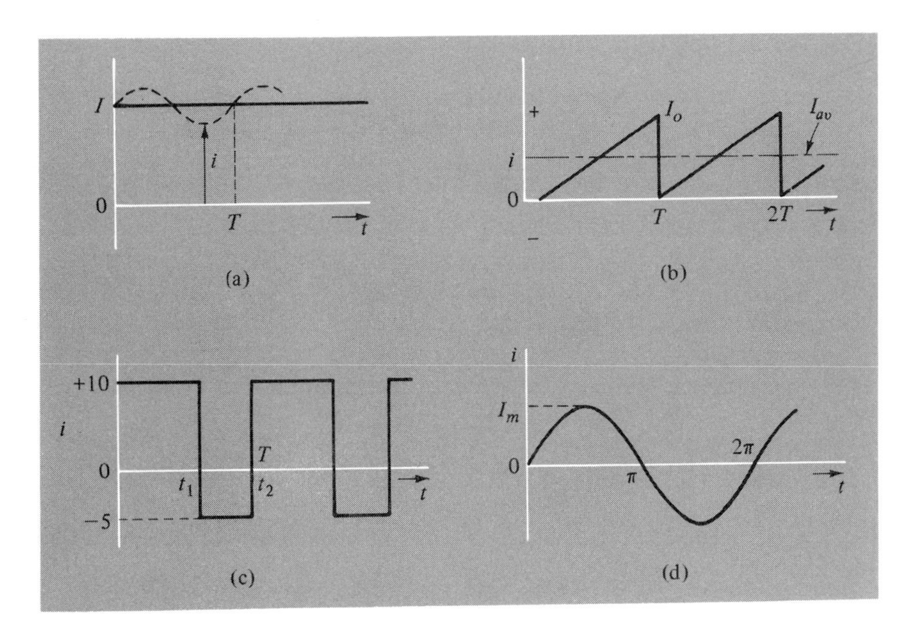

FIGURE 4.6
(a) Dc; (b) periodic triangular wave; (c) pulsed wave; (d) sinusoidal wave.

The area is given by the integral of the function over the period, and

$$I_{\text{av}} = \frac{1}{T}\int_0^T i\,dt \tag{4.50}$$

the current i being expressed as the appropriate function of time. An equivalent expression could be written for an average voltage value.

EXAMPLE. With the slope form of a straight-line equation for the ramp function of (b), we can write

$$i = I_o\frac{t}{T} \tag{4.51}$$

as the time variation of the current over the period from $t = 0$ to $t = T$. The average or dc value of the current is

$$I_{\text{av}} = \frac{1}{T}\int_0^T I_o\frac{t}{T}\,dt = \frac{I_o}{T}\left[\frac{t^2}{2T}\right]_0^T = \frac{I_o}{2} \tag{4.52}$$

The pulsed periodic wave form of (c) in Fig. 4.6 has positive and negative values, and the average value may or may not be zero.

EXAMPLE. We can use Eq. 4.50 over the several time segments of the figure in (c) to give

$$I_{\text{av}} = \frac{1}{T}\int_0^{t_1} 10\,dt + \frac{1}{T}\int_{t_1}^{t_2} (-5)\,dt$$

Taking $t_1 = 0.010$ s and $t_2 = 0.015$ s, then $T = 0.015$ s, and we can write

$$I_{av} = \frac{1}{0.015} 10t \Big]_0^{0.010} - \frac{1}{0.015} 5t \Big]_{0.010}^{0.015}$$

$$= 6.67 - 1.67 = 5.0 \text{ A}$$

In Chapter 1 instantaneous power was shown to be

$$p = vi \qquad \text{W} \tag{4.53}$$

and when power is supplied to a resistor the instantaneous power is

$$p = Ri^2 \qquad \text{and} \qquad p = \frac{v^2}{R} \tag{4.54}$$

Because of thermal storage, it is the *average value* of power which determines the temperature rise of our electrical devices. Power is not linear with voltage or current, and average power cannot be found by the use of average values of varying voltage or current waves. In fact the average value of the sine variation of (d) in Fig. 4.6 is zero, yet a sine current develops heat in a resistor.

However, it is possible to assign an effective value to a voltage or current wave which can satisfy the power equations; we are also able to design instruments which will measure this effective value directly. The average power dissipated in a resistor R is

$$P_{av} = \frac{1}{T}\int_0^T Ri^2\, dt = R\left[\frac{1}{T}\int_0^T i^2\, dt\right] \tag{4.55}$$

This represents R times the average square of the varying current. An *effective current* I_e is the equivalent steady value in R which will produce the same average power as produced by the varying current i and given by Eq. 4.55. That is,

$$P_{av} = RI_e^2 = R\left[\frac{1}{T}\int_0^T i^2\, dt\right]$$

and therefore the effective value of the varying current i is given by

$$I_e = I_{rms} = \sqrt{\frac{1}{T}\int_0^T i^2\, dt} \tag{4.56}$$

The subscript rms describes the mathematical operations indicated by Eq. 4.56 as the square root of the average of the squares of the current ordinates. We speak of *effective* current or voltage, or *root-mean-square* current or voltage, as characteristic values of any varying current or voltage wave.

The sinusoid of (d) in Fig. 4.6 is our most common ac wave form. Calculation of power with the sinusoidal wave form will be considered in the next section.

EXAMPLE. Let us determine the effective value of the current wave in Fig. 4.6(c). We can write the average of the squares by squaring the variables and introducing the root:

$$I_{rms} = \sqrt{\frac{1}{T}\int_0^{t_1} 10^2\, dt + \frac{1}{T}\int_{t_1}^{t_2} (-5)^2\, dt}$$

$$= \sqrt{\frac{1}{0.015}100t\Big]_0^{0.01} + \frac{1}{0.015}25t\Big]_{0.01}^{0.015}}$$

$$= \sqrt{66.67 + 8.33} = \sqrt{75} = 8.66 \text{ A}$$

The effective value of a wave is always equal to or greater than the average value.

With the wave of Fig. 4.6(c) the average power input to R is $I_{rms}^2 R$. To prove this, write

$$P_{av} = \frac{1}{T}\int_0^T vi\, dt = \frac{1}{T}\int_0^{t_1} (10R)(10)\, dt + \frac{1}{T}\int_{t_1}^{t_2} (-5R)(-5)\, dt$$

$$= R\left[\frac{1}{T}\int_0^{t_1} 100\, dt + \frac{1}{T}\int_{t_1}^{t_2} 25\, dt\right]$$

By comparison, this is seen to be $I_{rms}^2 R$.

4.8 Power with sinusoidal currents and voltages

The effective value of a sinusoidal current described by $i = I_m \sin \omega t$ can be found by use of Eq. 4.56 as

$$I_{rms} = \sqrt{\frac{1}{2\pi}\int_0^{2\pi} I_m^2 \sin^2 \omega t\, d\omega t}$$

after a change of variable from t to ωt and making $T = 2\pi$. Then

$$I_{rms} = \sqrt{\frac{I_m^2}{2\pi}\left[\frac{\omega t}{2} - \frac{\sin 2\omega t}{4}\right]_0^{2\pi}} = \sqrt{\frac{I_m^2}{2}}$$

$$= \frac{I_m}{\sqrt{2}} \cong 0.707 I_m \tag{4.57}$$

For a *sinusoidal function* the rms or effective value is $1/\sqrt{2}$ times the maximum value of the wave. *This relation applies only for the sinusoidal form*; other wave forms must be evaluated by use of Eq. 4.56.

The ratio of the voltage phasor to the current phasor is the impedance:

$$\mathbf{Z} = \frac{\mathbf{V}_m}{\mathbf{I}_m} = \frac{\mathbf{V}_m/\sqrt{2}}{\mathbf{I}_m/\sqrt{2}} = \frac{\mathbf{V}_{\text{rms}}}{\mathbf{I}_{\text{rms}}} = \frac{\mathbf{V}}{\mathbf{I}}$$

Scaling the voltage and current by the same factor, $\sqrt{2}$, does not alter this result, and we shall scale phasor voltages and currents to the rms values of the corresponding sinusoidal waves, indicated as $\mathbf{V}$ and $\mathbf{I}$ above.

Since we have instruments that directly measure voltage and current in rms (Section 4.15), we employ rms values to specify our ac currents and voltages. When we say that the laboratory supply voltage is 117 Vac in the United States, we mean the rms value. If any other function of the wave is meant, it will be specifically identified.

With $\mathbf{Z} = |\mathbf{Z}|\underline{/\theta}$ in a one-port ac circuit, we write

$$\mathbf{I} = \frac{\mathbf{V}}{|\mathbf{Z}|\underline{/\theta}} = \frac{\mathbf{V}}{|\mathbf{Z}|}\underline{/-\theta}$$

In general, the ac current may lead the voltage by an angle $+\theta$ or lag by an angle $-\theta$. Figure 4.7 illustrates the current and voltage relations for several values of θ, and shows the vi product for instantaneous power. The power is positive and into the one-port or negative and back to the source. Since the resistance cannot return energy, the return energy must be derived from the collapsing electric and magnetic fields in the C and L elements, as the current or voltage fall from a maximum to zero.

We can obtain an average of the instantaneous power,

$$P_{\text{av}} = \frac{1}{T}\int_0^T vi\,dt = \frac{1}{2\pi}\int_0^{2\pi} |V_m|\cos\omega t|I_m|\cos(\omega t - \theta)\,d\omega t \tag{4.58}$$

for cosine voltage and current and θ as the lag angle. Expanding $\cos(\omega t - \theta)$, the above becomes

$$P_{\text{av}} = \frac{1}{2\pi}\int_0^{2\pi} |V_m||I_m|\cos\omega t(\cos\omega t\cos\theta + \sin\omega t\sin\theta)\,d\omega t$$

Now we have the identities

$$\cos^2\omega t = \frac{1 + \cos 2\omega t}{2}$$

$$\cos\omega t\sin\omega t = \frac{\sin 2\omega t}{2}$$

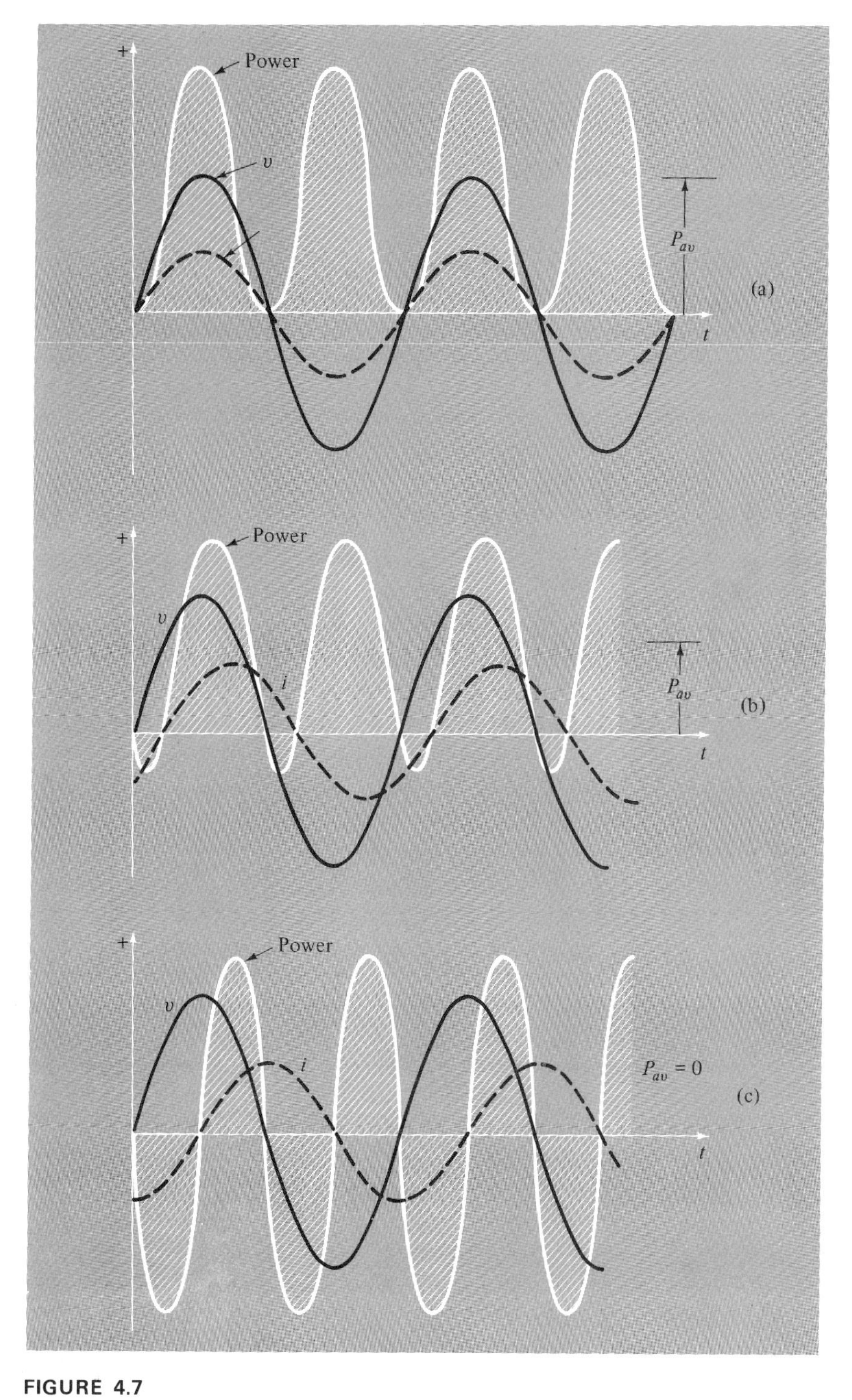

FIGURE 4.7

Voltage, current, and power: (a) $\theta = 0°$; (b) $-90° < \theta < 0°$; (c) $\theta = -90°$ (voltage reference).

and after collecting terms,

$$P_{av} = \frac{1}{2\pi}\int_0^{2\pi} \frac{|V_m||I_m|}{2}[\cos\theta + (\cos 2\omega t \cos\theta + \sin 2\omega t \sin\theta)]\, d\omega t \tag{4.59}$$

$$= \frac{1}{2\pi}\int_0^{2\pi} \left[\frac{|V_m||I_m|}{2}\cos\theta + \frac{|V_m||I_m|}{2}\cos(2\omega t - \theta)\right] d\omega t \tag{4.60}$$

The second term in the brackets represents instantaneous power at the second harmonic frequency, oscillating back and forth between the source and the L or C elements. However, the integral of $\cos(2\omega t - \theta)$ over 2π contributes zero to the average power. Therefore, integration of the first term leads to

$$\begin{aligned} P_{av} &= \frac{|V_m||I_m|}{2}\cos\theta = \frac{|V_m|}{\sqrt{2}}\frac{|I_m|}{\sqrt{2}}\cos\theta \\ &= |V_{rms}||I_{rms}|\cos\theta \end{aligned} \tag{4.61}$$

Average power in an ac sinusoidal circuit is written here in a form suited to the rms measurement of alternating currents and voltages. More importantly, it is written in a form suited to our phasor concepts, involving the magnitudes $|\mathbf{V}|$ and $|\mathbf{I}|$ and the cosine of the angle between the phasors.

The term cosine θ is called the *power factor* of the circuit. For a resistance, $\mathbf{V} = R\mathbf{I}$ and $\theta = 0°$. The power factor is unity; this is the condition of (a) in Fig. 4.7. For an ideal inductance $\mathbf{Z} = j\omega L$ and with an rms current $\mathbf{I}$ through the inductance,

$$\mathbf{V}_L = j\omega L\mathbf{I} = \omega L\mathbf{I}\,\underline{/90°}$$

As shown in (c) in Fig. 4.7, the inductor voltage leads the inductor current by 90°, the power factor $\cos\theta$ is zero, and an ideal inductance takes zero average power. The figure shows that energy oscillates into and out of the inductance once each half cycle or at double frequency, in accordance with the second term of Eq. 4.60. Energy oscillates in a similar fashion in a capacitance.

In (b) in Fig. 4.7, we have an RL circuit in which θ lies between 0° and 90°; the positive instantaneous energy into the one-port exceeds the negative energy returned to the source. The average power input is dissipated in the circuit resistance.

We have demonstrated that use of rms values permits average power to be determined as I^2R. At the input of a one-port, we have $P_{av} = |\mathbf{V}||\mathbf{I}|\cos\theta$, as power delivered to the one-port circuit.

4.9 Circuit solutions by use of phasors

We are ready to demonstrate the solution of simple circuits by the phasor methods just developed. Consider the circuit of Fig. 4.8(a) in which the source is $v = 28 \sin 10^3 t$. Since $V_m =$

28 V, the rms voltage is 20 V, with its reference phase angle made zero arbitrarily. The phasor **V** appears on the real axis in Fig. 4.8(d).

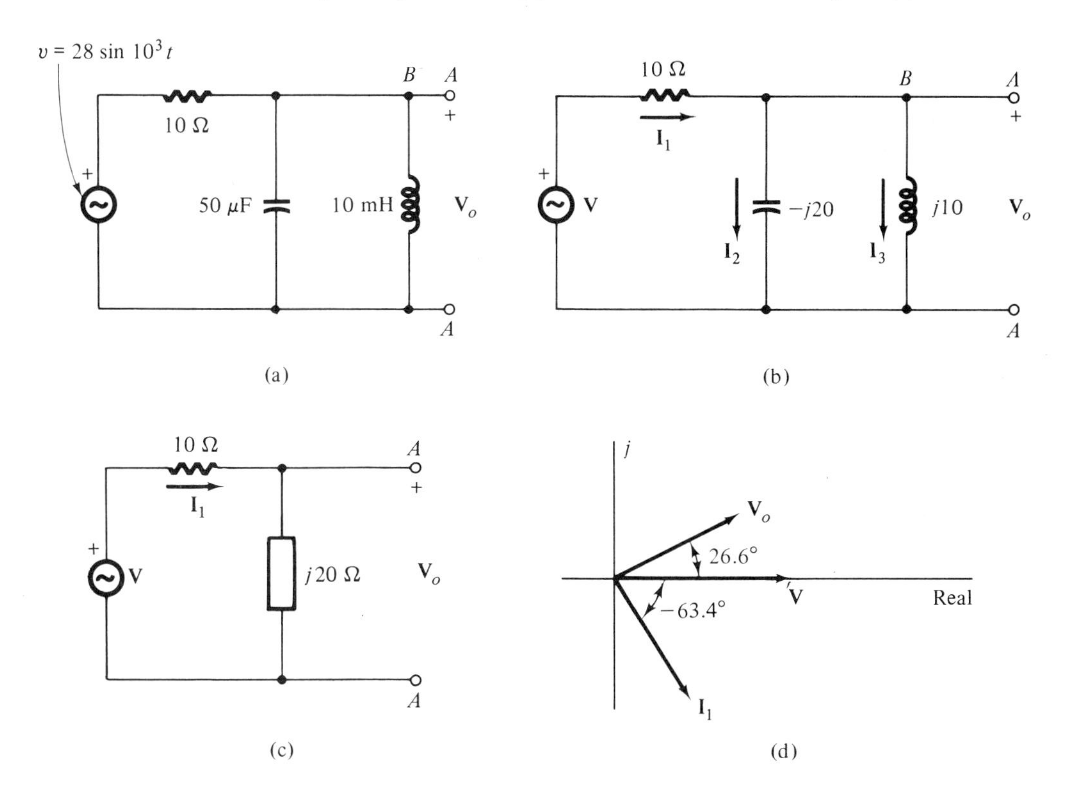

FIGURE 4.8
Examples.

With $\omega = 2\pi f = 10^3$,

$$\mathbf{Z}_L = j\omega L = j10^3 \times 10 \times 10^{-3} = j10\,\Omega$$

$$\mathbf{Z}_C = \frac{-j}{\omega C} = \frac{-j}{10^3 \times 50 \times 10^{-6}} = -j20\,\Omega$$

The circuit is redrawn in Fig. 4.8(b) using phasor voltages and currents and complex impedances in ohms.

Combining $\mathbf{Z}_L$ and $\mathbf{Z}_C$ in parallel yields

$$\mathbf{Z}_p = \frac{\mathbf{Z}_L\mathbf{Z}_C}{\mathbf{Z}_L + \mathbf{Z}_C} = \frac{j10(-j20)}{j10 - j20} = \frac{200}{-j10} = j20\,\Omega \tag{4.62}$$

and the equivalent circuit becomes that at (c) in Fig. 4.8. From Ohm's law,

$$\mathbf{I}_1 = \frac{\mathbf{V}}{\mathbf{Z}} = \frac{20\angle 0^\circ}{10 + j20} = \frac{20\angle 0^\circ}{22.3\angle 63.4^\circ} = 0.898\angle -63.4^\circ \tag{4.63}$$

As a phasor, current $\mathbf{I}_1$ is shown in Fig. 4.8(d).

The voltage $\mathbf{V}_o$ at the A,A terminals is

$$\begin{aligned}\mathbf{V}_o = \mathbf{I}_1\mathbf{Z}_p &= 0.898\underline{/-63.4^\circ} \times 20\underline{/90^\circ} \\ &= 17.96\underline{/26.6^\circ}\end{aligned} \tag{4.64}$$

That is, $\mathbf{V}_o$ leads V by 26.6°. The voltage $\mathbf{V}_o$ could also be found by use of the voltage-divider relation of Eq. 2.14 as

$$\begin{aligned}\mathbf{V}_o &= \frac{\mathbf{Z}_p}{R + \mathbf{Z}_p}\mathbf{V} = \frac{j20}{10 + j20} \times 20 \\ &= \frac{400\underline{/90^\circ}}{22.3\underline{/63.4^\circ}} = 17.96\underline{/26.6^\circ}\text{ V}\end{aligned} \tag{4.65}$$

Knowing the voltage V_o, the currents I_2 and I_3 can be found:

$$\mathbf{I}_2 = \frac{\mathbf{V}_o}{\mathbf{Z}_C} = \frac{17.96\underline{/26.6^\circ}}{20\underline{/-90^\circ}} = 0.898\underline{/116.6^\circ}\text{ A}$$

$$\mathbf{I}_3 = \frac{\mathbf{V}_o}{\mathbf{Z}_L} = \frac{17.96\underline{/26.6^\circ}}{10\underline{/+90^\circ}} = 1.796\underline{/-63.4^\circ}\text{ A}$$

We could also have used the current-divider relation of Eq. 2.13, leading to

$$\mathbf{I}_2 = \frac{\mathbf{Z}_L}{\mathbf{Z}_L + \mathbf{Z}_C}\mathbf{I}_1 = \frac{j10}{j10 - j20} \times 0.898\underline{/-63.4^\circ} = 0.898\underline{/116.6^\circ}\text{ A}$$

$$\mathbf{I}_3 = \frac{\mathbf{Z}_C}{\mathbf{Z}_L + \mathbf{Z}_C}\mathbf{I}_1 = \frac{-j20}{j10 - j20} \times 0.898\underline{/-63.4^\circ} = 1.796\underline{/-63.4^\circ}\text{ A}$$

We have a curious situation: The current $\mathbf{I}_1$ divides into two parts, each equal to or greater than $\mathbf{I}_1$ in magnitude. Checking by use of Kirchhoff's current law that $\mathbf{I}_1 - \mathbf{I}_2 - \mathbf{I}_3 = 0$ at B,

$$0.898\underline{/-63.4^\circ} - 0.898\underline{/116.6^\circ} - 1.796\underline{/-63.4^\circ} = 0$$

but a negative current at 116.6° equals a positive current at 116.6° + 180° = 296.6° = −63.4° or

$$0.898\underline{/-63.4^\circ} + 0.898\underline{/-63.4^\circ} - 1.796\underline{/-63.4^\circ} = 0 \tag{4.66}$$

and our work is proved.

The power output from the generator is

$$P_{av} = |V_1||I_1| \cos\theta = 20 \times 0.898 \cos(-63.4^\circ)$$
$$= 17.96 \times 0.4478 = 8.05 \text{ W} \tag{4.67}$$

Power is not dissipated in a perfect inductance or a perfect capacitance, and so this value of power must appear in the resistance element. The resistive power can be independently found as

$$P_{av} = RI^2 = 10 \times 0.898^2 = 8.05 \text{ W} \tag{4.68}$$

and we have a check on the result.

The generator voltage has been used as the phasor reference angle in this problem; a choice of reference must always be made and should be

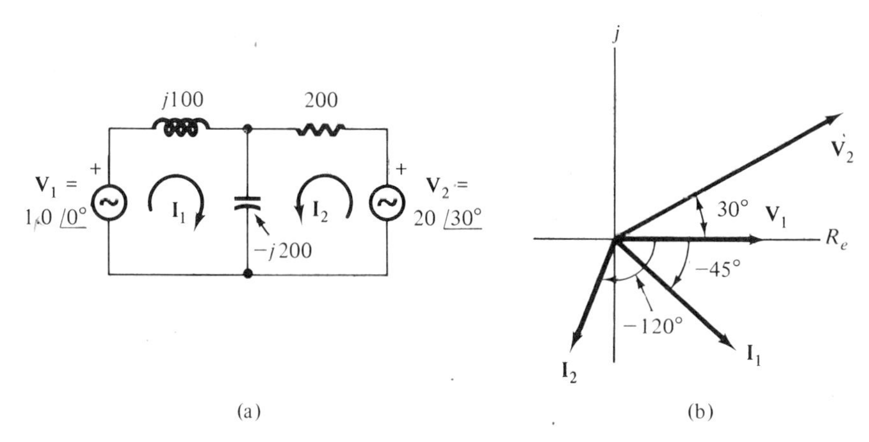

FIGURE 4.9
Examples.

stated. The point can be illuminated by writing the loop voltage equations for the circuit of Fig. 4.9, with $\mathbf{V}_1 = 10\underline{/0^\circ}$ as the reference, and

$$10\underline{/0^\circ} - j100\mathbf{I}_1 - (-j200)(\mathbf{I}_1 + \mathbf{I}_2) = 0$$

$$10\underline{/0^\circ} = -j100\mathbf{I}_1 - j200\mathbf{I}_2$$

Now

$$\mathbf{V}_2 = 20\underline{/30^\circ} = 17.3 + j10$$

and around the second loop we have

$$17.3 + j10 + j200\mathbf{I}_1 - (200 - j200)\mathbf{I}_2 = 0$$

From the first equation we obtain

$$\mathbf{I}_1 = -2\mathbf{I}_2 + j0.1$$

and substitution for $\mathbf{I}_1$ in the second equation gives

$$\mathbf{I}_2 = \frac{-2.7 + j10}{200(1 + j1)} = \frac{10.4\underline{/105^\circ}}{283\underline{/45^\circ}} = 0.0366\underline{/60^\circ}\ \text{A} \tag{4.69}$$

Using this value in the expression for $\mathbf{I}_1$, we obtain

$$\begin{aligned}\mathbf{I}_1 &= -0.0732\underline{/60^\circ} + j0.1 \\ &= -0.0366 - j0.0634 + j0.1 \\ &= -0.0366 + j0.0366 = 0.0517\underline{/135^\circ}\ \text{A}\end{aligned} \tag{4.70}$$

These phasors are drawn in Fig. 4.9(b), and the angles refer to the angle of $\mathbf{V}_1$, assumed to be zero. The meaning of the 135° angle of $\mathbf{I}_1$ should be explained by the student, especially as it affects the power output of $\mathbf{V}_1$.

EXAMPLE. We wish to determine the voltage $\mathbf{V}_o$ and the current $\mathbf{I}$ and the power from the source in the circuit of Fig. 4.10(a); two general methods are possible.

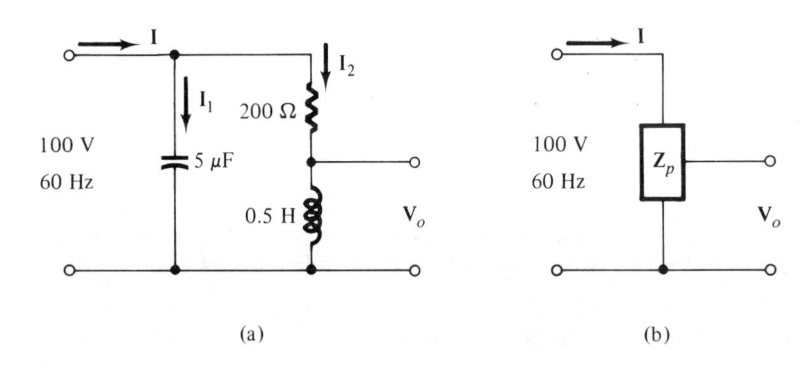

FIGURE 4.10
Circuit transformation.

Method I. We can find the current **I** from the source by consolidating the two branches into a single impedance. The power is then given by $|V||I|\cos\theta$. The second circuit of R and L is across **V**, and $\mathbf{V}_0$ can be found by use of the voltage division factor. That is,

$$\omega = 2\pi \times 60 = 377\ \text{r/s}$$

$$\mathbf{Z}_C = \frac{1}{j\omega C} = \frac{-j}{377 \times 5 \times 10^{-6}} = -j531\,\Omega$$

$$\mathbf{Z}_L = j\omega L = j \times 377 \times 0.5 = j188\,\Omega$$

$$\mathbf{Z}_2 = R + j\omega L = 200 + j188 = 274\underline{/43^\circ}\,\Omega$$

$$\mathbf{Z}_p = \frac{\mathbf{Z}_C\mathbf{Z}_2}{\mathbf{Z}_C + \mathbf{Z}_2} = \frac{531\underline{/-90^\circ} \times 274\underline{/43^\circ}}{-j531 + 200 + j188} = \frac{1.46 \times 10^5\underline{/-47^\circ}}{397\underline{/-60^\circ}}$$

$$= 368\underline{/13^\circ}\,\Omega$$

Then

$$\mathbf{I} = \frac{\mathbf{V}}{\mathbf{Z}_p} = \frac{100\underline{/0^\circ}}{368\underline{/13^\circ}} = 0.272\underline{/-13^\circ}\text{ A}$$

This is the generator output current into the two impedances in parallel. The generator power output is

$$\begin{aligned} P_{av} &= |\mathbf{V}||\mathbf{I}|\cos\theta = 100 \times 0.272\cos(-13^\circ) \\ &= 27.2 \times 0.974 = 26.5\text{ W} \end{aligned} \tag{4.71}$$

The voltage $\mathbf{V}_o$ can then be found by use of the voltage division factor in branch 2. That is,

$$\mathbf{V}_o = \frac{jX_L}{R + jX_L}\mathbf{V} = \frac{j188}{200 + j188} \times 100 = \frac{18{,}800\underline{/90^\circ}}{274\underline{/43^\circ}}$$

$$= 68.6\underline{/47^\circ}\text{ V}$$

Method II. The currents $\mathbf{I}_1$ and $\mathbf{I}_2$ in each branch can be directly found from the branch impedances, and the sum of these currents is $\mathbf{I}$. Power is then $|\mathbf{V}||\mathbf{I}|\cos\theta$ or I_2^2R. The voltage $\mathbf{V}_o$ is simply given by $\mathbf{Z}_L\mathbf{I}_2$.

Using the previously calculated impedances,

$$\mathbf{I}_1 = \frac{\mathbf{V}}{-jX_C} = \frac{100\underline{/0^\circ}}{531\underline{/-90^\circ}} = 0.188\underline{/90^\circ}\text{ A} = j0.188\text{ A}$$

$$\mathbf{I}_2 = \frac{\mathbf{V}}{R + jX_L} = \frac{100\underline{/0^\circ}}{274\underline{/43^\circ}} = 0.365\underline{/-43^\circ} = 0.266 - j0.248\text{ A}$$

$$\begin{aligned} \mathbf{I} = \mathbf{I}_1 + \mathbf{I}_2 &= j0.188 + 0.266 - j0.248 = 0.266 - j0.060 \\ &= 0.272\underline{/-13^\circ}\text{ A} \end{aligned}$$

The power from the generator is then

$$P_{av} = |\mathbf{V}||\mathbf{I}|\cos\theta = 100 \times 0.272 \times 0.974 = 26.5\text{ W} \tag{4.72}$$

The circuit power may also be found from $I_2^2 R$, since we have the only resistance in branch 2. That is,

$$P_{av} = I_2^2 R = 0.365^2 \times 200 = 26.7 \text{ W}$$

which checks the source output to the circuit.

Voltage $\mathbf{V}_o$ follows easily as

$$\begin{aligned} \mathbf{V}_o = \mathbf{I}_2 \mathbf{Z}_L &= 0.365\underline{/-43^\circ} \times 188\underline{/90^\circ} \\ &= 68.6\underline{/47^\circ} \text{ V} \end{aligned} \tag{4.73}$$

as by method I. It would appear that method II involves less calculation, since the calculation of a parallel impedance is not involved. Method II will usually be preferable when parallel impedances are included. The slight differences in magnitudes between method I and method II are a result of the frequent rounding of numbers.

4.10 Scaling the circuit variables

We may convert the magnitudes of our variables to more suitable numerical ranges. This is important in preparing problems for simulation on the analog computer to bring frequencies and voltage and current levels into the ranges of the computer instrumentation. After a solution is obtained, it is translated back to the original problem levels. The process is known as *scaling*.

The voltage or current may be converted to new ranges of v' and i' by use of the real constants k_1 and k_2, giving

$$v' = k_1 v \tag{4.74}$$

$$i' = k_2 i \tag{4.75}$$

Impedances are then scaled as

$$\begin{aligned} Z' &= \frac{k_1}{k_2} Z = pZ \\ R' &= pR \end{aligned} \tag{4.76}$$

We may scale time to a new variable t' as

$$t' = \frac{t}{q}$$

from which

$$\omega' = q\omega \tag{4.77}$$

Since

$$\omega' L' = p\omega L$$

it follows that

$$L' = \frac{p}{q} L \tag{4.78}$$

Similarly,

$$\frac{1}{\omega' C'} = \frac{p}{\omega C}$$
$$C' = \frac{1}{pq} C \tag{4.79}$$

EXAMPLE. A circuit operating at 159,000 Hz ($\omega = 10^6$ r/s) has series elements of $R = 5000\ \Omega$, $L = 1$ mH, and $C = 1000$ pF. Scale to a frequency of $\omega = 1$ r/s and an inductance of 1 H, with applied voltage of 10 V.

Scaling ω to $\omega' = 1$, we have

$$q = 10^{-6}$$

With L' as 1 H,

$$1 = \frac{p}{q} L = \frac{p}{10^{-6}} \times 10^{-3}$$

and

$$p = 10^{-3}$$

Therefore R will scale to

$$R' = pR = 10^{-3} \times 5000$$
$$= 5\ \Omega$$

and C becomes

$$C' = \frac{1}{pq} C = \frac{1}{10^{-3} \times 10^{-6}} \times 1000 \times 10^{-12}$$
$$= 1\ \text{F}$$

Either current or voltage may be arbitrarily scaled, since

$$\frac{V'}{I'} = p\frac{V}{I}$$

following the scaling of impedance. However, it seems possible to accept 10 V as a normal value, and therefore the current is scaled:

$$I' = \frac{1}{p} I$$

Therefore

$$I' = 10^3 I$$

That is, 1 mA in the actual circuit becomes 1 A in the scaled model, which now has

$$R = 5\ \Omega$$

$$L = 1\ \text{H}$$

$$C = 1\ \text{F}$$

$$\omega = 1\ \text{r/s}$$

EXAMPLE. A series circuit is drawn with $R = 1\ \Omega$, $C = 0.5$ F, $L = 2$ H, and $\omega = 1$ r/s. What should the circuit parameters be if ω is scaled to 10,000 r/s and the resistance is scaled to 500 Ω?

From Eq. 4.77 we have

$$\frac{\omega'}{\omega} = \frac{1}{10^4} = q = 10^{-4}$$

From Eq. 4.76,

$$\frac{R'}{R} = \frac{1}{500} = p$$

$$p = 2 \times 10^{-3}$$

Then from Eq. 4.78,

$$L' = 2 = \frac{p}{q} L = \frac{2 \times 10^{-3}}{10^{-4}} L$$

$$L = \frac{2}{20} = 0.1\ \text{H}$$

and from Eq. 4.79,

$$C' = 0.5 = \frac{1}{pq} C = \frac{1}{2 \times 10^{-3} \times 10^{-4}} C$$

$$C = 0.5 \times 2 \times 10^{-7} = 1 \times 10^{-7}\ \text{F}$$

$$= 0.1\ \mu\text{F}$$

For operation at $\omega = 10{,}000$ r/s, we need

$$R = 500\ \Omega$$

$$L = 0.10\ \text{H}$$

$$C = 0.1\ \mu\text{F}$$

4.11 Network definitions and topology

Kirchhoff's circuit laws, when used with phasor representation of alternating voltages and currents, provide basic methods for ac network analysis. To systematize some of the procedures, we start with a general network, drawn in Fig. 4.11, and certain definitions.

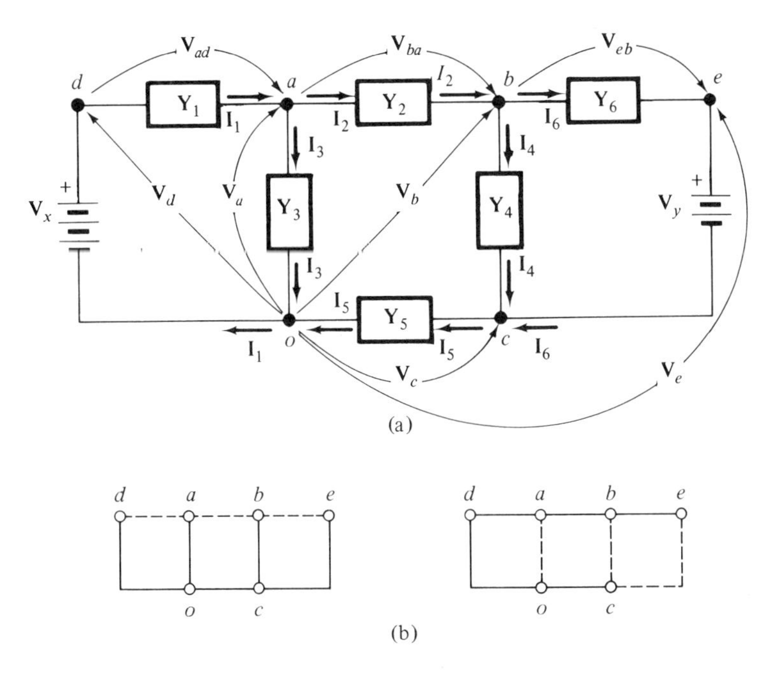

FIGURE 4.11
(a) General network; (b) two possible trees.

BRANCH Every network element or source is called a branch; thus $\mathbf{V}_x$, $\mathbf{V}_y$ and $\mathbf{Y}_1 \cdots \mathbf{Y}_6$ of Fig. 4.11 are branches.

NODE A junction where two or more branches connect, as *a*, *b*, *c*, *d*, *e*, and *o* of Fig. 4.11.

TREE Any combination of branches which connects to all nodes but forms no closed paths is a tree; two trees are illustrated in Fig. 4.11(b). Other trees could also have been formed.

CHORD Applies to any branch not contained in the chosen tree. The chords for the indicated trees are shown dashed in Fig. 4.11(b).

The circuit laws are used to write systems of equations which can be solved if the number of equations equals or exceeds the number of unknowns. The manner of connection of the circuit elements is of primary importance in determining the needed equations, and the study of network geometry is called *network topology*. Only a brief introduction is offered here.

It is possible to designate six currents in Fig. 4.11, labeled $\mathbf{I}_1 \cdots \mathbf{I}_6$. By use of Kirchhoff's current law,

$$-\mathbf{I}_1 + \mathbf{I}_2 + \mathbf{I}_3 = 0$$

$$-\mathbf{I}_2 + \mathbf{I}_4 + \mathbf{I}_6 = 0$$

$$-\mathbf{I}_4 - \mathbf{I}_6 + \mathbf{I}_5 = 0$$

Manipulation of these equations yields

$$+\mathbf{I}_1 - \mathbf{I}_3 - \mathbf{I}_5 = 0$$

and this is a summation at the fourth node. If three of the currents are known, the other three can be determined by linear combination of the three known currents. Therefore, there are three independent variables, but we are free to choose which three. Similarly, there are two independent voltages, the others forming subsets of the set of all voltages.

Looking at the trees in Fig. 4.11(b), we see that the first branch used to draw a tree has two nodes and that each added branch adds one more node, so the number of branches used to draw a tree is $N - 1$, where N is the number of nodes. In Fig. 4.11, N is 6.

Let B equal the total number of branches in the network; each tree must have the same number of branches. Since there are $N - 1$ branches in the tree, the number of chords not included in the tree is $B - (N - 1) = B - N + 1$. As we add each chord to the tree, we form a closed path, or *loop*. Since we have $B - N + 1$ chords, we can form $B - N + 1$ *fundamental loops*. The Kirchhoff voltage equations written around these fundamental loops must be independent, since each equation will contain a branch voltage across the chord that is not contained in any other equation. Therefore, there are $B - N + 1$ independent Kirchhoff voltage equations.

It may be possible to have a tree in which two chords are used to close a loop. However, a voltage equation covering two chords is a combination of the equations written separately with each chord, and $B - N + 1$ remains as the number of possible independent voltage equations.

In nodal analysis using Kirchhoff's current law, one node must be chosen as a reference, or *ground*; in Fig. 4.11, node *o* has been arbitrarily selected. It is then possible to write $N - 1$ nodal equations in terms of the potentials from the reference to each node and the branch admittances or impedances.

In more general terms, the branches connecting one portion of a network to another may be cut, and Kirchhoff's current law then requires the sum of the currents crossing the cut boundary to be zero. A *cut set* of branches might be formed in Fig. 4.11 by cutting through $\mathbf{Y}_1$, $\mathbf{Y}_3$, and $\mathbf{Y}_5$ and so $\mathbf{I}_1 - \mathbf{I}_3 - \mathbf{I}_5 = 0$. A second cut set might be formed by cutting through $\mathbf{Y}_2$ and $\mathbf{Y}_5$, yielding $\mathbf{I}_2 - \mathbf{I}_5 = 0$. These equations will be independent, since the second includes a branch not included in the first. Of course, the writing of a current summation at a node represents the formation of a cut set at that node. Since there are $N - 1$ tree branches, $N - 1$ independent current summations may be written.

In the circuit studied, the voltages V_x and V_y are ideal voltage sources, and there is no stated relation between V_x and I_1 or V_y and I_6. To analyze a network containing an independent voltage branch, a new tree must be drawn which includes the voltage source in one of the tree branches with another element. A tree might then appear as in Fig. 4.12, and two nodes

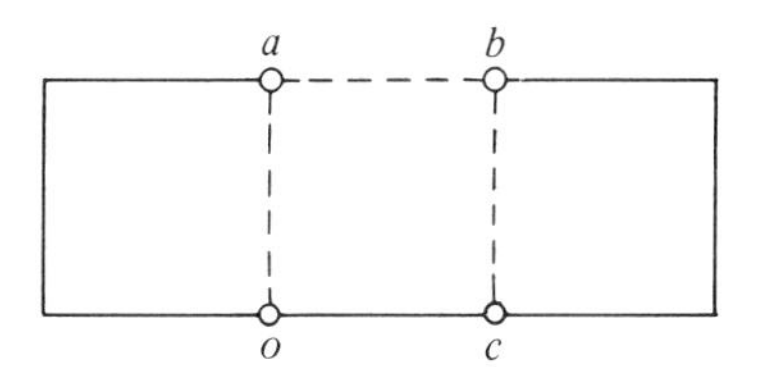

FIGURE 4.12
Tree modified to include voltage sources.

have disappeared, one for each voltage source. Thus we must modify the equation requirement for *the current law* by stating that *the number of independent equations is $N - 1$ minus the number of branches containing independent voltage sources only*. For the circuit of Fig. 4.12 this is $6 - 1 - 2 = 3$, written at *a,b,c*.

Similarly, we might analyze a tree containing ideal independent current sources and conclude that the independent loop equations, written by Kirchhoff's *voltage law*, would number $B - N + 1$ *minus the number of branches containing independent current sources only*. For Fig. 4.12 the result is $8 - 6 + 1 = 3$, and these equations would be written around the three loops.

A problem arises in absorbing a node with a circuit such as that of Fig. 4.13(a). The voltage source has no single branch in series with it. However, the situation can be resolved if node d is split as in Fig. 4.13(b), with a voltage source V_1 in series with each. Nodes d and d' will have the same potential to o, or the equations

$$V_{eo} = V_1 - I_2 R_2$$
$$V_{fo} = V_1 - I_1 R_1$$

would apply to both circuits, at (a) and (b) in Fig. 4.13.

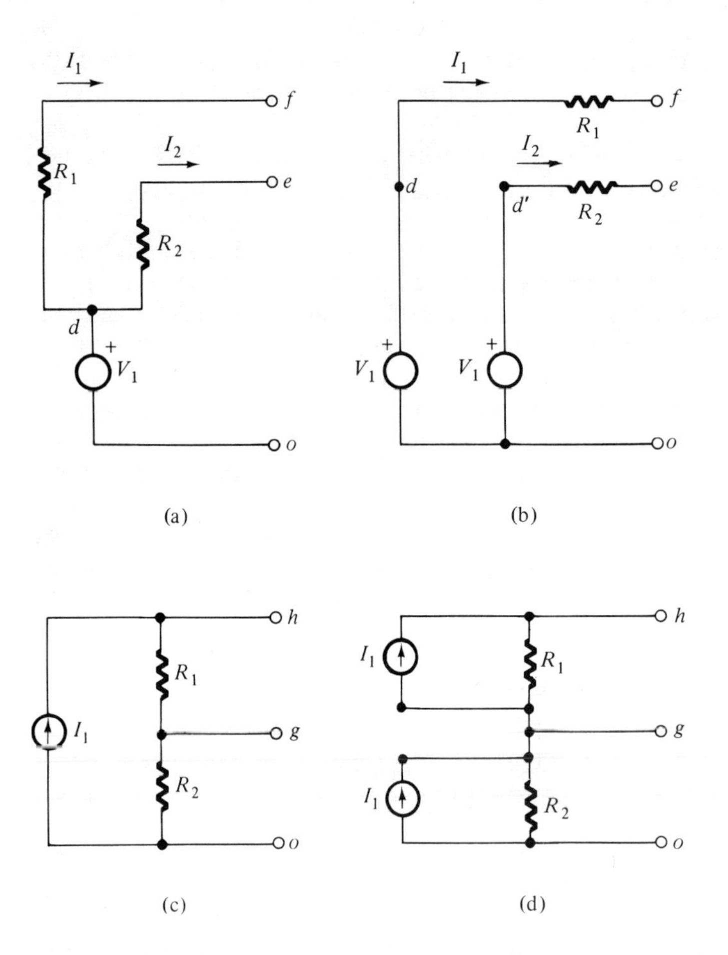

FIGURE 4.13

(a) Voltage source; (b) solution; (c) similar problem with a current source; (d) solution.

The new generators now have individual currents I_1 and I_2, and the generator current in the original circuit is $I_1 + I_2$.

A similar problem can arise with a current source as in Fig. 4.13(c), in which there is no isolated branch to include with the source. Here we may split the source as at (d) in Fig. 4.13 into two identical current sources. The same current summation exists at g, h, and o but each current source is now in parallel with a branch, and the solution can continue in the original manner.

The final choice of method of solution rests largely on which method requires the least number of equations. In the simpler electronic circuits the choice of the easiest method can usually be made by inspection of the circuit. If parallel branches predominate, the current law will be most productive, and if series branches exist for the most part, the voltage law will be the easiest to apply.

4.12

Formulation of network equations

We now apply the methods of the previous section to the network of Fig. 4.14.

LOOP METHOD The circuit at (a) in Fig. 4.14 has four nodes, A, B, C, and o, the latter representing the common line as reference. There are six branches and no current sources, so

$$B - N + 1 = 6 - 4 + 1 = 3$$

and three independent loop equations may be written.

The first step is to assign positive reference directions for the current

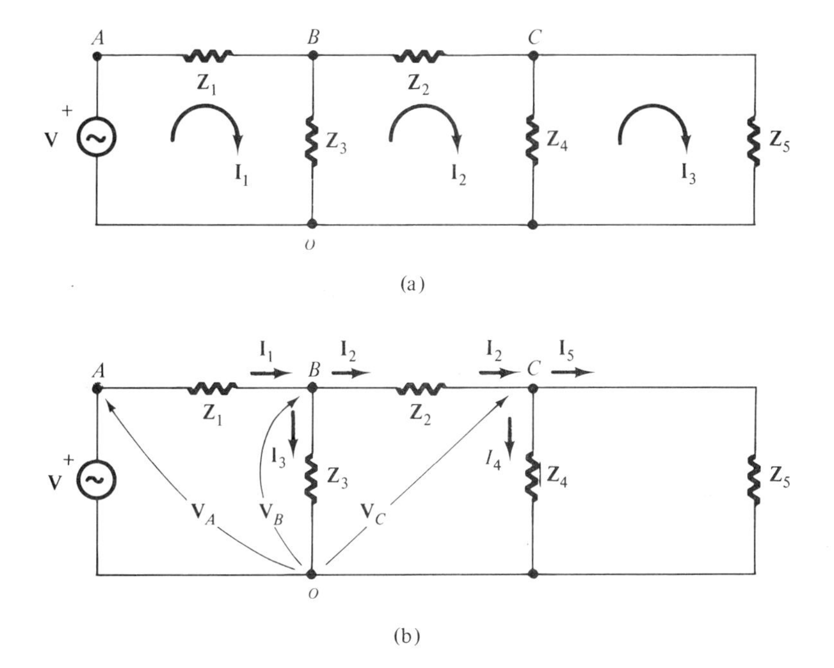

FIGURE 4.14
(a) Loop analysis; (b) nodal analysis.

around each selected circuit loop. While arbitrary, it is helpful to always assume positive currents as traversing the loops in the same direction, clockwise in this case. The chosen phasor currents are designated $\mathbf{I}_1$, $\mathbf{I}_2$, and $\mathbf{I}_3$ and are shown by the arrows.

In writing the loop equations, it is usual practice to travel around the loop in the direction of the assigned current; if the traverse is reversed, the equation simply appears as multiplied by -1. The equations for Fig. 4.14 are

$$\mathbf{V} = (\mathbf{Z}_1 + \mathbf{Z}_3)\mathbf{I}_1 - \mathbf{Z}_3\mathbf{I}_2$$

$$0 = -\mathbf{Z}_3\mathbf{I}_1 + (\mathbf{Z}_2 + \mathbf{Z}_3 + \mathbf{Z}_4)\mathbf{I}_2 - \mathbf{Z}_4\mathbf{I}_3 \tag{4.80}$$

$$0 = -\mathbf{Z}_4\mathbf{I}_2 + (\mathbf{Z}_4 + \mathbf{Z}_5)\mathbf{I}_3$$

Simultaneous solution leads to values for $\mathbf{I}_1$, $\mathbf{I}_2$, and $\mathbf{I}_3$.

NODE METHOD The circuit is redrawn at (b) in Fig. 4.14. The number of nodes is four, as indicated at A, B, C, and o, but there is a voltage source so the number of independent nodal equations is

$$(N - 1) - 1 = 4 - 1 - 1 = 2$$

Branch currents should be arbitrarily assigned by symbol and the assumed positive direction shown by arrows. The reference node is conveniently chosen as the common lead o, and the voltages to nodes A, B, and C are designated $\mathbf{V}_A$, $\mathbf{V}_B$, and $\mathbf{V}_C$. However, $\mathbf{V}_A = \mathbf{V}$ since it is across the voltage source, and this node was eliminated in determining the possible number of circuit equations. Thus $\mathbf{V}_B$ and $\mathbf{V}_C$ are the remaining unknown voltages, and the two nodal equations will be written from the cut sets at nodes B and C.

Currents enter and leave these nodes, and while choice of the positive direction is arbitrary, it is usual practice to designate the leaving current as positive. Then at B,

$$\mathbf{I}_2 + \mathbf{I}_3 - \mathbf{I}_1 = 0 \tag{4.81}$$

At C,

$$\mathbf{I}_4 + \mathbf{I}_5 - \mathbf{I}_2 = 0 \tag{4.82}$$

The currents can be developed in terms of the node voltages as

$$\mathbf{I}_1 = \frac{\mathbf{V} - \mathbf{V}_B}{\mathbf{Z}_1}$$

$$\mathbf{I}_2 = \frac{\mathbf{V}_B - \mathbf{V}_C}{\mathbf{Z}_2}$$

$$\mathbf{I}_3 = \frac{\mathbf{V}_B}{\mathbf{Z}_3}$$

$$\mathbf{I}_4 = \frac{\mathbf{V}_C}{\mathbf{Z}_4}$$

$$\mathbf{I}_5 = \frac{\mathbf{V}_C}{\mathbf{Z}_5}$$

Substitution of these values into Eq. 4.81 leads to

$$\frac{\mathbf{V}_B - \mathbf{V}_C}{\mathbf{Z}_2} + \frac{\mathbf{V}_B}{\mathbf{Z}_3} - \frac{\mathbf{V} - \mathbf{V}_B}{\mathbf{Z}_1} = 0 \tag{4.83}$$

$$\frac{\mathbf{V}}{\mathbf{Z}_1} = \left(\frac{1}{\mathbf{Z}_1} + \frac{1}{\mathbf{Z}_2} + \frac{1}{\mathbf{Z}_3}\right)\mathbf{V}_B - \frac{\mathbf{V}_C}{\mathbf{Z}_2} \tag{4.84}$$

or

$$\mathbf{Y}_1\mathbf{V} = (\mathbf{Y}_1 + \mathbf{Y}_2 + \mathbf{Y}_3)\mathbf{V}_B - \mathbf{Y}_2\mathbf{V}_C \tag{4.85}$$

From Eq. 4.82,

$$\frac{\mathbf{V}_C}{\mathbf{Z}_4} + \frac{\mathbf{V}_C}{\mathbf{Z}_5} - \frac{\mathbf{V}_B - \mathbf{V}_C}{\mathbf{Z}_2} = 0$$

$$0 = -\frac{\mathbf{V}_B}{\mathbf{Z}_2} + \left(\frac{1}{\mathbf{Z}_2} + \frac{1}{\mathbf{Z}_4} + \frac{1}{\mathbf{Z}_5}\right)\mathbf{V}_C \tag{4.86}$$

or

$$0 = -\mathbf{Y}_2\mathbf{V}_B + (\mathbf{Y}_2 + \mathbf{Y}_4 + \mathbf{Y}_5)\mathbf{V}_C \tag{4.87}$$

Equations 4.84 and 4.86 or 4.85 and 4.87 can be solved simultaneously for values of $\mathbf{V}_B$ and $\mathbf{V}_C$.

For the simple circuit used, the node method led to the solution of two equations and the loop method required solution of three equations. If the energy source had been a current source, the situation would have been reversed. In general, it is possible to achieve either situation, since we shall show in Section 6.7 that voltage and current sources may be developed as interchangeable equivalents.

4.13 Matrix solution of networks

We can organize the circuit equations for elaborate networks by the use of matrices and formalize the procedures for their solution by use of matrix algebra. Such organization is also helpful in network solution by digital computer.

Equations 4.80 might have been written in matrix form as

$$\begin{bmatrix} \mathbf{V} \\ 0 \\ 0 \end{bmatrix} = \begin{bmatrix} \mathbf{Z}_1 + \mathbf{Z}_3 & -\mathbf{Z}_3 & 0 \\ -\mathbf{Z}_3 & \mathbf{Z}_2 + \mathbf{Z}_3 + \mathbf{Z}_4 & -\mathbf{Z}_4 \\ 0 & -\mathbf{Z}_4 & \mathbf{Z}_4 + \mathbf{Z}_5 \end{bmatrix} \begin{bmatrix} \mathbf{I}_1 \\ \mathbf{I}_2 \\ \mathbf{I}_3 \end{bmatrix} \tag{4.88}$$

With the assumed currents, the matrix of impedances is symmetrical about the major diagonal, and the elements on this diagonal are positive; all others are negative or zero. The signs are the result of choosing to traverse all loops in a common direction.

The terms on the diagonal are total impedances in each loop, and the terms off the diagonal are the impedances common to the two loops, indicated by the subscript.

In general, a system of linear equations can be written:

$$\begin{aligned} \mathbf{I}_1 &= a_{11}\mathbf{V}_1 + a_{12}\mathbf{V}_2 + a_{13}\mathbf{V}_3 \\ \mathbf{I}_2 &= a_{21}\mathbf{V}_1 + a_{22}\mathbf{V}_2 + a_{23}\mathbf{V}_3 \\ \mathbf{I}_3 &= a_{31}\mathbf{V}_1 + a_{32}\mathbf{V}_2 + a_{33}\mathbf{V}_3 \end{aligned} \tag{4.89}$$

In matrix arrangement these become

$$\begin{bmatrix} \mathbf{I}_1 \\ \mathbf{I}_2 \\ \mathbf{I}_3 \end{bmatrix} = \begin{bmatrix} a_{11} & a_{12} & a_{13} \\ a_{21} & a_{22} & a_{23} \\ a_{31} & a_{32} & a_{33} \end{bmatrix} \begin{bmatrix} \mathbf{V}_1 \\ \mathbf{V}_2 \\ \mathbf{V}_3 \end{bmatrix} \tag{4.90}$$

and as a matrix equation

$$[\mathbf{I}] = [\mathbf{A}]\,[\mathbf{V}] \tag{4.91}$$

An advantage of matrix notation is the visible separation of the variables. The column matrix on the right represents the chosen independent variables and that on the left contains the dependent variables. The impedance matrix between represents the effect of the network configuration in relating the independent variables to the dependent variables.

The general term in $[\mathbf{A}]$ is a_{mn}, and for a square matrix, m and n range over the same values. It is possible to solve Eq. 4.91 for the independent variables, providing that the determinant $|\mathbf{A}|$ exists. With a rectangular matrix, m and n do not range over the same values, a determinant does not exist, and a complete set of equations has not been written.

A matrix of network parameters serves as a network model and is especially suited to digital computer solution of networks.

4.14 Fourier analysis of nonsinusoidal periodic waves

We have been working largely with sinusoidal wave forms but have not overlooked the fact that nonsinusoidal periodic waves such as those of Fig. 4.15 are frequently encountered. Fourier (1768–1830) showed that a periodic function can be represented by a series made up of an average value, plus sine and cosine terms of the fundamental

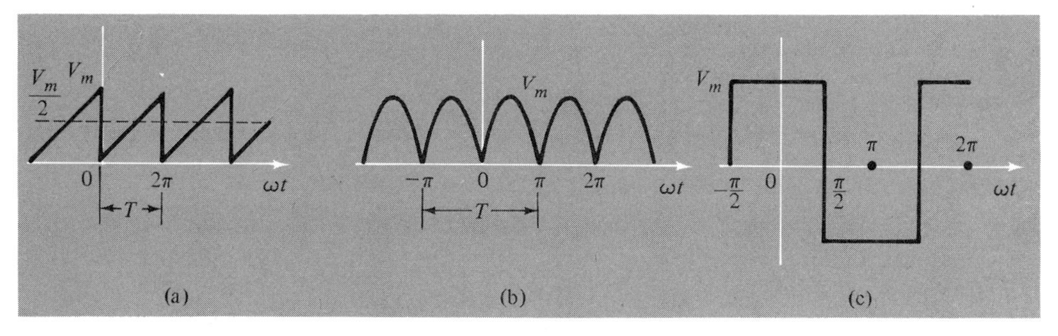

FIGURE 4.15

(a) Triangular wave; (b) rectified sine wave; (c) square wave.

frequency and all harmonic frequencies. A Fourier series for a periodic function of t is

$$f(t) = \frac{a_o}{2} + \sum_{n=1}^{\infty} (a_n \cos n\omega t + b_n \sin n\omega t) \tag{4.92}$$

where $\omega t = 2\pi ft = 2\pi t/T$, with period T measured in seconds. The component of angular frequency with $n = 1$ is called the *fundamental*; all higher multiples of frequency are called *harmonics*; a_n and b_n are the Fourier coefficients.

Any single-valued function $f(t)$, continuous except for a finite number of finite discontinuities in an interval of 2π and having only a finite number of maxima and minima in this interval, may be expressed as a convergent Fourier series. These limitations are satisfied by functions to be encountered in our circuit studies.

The coefficients can be evaluated by algebraic manipulation and the use of trigonometric identities. Expansion of Eq. 4.92 gives

$$f(t) = \frac{a_o}{2} + a_1 \cos \omega t + a_2 \cos 2\omega t + \cdots + a_n \cos n\omega t + \cdots + b_1 \sin \omega t$$
$$+ b_2 \sin 2\omega t + \cdots + b_n \sin n\omega t + \cdots \tag{4.93}$$

If both sides of the above are multiplied by $d\omega t$ and integrated over the period T, or 0 to 2π in radian measure, all sine and cosine terms on the right go to zero. We have

$$\int_0^{2\pi} f(t)\, d\omega t = \frac{a_o}{2}\omega t \Big]_0^{2\pi}$$

which can be rearranged as the integral

$$\frac{a_o}{2} = \frac{1}{2\pi}\int_0^{2\pi} f(t)\, d\omega t \tag{4.94}$$

which is the average value of the function $f(t)$; we also call this the dc value of the wave.

If both sides of Eq. 4.93 are multiplied by cos $n\omega t\, d\omega t$ and integrated over the period 0 to 2π, we have

$$\begin{aligned}\int_0^{2\pi} f(t)\cos n\omega t\, d\omega t &= \int_0^{2\pi} \frac{a_o}{2}\cos n\omega t\, d\omega t + \int_0^{2\pi} a_1 \cos \omega t \cos n\omega t\, d\omega t\\
&+ \int_0^{2\pi} a_2 \cos 2\omega t \cos n\omega t\, d\omega t + \cdots\\
&+ \int_0^{2\pi} a_n \cos^2 n\omega t\, d\omega t + \cdots\\
&+ \int_0^{2\pi} b_1 \sin \omega t \cos n\omega t\, d\omega t\\
&+ \int_0^{2\pi} b_2 \sin 2\omega t \cos n\omega t\, d\omega t + \cdots\\
&+ \int_0^{2\pi} b_n \sin n\omega t \cos n\omega t\, d\omega t + \cdots\end{aligned}$$

All integrals on the right go to zero except that containing $\cos^2 n\omega t$, and we have

$$\begin{aligned}\int_0^{2\pi} f(t)\cos n\omega t\, d\omega t &= \int_0^{2\pi} a_n \cos^2 n\omega t\, d\omega t\\
&= \pi a_n\end{aligned}$$

We then write the definition for the a coefficients as

$$a_n = \frac{1}{\pi}\int_0^{2\pi} f(t)\cos n\omega t\, d\omega t \tag{4.95}$$

Similarly, we can multiply Eq. 4.93 by sin $n\omega t\, d\omega t$ and integrate over one period from 0 to 2π. Then

$$\begin{aligned}\int_0^{2\pi} f(t)\sin n\omega t\, d\omega t &= \int_0^{2\pi} \frac{a_o}{2}\sin n\omega t\, d\omega t + \int_0^{2\pi} a_1 \cos \omega t \sin n\omega t\, d\omega t\\
&+ \int_0^{2\pi} a_2 \cos 2\omega t \sin n\omega t\, d\omega t + \cdots\\
&+ \int_0^{2\pi} a_n \cos n\omega t \sin n\omega t\, d\omega t + \cdots\\
&+ \int_0^{2\pi} b_1 \sin \omega t \sin n\omega t\, d\omega t\\
&+ \int_0^{2\pi} b_2 \sin 2\omega t \sin n\omega t\, d\omega t + \cdots\\
&+ \int_0^{2\pi} b_n \sin^2 n\omega t\, d\omega t + \cdots\end{aligned}$$

The integrals on the right again vanish except for the term containing $\sin^2 n\omega t$. Then we have

$$\int_0^{2\pi} f(t) \sin n\omega t \, d\omega t = \int_0^{2\pi} b_n \sin^2 n\omega t \, d\omega t$$
$$= \pi b_n$$

We write the definition for the b coefficients as

$$b_n = \frac{1}{\pi} \int_0^{2\pi} f(t) \sin n\omega t \, d\omega t \tag{4.96}$$

It is not necessary that the limits of integration be 0 to T or 0 to 2π. The only requirement is that one period of the function be covered by the integration, and so the integration can take place from $-\pi$ to $+\pi$ or from t_1 to $T + t_1$, where t_1 is an arbitrary time.

An even function has the property that $f(-t) = f(t)$, and an odd function has the property that $f(-t) = -f(t)$. As examples, a cosine is an even function and a sine is an odd function. *If $f(t)$ is an odd function*, then we write

$$a_n = \frac{1}{\pi} \int_0^{2\pi} f(t) \cos n\omega t \, d\omega t = \frac{1}{\pi} \int_{-\pi}^{+\pi} f(t) \cos n\omega t \, d\omega t$$

The integrand is odd because of $f(t)$, and the integral of an odd function over equal positive and negative limits is zero. Therefore, if $f(t)$ is odd, its a_n coefficients will be zero, and the series for an odd function will include only sine terms. Likewise, *if $f(t)$ is even*, then we write

$$b_n = \frac{1}{\pi} \int_0^{2\pi} f(t) \sin n\omega t \, d\omega t = \frac{1}{\pi} \int_{-\pi}^{+\pi} f(t) \sin n\omega t \, d\omega t$$

The integrand is an odd function because of $\sin n\omega t$, and its integral between $-\pi$ and $+\pi$ is zero. Therefore, if $f(t)$ is an even function, its b_n coefficients will be zero and the series for an even function will include only cosine terms.

As an example, consider (a) in Fig. 4.15. For this wave form,

$$f(\omega t) = \frac{V_m \omega t}{2\pi}$$

The wave is drawn with the origin located to make $f(\omega t)$ an odd function when the amplitude zero is placed at the wave minimum and an average value is present. Therefore

$$a_o = \frac{1}{2\pi} \int_{-\pi}^{\pi} \frac{V_m \omega t}{2\pi} \, d\omega t = \frac{V_m}{2} \tag{4.97}$$

We call this the *dc value* of a voltage or current wave.

Since $f(\omega t)$ is an odd function around the translated origin, $a_n = 0$, leaving the b_n coefficients. Then

$$b_1 = \frac{1}{\pi}\int_0^{2\pi} \frac{V_m \omega t}{2\pi} \sin \omega t \, d\omega t = -\frac{V_m}{\pi} \tag{4.98}$$

and the complete series for this triangular wave is

$$f(\omega t) = \frac{V_m}{2} - \frac{V_m}{\pi}\left(\sin \omega t + \frac{1}{2}\sin 2\omega t + \frac{1}{3}\sin 3\omega t + \cdots\right) \tag{4.99}$$

The function of (b) in Fig. 4.15 is a rectified sine wave and is an even function. Its series can be found to be

$$f(\omega t) = \frac{2V_m}{\pi} - \frac{4V_m}{\pi}\left(\frac{1}{3}\cos 2\omega t + \frac{1}{15}\cos 4\omega t + \frac{1}{35}\cos 6\omega t + \cdots\right) \tag{4.100}$$

The square wave, with the origin located as shown in Fig. 4.15(c), is an even function, and its series is

$$f(\omega t) = \frac{2V_m}{\pi}\left(\cos \omega t - \frac{1}{3}\cos 3\omega t + \frac{1}{5}\cos 5\omega t - \frac{1}{7}\cos 7\omega t + \cdots\right) \tag{4.101}$$

We can consider these series as representing the outputs of sinusoidal voltage generators connected in series, with voltage magnitudes specified by

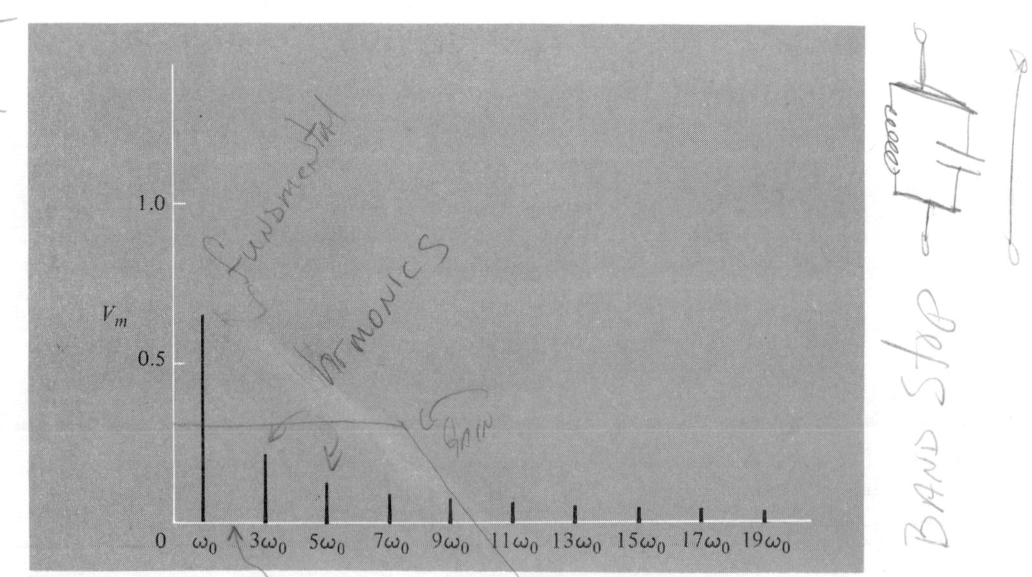

FIGURE 4.16
Frequency spectrum of the square wave of Fig. 4.15(c).

the coefficients. The result is displayed as a *frequency spectrum* in Fig. 4.16, in which the signal power is distributed among a set of harmonically related frequencies. Amplitude information for each frequency component is presented. The spectrum shown is for the square wave of Eq. 4.101. Since amplitudes are plotted, all terms appear positive; the negative signs of the series imply a 180° phase shift. The spectrum indicates that the amplitudes do not decline very fast after the third harmonic, and harmonics of high order are needed if a square wave is to be compounded by addition of these successive frequency terms.

The Fourier analysis of a wave shows that methods suited to linear circuit analysis with single frequencies can be applied to each of a number of frequencies and the superimposed result used to predict the response for a nonsinusoidal periodic wave.

A useful and concise Fourier expression can be written in exponential form. Substitute the exponential forms for the sine and cosine terms in Eq. 4.92, giving

$$\begin{aligned} f(\theta) &= \frac{a_o}{2} + \sum_{n=1}^{\infty}\left[a_n\frac{\varepsilon^{jn\theta}+\varepsilon^{-jn\theta}}{2} - jb_n\frac{\varepsilon^{jn\theta}-\varepsilon^{-jn\theta}}{2}\right] \\ &= \frac{a_o}{2} + \sum_{n=1}^{\infty}\frac{a_n - jb_n}{2}\varepsilon^{jn\theta} + \sum_{n=1}^{\infty}\frac{a_n + jb_n}{2}\varepsilon^{-jn\theta} \end{aligned} \tag{4.102}$$

If we change the summation from $+n$ to $-n$, we do not affect the magnitude of a_n in Eq. 4.95, but we do change the sign of b_n derived from Eq. 4.96. Therefore we can change the second summation and have

$$f(\theta) = \frac{a_o}{2} + \sum_{n=1}^{\infty}\frac{a_n - jb_n}{2}\varepsilon^{jn\theta} + \sum_{n=1}^{-\infty}\frac{a_n - jb_n}{2}\varepsilon^{jn\theta}$$

We can now include all terms in one summation:

$$f(\theta) = \sum_{n=-\infty}^{\infty}\frac{a_n - jb_n}{2}\varepsilon^{jn\theta} = \sum_{n=-\infty}^{\infty} c_n\varepsilon^{jn\theta} \tag{4.103}$$

This is the *exponential form of the Fourier series*, with c_n determining the amplitude of the components in the frequency spectrum, and with c_n a function of the a_n and b_n coefficients previously determined.

4.15 Polyphase systems

For reasons of distribution economy, and ease of electrical motor and generator design, most industrial electrical power is generated and distributed by *three-phase systems.* Polyphase sources also furnish the power for large electronic rectifiers supplying direct current. The ordinary domestic supply lines are derived from individual phases of a three-phase system.

A three-phase system is supplied by three sinusoidal voltages of equal magnitude and frequency but differing by 120° in phase from each other. A three-phase system can be analyzed as three phase-displaced single-voltage systems, with the currents and voltages combined in the lines and loads.

The three voltage generators can be connected in *Y* or *delta*, as indicated in Fig. 4.17. The voltages supplied to the three lines are $\mathbf{V}_{12}$, $\mathbf{V}_{23}$, and

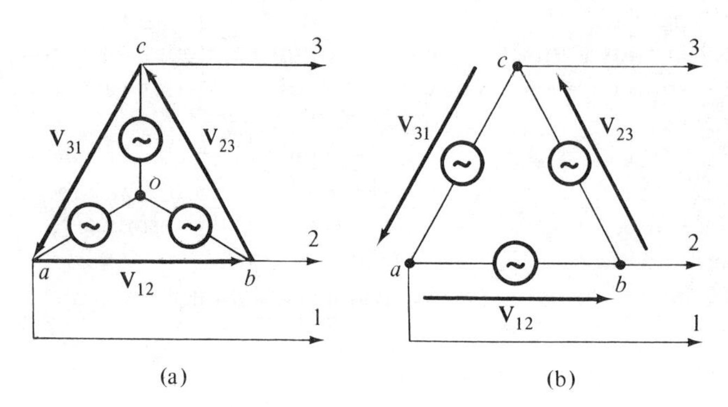

FIGURE 4.17
(a) *Y*-Connected generator; (b) delta-connected generator.

$\mathbf{V}_{31}$, and in (a) these are formed from the individual phase voltages V_p as

$$\begin{aligned}\mathbf{V}_{12} &= -\mathbf{V}_{oa} + \mathbf{V}_{ob} = -V_p(-0.866 - j0.5) + V_p(0.866 - j0.5) \\ &= \sqrt{3}V_p\angle 0^\circ\end{aligned} \tag{4.104}$$

$$\begin{aligned}\mathbf{V}_{23} &= -\mathbf{V}_{ob} + \mathbf{V}_{oc} = -V_p(0.866 - j0.5) + V_p(0 + j1) \\ &= \sqrt{3}V_p\angle 120^\circ\end{aligned} \tag{4.105}$$

$$\begin{aligned}\mathbf{V}_{31} &= -\mathbf{V}_{oc} + \mathbf{V}_{oa} = -V_p(0 + j1) + V_p(-0.866 - j0.5) \\ &= \sqrt{3}V_p\angle -120^\circ\end{aligned} \tag{4.106}$$

As phasors these line voltages are illustrated in Fig. 4.18(b). For the delta connection the line voltages are the same as the individual phase voltages.

The three-phase generators have equal voltage magnitudes, and when the three loads, Z_p in Fig. 4.18(c), are equal, the circuit is said to be *balanced*. Because of the symmetry of the balanced *Y*-connected circuit, points *o* and *o′* have the same potential with respect to the lines and are called the *neutrals*; in some four-wire three-phase systems a connection between the neutrals is actually made, and in others the neutral points are grounded. With equality of potentials at the neutrals, each generator appears to supply its own phase load, and computations on a per-phase basis follow.

With balanced loads,

$$\mathbf{I}_a = I_p\angle\theta^\circ$$

$$\mathbf{I}_b = I_p \angle 120° + \theta°$$

$$\mathbf{I}_c = I_p \angle -120° + \theta°$$

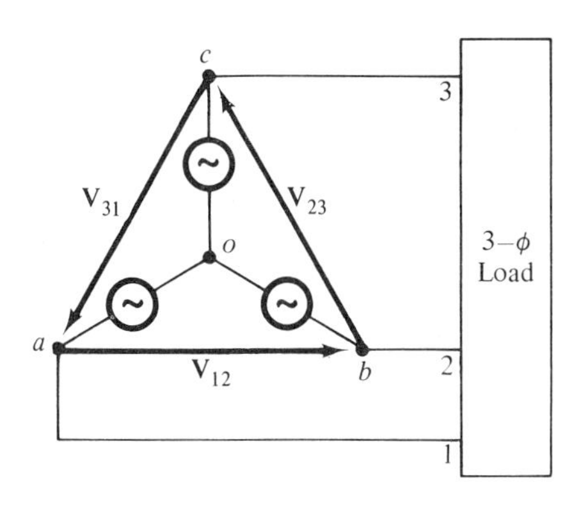

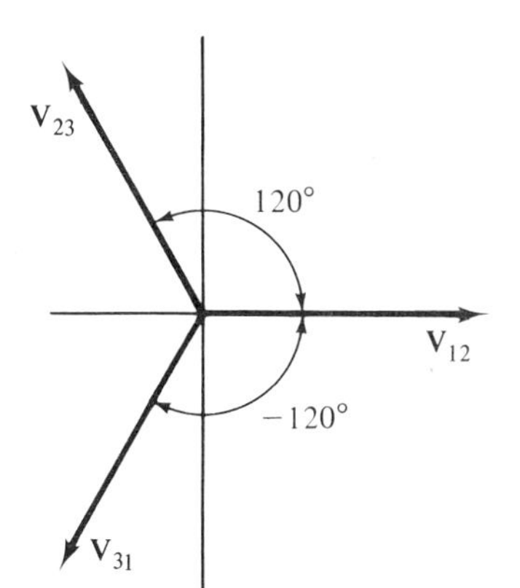

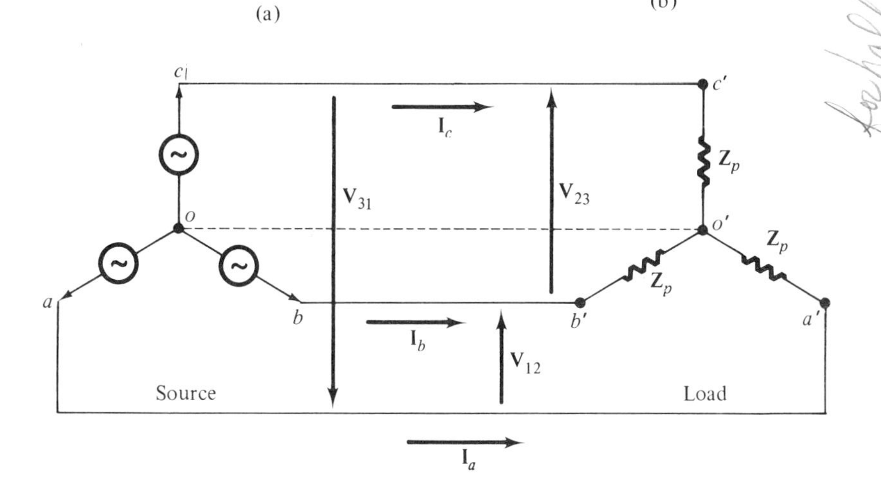

FIGURE 4.18

(a) Three-phase, *Y*-connected generator; (b) three-phase line voltages; (c) balanced three-phase system.

where I_p is the effective current in each phase and θ is the power factor angle of each phase load. The average power delivered to each phase load is $V_p I_p \cos\theta$, where V_p is the line to neutral voltage as defined in Eq. 4.104. The total average power is then

$$P = 3V_p I_p \cos\theta \tag{4.107}$$

The phase current for the Y connection is the same as the line current, or $\mathbf{I}_p = \mathbf{I}_L$, and $\mathbf{V}_L = \sqrt{3}\mathbf{V}_p$, so the total power expressed in terms of line currents and line voltages is

$$P = \sqrt{3} V_L I_L \cos\theta \tag{4.108}$$

the angle θ being that between phase voltage and phase current.

In a *delta-connected* system the line voltage and phase voltage are equal, but the line current is the sum of two phase currents, or $\mathbf{I}_p = \mathbf{I}_L/\sqrt{3}$. For total power in a delta-connected system, substitution of this current and $\mathbf{V}_L = \mathbf{V}_p$ in Eq. 4.107 gives

$$P = \sqrt{3} V_L I_L \cos\theta \tag{4.109}$$

so that the total average power is expressed in an identical manner for both Y and delta loads.

The statement concerning economy of the three-phase system can now be supported: A third conductor has been added, increasing the conductor material by 50 percent, but the power-carrying capacity of the circuit has increased 73 percent.

In single-phase power generation, there is a steady value of average power plus a power variation at sinusoidal double frequency. In three-phase circuits the pulsations in power cancel, and a three-phase generator is called upon to supply only a steady power component; vibration is also reduced. For some forms of generator prime movers, this makes polyphase generation an advantage.

The product $V_L I_L$ is called the *volt-amperes* (VA). Remembering that these phasors differ by an angle θ, we could write

$$\begin{aligned} \mathbf{VI} &= V_L I_L \varepsilon^{j\theta} \\ &= V_L I_L \cos\theta + j V_L I_L \sin\theta \\ &= P + jQ \end{aligned} \tag{4.110}$$

We can identify P as the average power in the circuit, but a new term Q has been introduced and this is called the *reactive power*, measured in *VAR*s (*v*olt-*a*mperes, *r*eactive).

The reactive component arises from energy which flows into an inductive or capacitive circuit; it represents the energy momentarily stored in the fields but which is returned to the circuit as the electric or magnetic fields collapse in each half cycle. A quadrature current is present as a result of this stored energy. Being in quadrature with the applied voltage, this current represents zero average power input, except for the losses in any resistance through which this current passes.

A power customer pays for energy, that is, watt-hours consumed. The reactive power which enters his motors and transformers is returned and does not add to his bill. However, the quadrature current adds vectorially to the in-phase power component of current, and this requires larger conductors in the generators and lines. A generator rated at 10,000 kW at unity power factor can supply 10,000 V and 1000 A, and the power company can earn a return on its investment due to the sale of 10,000 kW. However, if the power factor is only 0.5, the maximum current remains at 1000 A, the voltage is 10,000 V, but the power generated is only 5000 kW. The power company receives only one half of the return on its investment in the machine. Because of this undesirable effect of low power factor, a low power factor penalty is sometimes included in rate schedules to encourage customers to improve the power factor of their load.

Power factor correction equipment is advantageous in obtaining these lower power rates. Since most industrial loads take power at a lagging current because of inductance present in the motors and transformers, the correction equipment usually takes the form of capacitors in parallel with the load. Taking no net power themselves, their quadrature current is opposite in phase angle to that of the quadrature current taken by the lagging power factor load, and the net line current is reduced, or the overall power factor is raised.

EXAMPLE. An industry has an average load of 1000 kW, with a power factor of 0.75, taken with a Y connection from a three-phase line at 2300 V to neutral. This is to be corrected to a power factor of 0.95 with capacitors.

With

$$P = |V||I| \cos \theta = 1000 \text{ kW}$$

the volt-ampere load at 0.75 power factor is

$$|V||I| = \frac{1000}{0.75} = 1333 \text{ kVA}$$

The reactive volt-amperes, VAR, are

$$Q = 1333 \sin(\cos^{-1} 0.75) = 1333 \times 0.661 = 882 \text{ kVAR}$$

The uncorrected load calls for $P_L = 1000$ kW and $Q = 882$ kVAR.

At a power factor of 0.95, the total volt-amperes required to supply 1000 kW will be

$$|V||I| = \frac{1000}{0.95} = 1052 \text{ kVA}$$

from which

$$P = 1052 \times 0.95 = 1000 \text{ kW}$$

$$Q_L = 1052 \sin(\cos^{-1} 0.95) = 328 \text{ kVAR}$$

Thus the corrected load calls for $P = 1000$ kW and $Q_L = 328$ kVAR.

The volt-amperes taken by a lossless capacitor are

$$|V_c||I_c|\varepsilon^{-j90^\circ} = |V_c||I_c|(\cos 90^\circ - j\sin 90^\circ)$$
$$= 0 - jQ_C$$

The volt-amperes taken by two parallel loads is the sum of their individual volt-ampere requirements, so that the industry load plus the capacitor will be

$$\mathbf{VI} = P + jQ - jQ_C$$

At a power factor of 0.75, $Q = 882$ kVAR. At a power factor of 0.95 the total value is

$$Q = Q_L - Q_C$$
$$882 = 328 - Q_C$$

so that

$$Q_C = 882 - 328 = 554 \text{ kVAR}$$

This is the corrective kilovolt-ampere requirement.

With a Y load, one third of the kilovolt amperes, or 185 kVA, is required per phase. The current taken by one phase of the capacitor is

$$I_C = \frac{185{,}000}{2300} = 79 \text{ A}$$

The capacitor impedance is

$$|\mathbf{Z}_c| = \frac{1}{\omega C} = \frac{V}{I_C} = \frac{2300}{79} = 29.6\,\Omega$$

Therefore

$$C \text{ (per phase)} = \frac{1}{\omega Z_C} = \frac{1}{2\pi \times 60 \times 29.6} = 90\,\mu\text{F}$$

The cost of this capacitor, capitalized on an annual basis, may be less than the penalty previously paid in a higher energy rate for the low power factor condition, and the investment in the capacitor is then justified.

4.16

Instruments for average and rms measurement

An instrument for measurement of currents having steady values, or for measuring the average value of periodic currents, actually measures the force on the current in a fixed magnetic field. A coil of fine wire is suspended in a magnetic field produced by a permanent magnet, and the current to be measured is passed around the

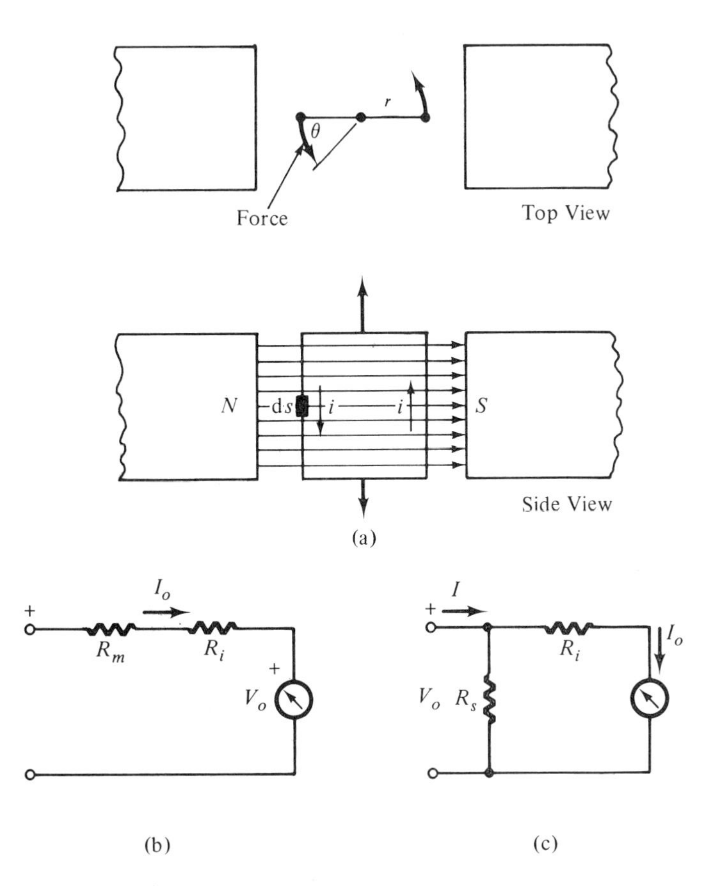

FIGURE 4.19
(a) Permanent-magnet moving-coil (PMMC) instrument; (b) as a voltmeter; (c) as an ammeter.

coil in the direction indicated in Fig. 4.19(a). The magnetic field of the current element $i\,ds$ interacts with the magnet flux of density B Wb/m^2, and the result is a force directed out of the page:

$$f = Bi\,ds$$

For a length l of the coil side and N turns, this becomes

$$f = NBli$$

The torque is exerted at radius r and produces a deflection θ; the torque is additive on both sides of the coil, so

$$T = 2NBrli$$

However $2r$ is the width of the coil and l the length of the coil, and $2rl$ represents the coil area A. Then

$$T = NABi \qquad \text{N-m} \tag{4.111}$$

This torque is opposed by that of spiral springs at the pivots, and the rotation θ becomes directly proportional to the current i,

$$\theta = \frac{NABi}{K} \tag{4.112}$$

where K is the spring constant. The angle of rotation is limited so that the coil remains in a region of constant flux density.

Inertial and damping torques are adjusted to produce a motion approximating the response of a slightly underdamped mechanical system. The low natural frequency of the coil system prevents it from following rapid current variations, and the instrument indicates the average or the dc component of the current applied. We speak of it as a *permanent-magnet moving-coil* (PMMC) *instrument* or as a dc instrument.

The designer provides sufficient torque to minimize the effect of friction, and a given instrument is designed for a maximum deflection current I_o; this may be 1 mA. The coil of the instrument will have some internal resistance R_i, and the voltage required for full deflection will be $V_o = R_i I_o$.

To allow measurement of higher voltages, a series multiplier resistance R_m is added as in Fig. 4.19(b). Since I_o passes through both R_m and R_i, we can write

$$\frac{V}{V_o} = \frac{R_m + R_i}{R_i}$$

where V is the desired maximum reading of the instrument as a *voltmeter*. The multiplier resistance is then

$$R_m = R_i\left(\frac{V}{V_o} - 1\right) \cong R_i\frac{V}{V_o} \tag{4.113}$$

A voltmeter is connected *across* a port and should not appreciably alter the current entering that port, for reasons of accuracy. Thus a voltmeter should have a high internal resistance $R_m + R_i$ compared to the resistance of the connected circuit. This criterion is provided for in the *sensitivity* or *ohms-per-volt* rating of a voltmeter, defined as

$$S = \frac{R_m + R_i}{V} = \frac{1}{I_o} \tag{4.114}$$

Reflecting the relative resistance of the instrument, the sensitivity figure indicates the effect of the voltmeter on the connected circuit.

The same basic movement can be used to measure currents larger than I_o by paralleling the instrument with a resistive *shunt*, such as R_s in Fig. 4.19(c). With full deflection voltage V_o across the meter terminals with current I_o, it is possible to write

$$R_i I_o = R_s(I - I_o)$$

where I is the current to be measured at full deflection with the shunted movement. Then

$$R_s = \frac{R_i I_o}{I - I_o} = \frac{V_o}{I - I_o} \cong \frac{V_o}{I} \tag{4.115}$$

Such an instrument is called an *ammeter*.

In the PMMC instrument the torque reverses when the current reverses, and the coil will not deflect for ac currents. If the field of the permanent magnet is replaced with a field created by a coil through which alternating current passes, the field will reverse as the current reverses and a unidirectional torque will be obtained.

The inductance of the fixed field coils in Fig. 4.20 is L_1. The rotating

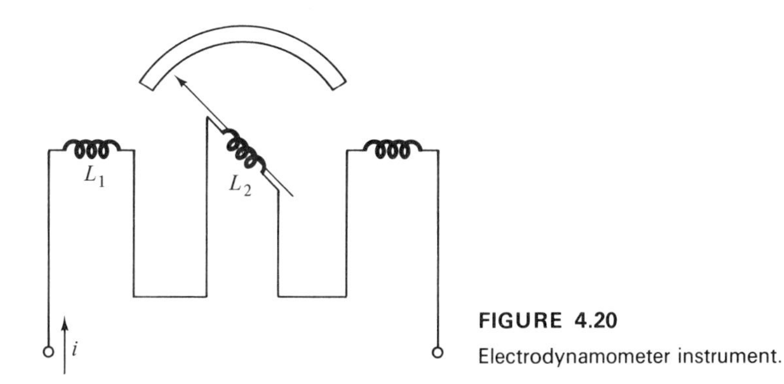

FIGURE 4.20
Electrodynamometer instrument.

coil has inductance L_2 and is connected in series with the fixed coils, so that current and flux reverse together. The energy in the system is

$$w = \frac{L_1 i_1^2}{2} + \frac{L_2 i_2^2}{2} + M i_1 i_2 \tag{4.116}$$

where M is the mutual inductance between the fixed and moving coils. The rate of doing work is dw/dt and

$$\frac{dw}{dt} = T\frac{d\theta}{dt}$$

in a rotational system, where θ is the angle of deflection of the moving coil. Therefore

$$T = \frac{dw}{d\theta}$$

and

$$T = \frac{i_1^2}{2}\frac{dL_1}{d\theta} + \frac{i_2^2}{2}\frac{dL_2}{d\theta} + i_1 i_2 \frac{dM}{d\theta} \tag{4.117}$$

With the magnetic path consisting only of air or nonmagnetic material,

$$\frac{dL_1}{d\theta} = \frac{dL_2}{d\theta} = 0$$

since there is no change in the flux relationship. However, the mutual inductance does change with moving coil rotation, and so

$$T = i_1 i_2 \frac{dM}{d\theta}$$

Since the currents are identical in the coils,

$$T = i^2 \frac{dM}{d\theta} \tag{4.118}$$

With springs provided to balance this torque and with the mechanical time constant long with respect to the periods of current variation, we have an instrument in which

$$\frac{1}{t_0}\int_0^{t_0} i^2 \frac{dM}{d\theta}\, dt = K\theta \tag{4.119}$$

where t_0 is the time of a period. The deflection is proportional to the average of the square of the current. The mutual inductance M varies as the cosine of the rotation, so that $dM/d\theta$ varies as the sine of the angle, but this variation may be calibrated into the scale of the instrument. The square root may also be extracted by suitable marking of the scale, and we then have an instrument capable of indicating rms values of applied voltages and currents. This is the *electrodynamometer* instrument.

Such an instrument cannot readily be shunted as an ammeter, because the resistive shunt and inductive movement would not accurately divide the input current at all frequencies of interest. Ammeters for large currents are wound with larger wire and carry the measured current directly in the moving coil, or are coupled through a *current transformer*, to receive a fixed fraction of the circuit current. Voltmeters can employ series multiplier resistors; the value of R_o is made large with respect to the coil impedance so that the current is relatively independent of frequency. However, the usual electrodynamometer instruments are somewhat limited in useful frequency range.

Rectifier ac instruments convert the applied alternating current to direct current, which is measured in a PMMC movement. The dc current represents an average value of half a wave, but the scale is marked in rms values. If the wave is distorted and nonsinusoidal, particularly if even-order harmonics are present, considerable inaccuracy is possible.

Average power can also be measured by use of the electrodynamometer movement connected as a *wattmeter*. The fixed coils carry the current to be measured, and the rotating coil current is made proportional to the applied voltage by supplying it through a fixed resistor. The torque is proportional to the product vi by Eq. 4.118, and the average power is indicated by the rotation of the moving coil.

4.17 Polyphase power measurement

Two wattmeters, connected as in Fig. 4.21, can be used to measure total three-phase average power. The power values measured by the instruments are

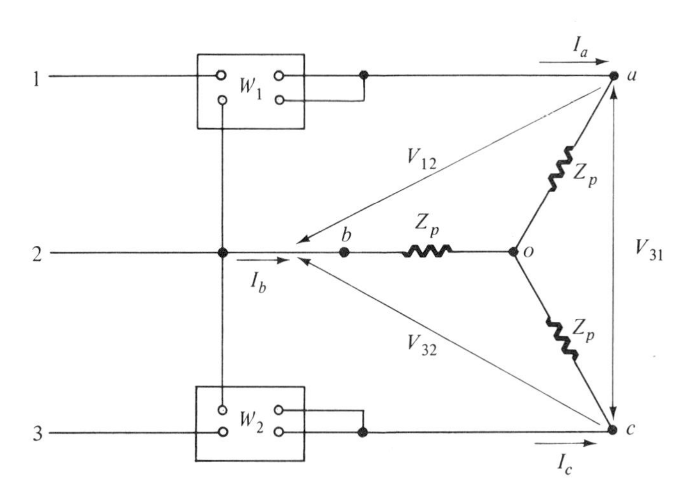

FIGURE 4.21

Polyphase wattmeter connection.

$$W_1 = V_{12} I_a \cos \lambda_1 \tag{4.120}$$

$$W_2 = V_{32} I_c \cos \lambda_2 \tag{4.121}$$

where λ_1 and λ_2 are the phase angles between V_{12} and I_a and V_{32} and I_c. For a given load I_a and V_{ao} are at an angle θ; likewise, I_c and V_{co} are at an angle $-\theta$. However, V_{12} and V_{ao} differ by 30°, as do V_{32} and V_{co}. Therefore

$$\lambda_1 = 30° + \theta$$

$$\lambda_2 = 30° - \theta$$

for balanced phase loads, where θ is the power factor angle of each load.

Addition of Eqs. 4.120 and 4.121 gives

$$W_1 + W_2 = V_{12}I_a\cos(30^\circ + \theta) + V_{32}I_c\cos(30^\circ - \theta)$$
$$= V_{12}I_a(\cos 30^\circ \cos\theta - \sin 30^\circ \sin\theta) \qquad (4.122)$$
$$+ V_{32}I_c(\cos 30^\circ \cos\theta + \sin 30^\circ \sin\theta)$$

For a balanced load and balanced voltage source, $V_{12} = V_{32} = V_L$ and $I_a = I_c = I_L$, magnitudes being used. Therefore

$$W_1 + W_2 = \sqrt{3}V_L I_L \cos\theta \qquad (4.123)$$

and the sum of the two readings is the total three-phase average power delivered to the load.

An equivalent development could be used with a delta-connected load, leading to the same expression. In fact the result is correct for any three-phase load, balanced or unbalanced.

4.18 The bridge circuit for parameter measurement

Values of the R, L, and C parameters are usually measured in circuits known as *bridges*. The Wheatstone form for measuring resistances with direct current is shown in Fig. 4.22(a). Current

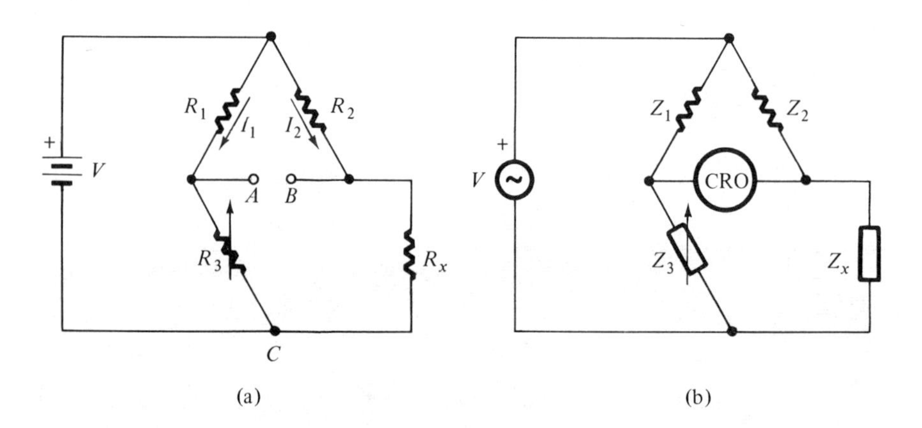

FIGURE 4.22
(a) Wheatstone resistance bridge; (b) impedance bridge.

I_1 is through R_1 and R_3 and voltage V_{AC} between A and C is given by voltage division as

$$V_{AC} = \frac{R_3}{R_1 + R_3}V$$

Likewise, current I_2 is through R_2 and R_x and voltage V_{BC} is

$$V_{BC} = \frac{R_x}{R_2 + R_x} V$$

For the *balance condition* of the bridge there is to be zero voltage between A and B, a condition usually indicated by zero deflection of a PMMC galvanometer at those terminals. That is,

$$V_{AC} - V_{BC} = 0$$

$$\frac{R_3}{R_1 + R_3} - \frac{R_x}{R_2 + R_x} = 0$$

from which the balance equation is obtained as

$$R_x = \frac{R_2}{R_1} R_3 \tag{4.124}$$

Resistors R_1 and R_2 are called the *ratio arms* and their ratio is often made unity. Then R_x is directly measured as a function of R_3.

A bridge for measurement of impedances appears in Fig. 4.22(b). The indicator of the balance condition is most usually a cathode-ray oscilloscope, and for zero voltage across its terminals we must have

$$|Z_x| \underline{/\theta_x} = \frac{|Z_2| \underline{/\theta_2}}{|Z_1| \underline{/\theta_1}} |Z_3| \underline{/\theta_3} \tag{4.125}$$

The achievement of simultaneous balance of magnitude and phase angle requires two bridge adjustments which interlock, and the balancing of an ac bridge is more difficult than that of a dc bridge. The two conditions are

$$|Z_x| = \frac{|Z_2|}{|Z_1|} |Z_3|; \qquad \theta_x = \theta_2 + \theta_3 - \theta_1 \tag{4.126}$$

In certain circuits it is possible to make $|Z_1| = R_1$ and $|Z_2| = R_2$, and the phase angle requirement reduces to $\theta_x = \theta_3$.

Many forms of ac bridges have been devised for ease in special measurements.

4.19 General comments

In this chapter we have shown the logical development of methods for the solution of steady-state ac electrical circuits. Basically, we have been seeking methods for ac circuit solution which parallel the simplicity of Ohm's law, $V = IR$, and Joule's law, $P = I^2R$, for direct current.

In our initial approach we formulated circuit integrodifferential equations to account for the effect of time variation of current or voltage with inductive or capacitive circuit elements. To find the steady-state response, we chose $\varepsilon^{j\omega t}$ as the form of our forcing function. Each derivative was replaced with $j\omega$ and each integral with $1/j\omega$; the integrodifferential equation was accordingly transformed to an algebraic equation involving phasors. By the use of complex numbers for impedance we reached our first objective, a solution in the ac form of Ohm's law, $\mathbf{V} = \mathbf{IZ}$.

By defining a root-mean-square value for a periodic wave form we achieved a form of measurement of a varying current or voltage which satisfies the relation $P = I^2_{\text{rms}}R$.

We found that we did not need to write the integrodifferential equation at all, as we had also learned to do for the Laplace transform method, but could directly draw the circuit transformed to $j\omega$ impedances and phasors. With this simplicity we shall now proceed to further study of the properties of circuits.

PROBLEMS

4.1 Given the one-port circuits of Fig. 4.23, write the impedances as complex numbers if the applied frequency is 50 Hz. Give both rectangular and polar forms of $\mathbf{Z}$.

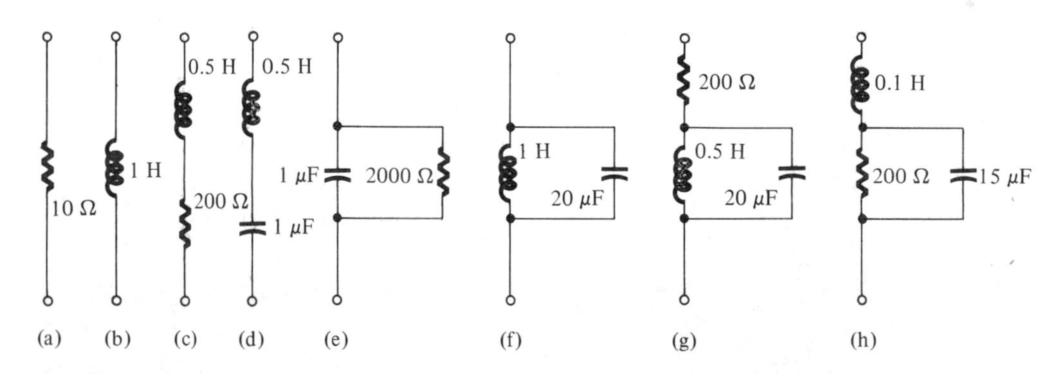

FIGURE 4.23

4.2 Given the complex numbers $\mathbf{A} = 3 + j2$, $\mathbf{B} = 4 - j3$, and $\mathbf{C} = -1 + j2$, express the results of the following operations in polar form:

(a) $\mathbf{AC} =$ (b) $\mathbf{B} - \mathbf{C} =$
(c) $(\mathbf{A} - \mathbf{B})(\mathbf{C} + j1) =$ (d) $(\mathbf{A} + \mathbf{B})/\mathbf{C} =$
(e) $\sqrt{\mathbf{B} + \mathbf{C}} =$ (f) $\mathbf{BC}/(\mathbf{B} + \mathbf{C}) =$

4.3 Using the complex numbers of Problem 4.2, give the results of the following operations in polar form:

(a) $(\mathbf{A} + \mathbf{B})^* =$ (b) $|\mathbf{B} - \mathbf{A}| =$
(c) $|\mathbf{B}|^2 =$ (d) $\mathbf{CB}/(\mathbf{C} + \mathbf{B}) =$

4.4 The voltmeter in Fig. 4.24(a) reads rms magnitudes. The following readings are made on the circuit:

$$|V_{ae}| = 15 \text{ V}$$

$$|V_{ab}| = 9 \text{ V}$$

Graphically determine what the voltmeter should read across V_{bc}.

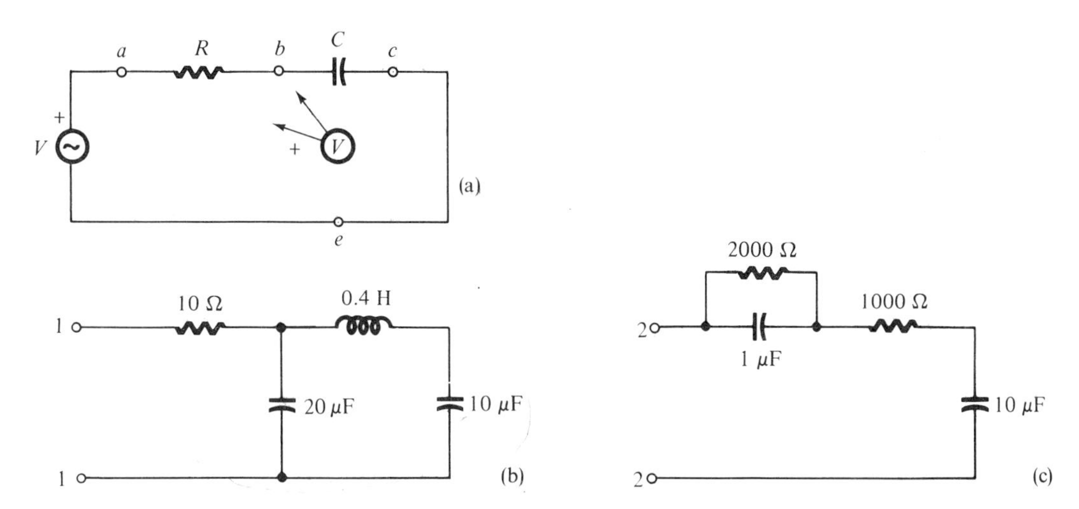

FIGURE 4.24

4.5 Find the input impedance at terminals 1,1 in Fig. 4.24(b) when driven at (a) $f = 5$ Hz and (b) $f = 60$ Hz.

4.6 Determine the impedance at terminals 2,2 in Fig. 4.24(c) for (a) $\omega = 20$ r/s and (b) $\omega = 1000$ r/s.

4.7 Calculate the polar form of the following expressions:

(a) $\dfrac{(10 + j12)(3 + j6)}{(9 + j3)(12 - j7)}$

(b) $\dfrac{(8 + j7)(3 - j5)}{10\angle 135^\circ}$

(c) $\dfrac{12 - j7}{10\angle 77^\circ \times 22\angle 130^\circ}$

4.8 Given phasors $\mathbf{A} = 5\angle 0^\circ$, $\mathbf{B} = |B|\angle -105^\circ$, and $\mathbf{C} = |C|\angle 95^\circ$, find the magnitudes of $\mathbf{B}$ and $\mathbf{C}$ if $\mathbf{A} + \mathbf{B} + \mathbf{C} = 0$.

4.9 Phasors $\mathbf{A} = 10\varepsilon^{j30^\circ}$ and $\mathbf{B} = 20\varepsilon^{j50^\circ}$ are used in the following operations; find the results and express them in polar form:

(a) $\mathbf{A} + \mathbf{B} =$

(b) $\mathbf{A}/\mathbf{B} =$

(c) $\sqrt{\mathbf{AB}} =$

4.10 As time functions two currents are $i_1 = 10 \sin(377t + 20°)$ and $i_2 = 20 \sin(377t + 77°)$. They add at node A to form i_3. (a) Express i_3 as a function of time. (b) Find the rms value of i_3.

4.11 For the network in the box in Fig. 4.25(a), the source is $v = 100 \sin(10^3 t + 30°)$, and the input current is $i = 6 \sin(10^3 t + 50°)$. (a) Find the equivalent series impedance represented by the box. Is the network capacitive or inductive in effect? (b) Compute the average power supplied by the generator to the network. (c) Check (b) using the average of the instantaneous $i^2 R$; also check (b) by using $I_{rms}^2 R$.

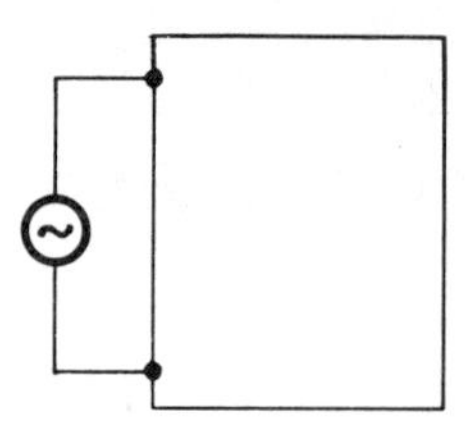

FIGURE 4.25

4.12 For the voltage and current variations shown in Fig. 4.26(a), over one period, determine the (a) average current and (b) average power supplied to a resistor of 20 Ω.

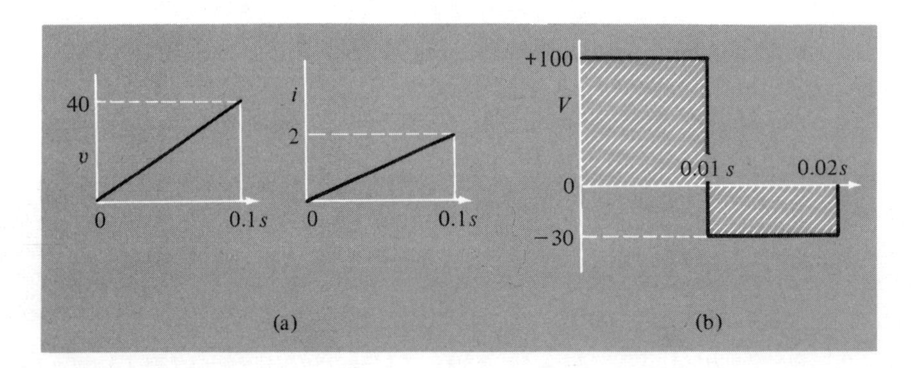

FIGURE 4.26

4.13 (a) For the wave forms of Fig. 4.26(a), calculate the rms voltage and current values. (b) Determine the power delivered to a 20-Ω resistor by use of these rms values.

4.14 Calculate the dc and the effective voltage values for the wave form of (b) in Fig. 4.26. Find the average power delivered to a 20-Ω resistor.

4.15 Determine the dc and the rms voltage values for the wave form of (a) in Fig. 4.27. Find the average power delivered if this voltage is connected across a 10-Ω resistor.

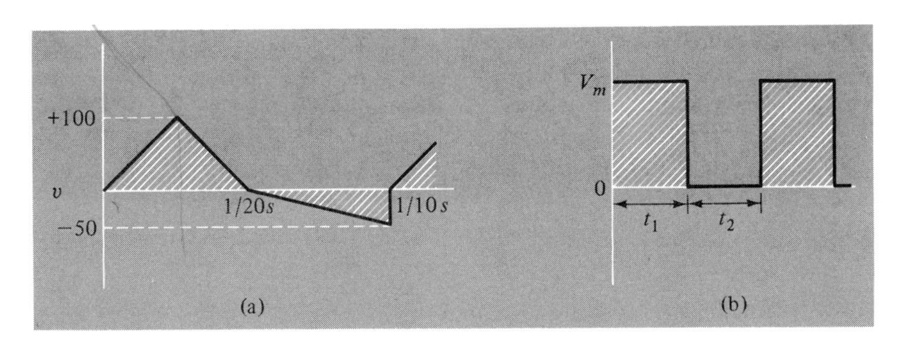

FIGURE 4.27

4.16 The rectangular wave of Fig. 4.27(b) causes a PMMC voltmeter to read 240 V and an electrodynamometer voltmeter to read 300 V. The period of the wave is $t_1 + t_2 = 0.05$ s. Find V_m, t_1, and t_2.

4.17 An ac voltage of 120 V rms is applied to the network of Fig. 4.28(a). Determine the average power dissipated in all resistors and show that this is equal to the power input from the source.

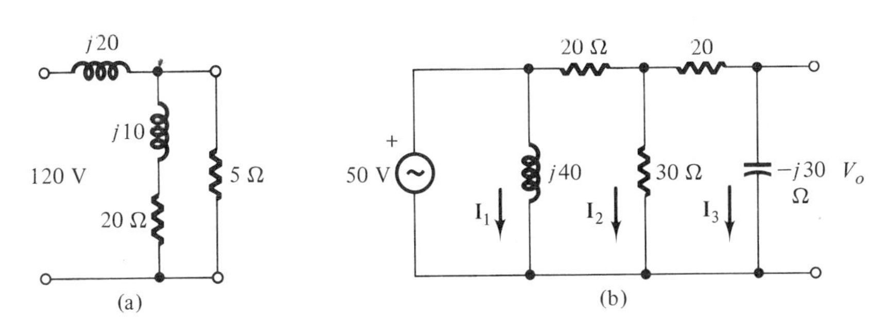

FIGURE 4.28

4.18 In Fig. 4.28(b), find $\mathbf{I}_1$, $\mathbf{I}_2$, $\mathbf{I}_3$, and $\mathbf{V}_o$.

4.19 In Fig. 4.29(a), $\mathbf{Z}_1 = 22.3\angle 27°\ \Omega$ and $\mathbf{Z}_2 = 35.8\angle -56°\ \Omega$. Find the power supplied by the 100-V ac generator.

4.20 A basic PMMC instrument movement indicates full scale with a 1.0-mA current; its resistance is 80 Ω. A series resistor of 199,920 Ω

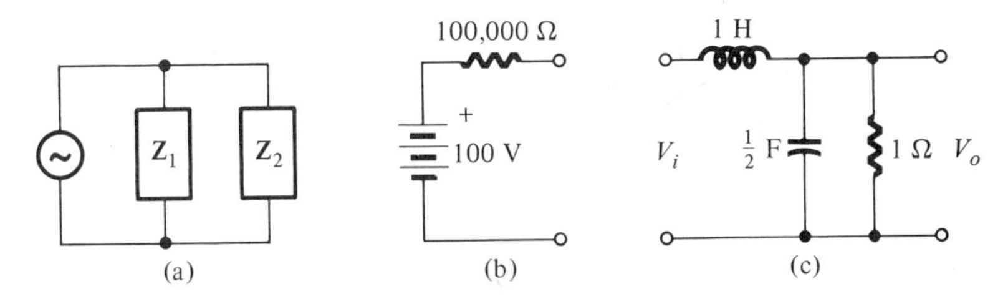

FIGURE 4.29

is added to make it a voltmeter. (a) What is the full-scale reading of the resultant instrument, in volts. (b) What does the instrument indicate when connected across the port of Fig. 4.29(b)? (c) Why does it not read 100 V?

4.21 Scale the circuit of Fig. 4.29(c) to provide a shunt resistance of 10,000 Ω, replacing the 1 Ω resistance.

4.22 You have a box with one port. Voltages of 100 V, dc, and 100 V, 100 Hz, are separately applied to the port, and the input current and average power readings are taken as

	Volts	*Amps*	*Watts*
At dc	100	1.0	100
At 100 Hz	100	0.47	22

Find the resistance and reactance values for a series circuit which would be equivalent to the box. How can you determine what kind of reactance is present?

4.23 Find the rms value of the voltage-time function of Fig. 4.15(b) as a function of V_m; repeat for the wave at (c).

4.24 Using the Laplace transform, write V_o of Fig. 4.30(a). Using $j\omega = s$, transform the voltage expression to a function of ω.

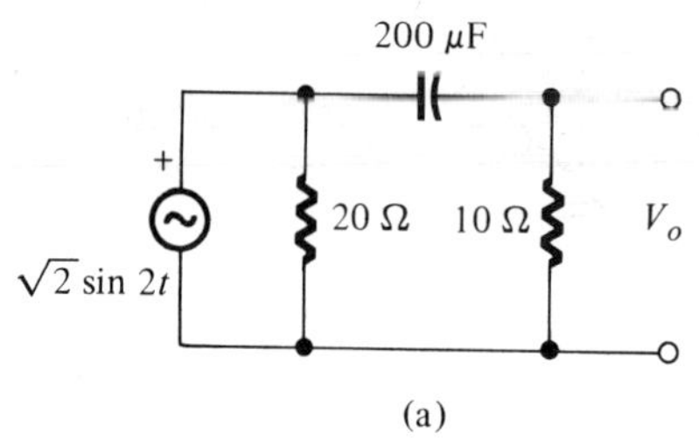

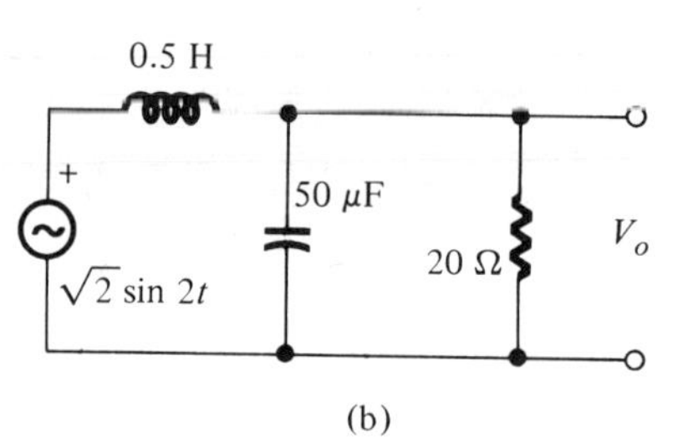

FIGURE 4.30

4.25 For the circuit of Fig. 4.30(b), find V_o.

4.26 For the circuit of Fig. 4.31(a), find (a) the current and power supplied by each generator and (b) the current in the 10 Ω resistance.

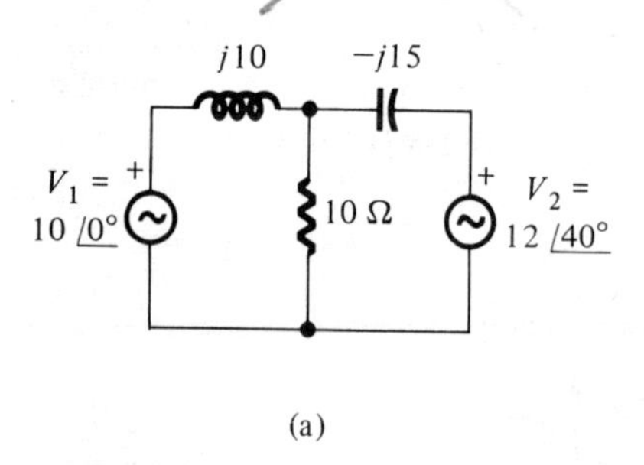

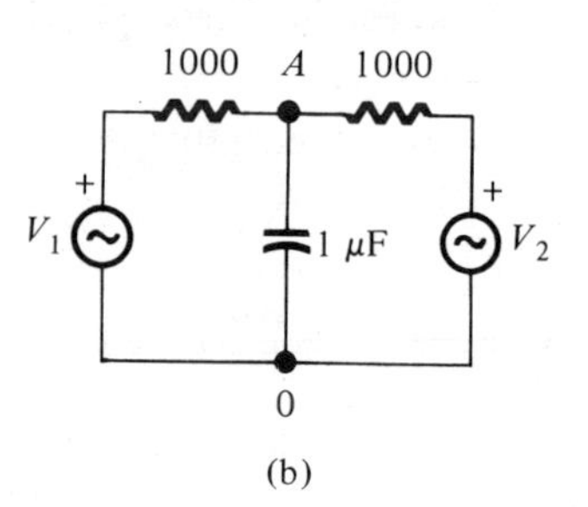

FIGURE 4.31

4.27 In Fig. 4.31(b), with $V_1 = 14\sin 10^3 t$ and $V_2 = 10\cos 10^3 t$, find the voltage V_{Ao} as a function of time.

4.28 With $V = 10\angle 0°$ V in Fig. 4.32(a), find the average power supplied to resistor R_3 and the total power supplied to the network. The frequency of the generator is 100 Hz.

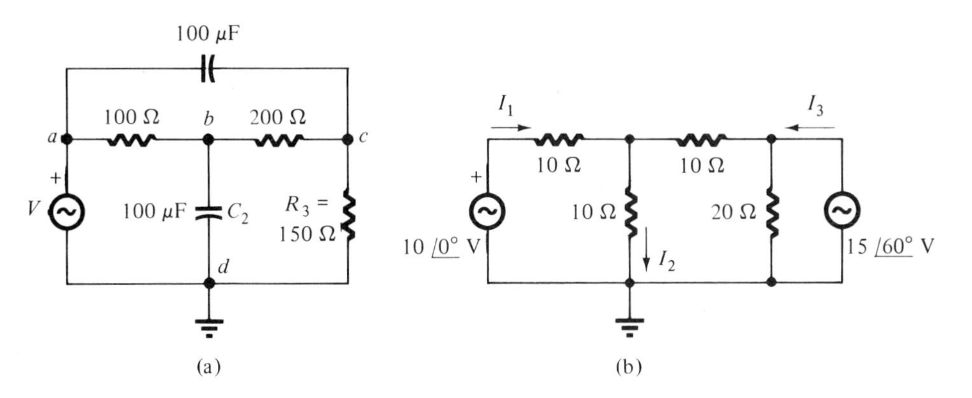

FIGURE 4.32

4.29 In Fig. 4.32(b), find the power supplied by each generator.

4.30 In Fig. 4.33(a), $V_1 = 30\angle 0°$ V and $V_2 = 50\angle 36°$ V, both at 60 Hz. Find the current and power supplied by each generator.

4.31 In Fig. 4.33(b), the current I is 5.7 A, as read by an electrodynamometer instrument, with frequency at 100 Hz. Find the value of R in ohms.

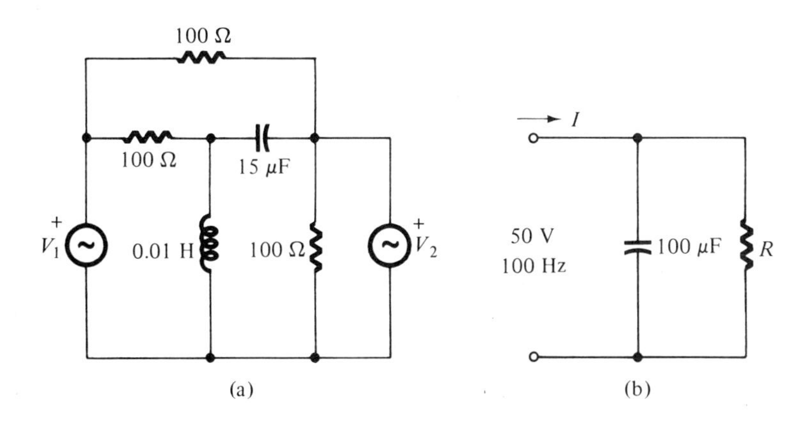

FIGURE 4.33

4.32 An inductor L, having a 30-Ω series resistance, is connected in parallel with 100 Ω of resistance. When the combination is connected to a source of 100 V, 16-Hz frequency, the circuit takes 150 W. What is the value of the inductance in henrys?

4.33 In Fig. 4.34(a), we want V_o to be 2.7 V (rms). Find the needed value of R.

4.34 For the circuit of Fig. 4.34(b), (a) determine the values in ohms, henrys, and farads for an equivalent series circuit at 60 Hz. (b) If 100 V at 60 Hz is applied at port 1,1, find the input power.

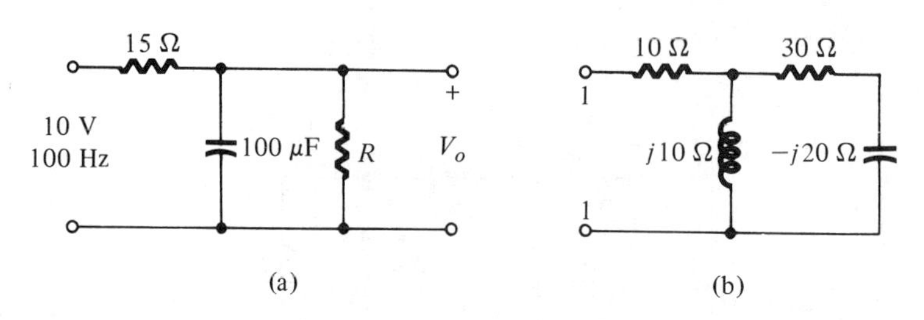

FIGURE 4.34

4.35 The generators A and B in Fig. 4.35(a) give outputs at 796 Hz. (a) Find the voltage V_o. (b) To what quantity does the angle on V_o refer? (c) Find the power output from each generator.

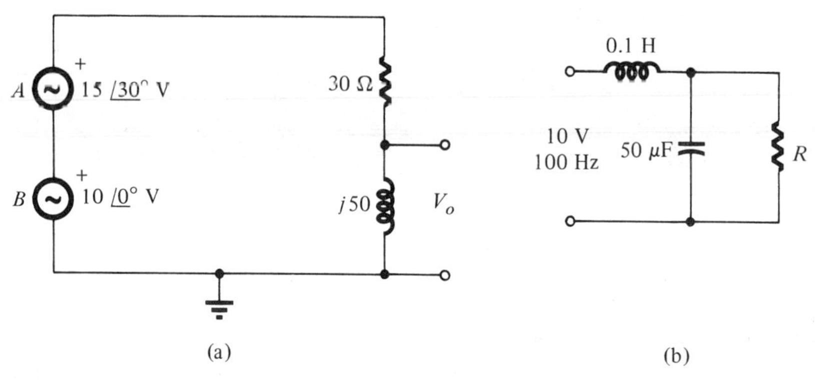

FIGURE 4.35

4.36 For the circuit of Fig. 4.35(b), we want 0.5 W delivered to the resistor R. What must be the value of R in ohms?

4.37 The power input to the circuit of Fig. 4.34(a) is measured as 5.25 W. Find the value of R in ohms.

4.38 In the circuit of Fig. 4.36(a), the generator voltages are $V_{Ao} = 20\angle 35^\circ$ V, $V_{Bo} = 27\angle 12^\circ$ V, and $V_{Co} = 18\angle -15^\circ$ V. (a) Find the voltage V_o across the load of $10 + j10\ \Omega$. (b) Compute the power output from each source at A, B, and C, respectively.

4.39 The ac bridge of Fig. 4.36(b) has $Z_1 = R_1 = 1000\ \Omega$ and $Z_4 = R_4 = 2000\ \Omega$, with a supply frequency of 159 Hz. Find the impedance Z_3 needed for balance of the bridge and its R, L, or C values and arrangement if the unknown impedance is Z_2 due to a 5-μF capacitor.

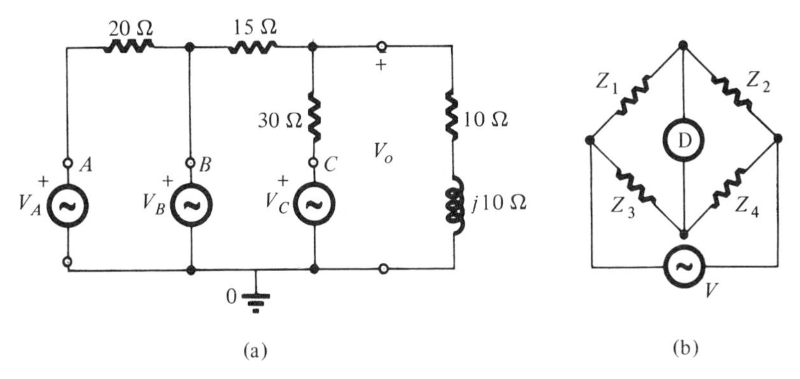

FIGURE 4.36

4.40 An ac motor is connected to a 120-V, 60-Hz source and under steady load takes 480 W with a lagging power factor of 0.8 ($\cos\theta = 0.8$). (a) With voltage as the reference angle, state the current through the motor, in polar form. (b) Draw a circuit of passive elements which is equivalent to this motor under load. Give the element values in ohms and henrys or farads.

4.41 A capacitance C is connected in parallel with an inductance whose internal series resistance is 10 Ω. A source of 10 V and 30 Hz is connected to the circuit. Find the value of C which will make the generator current a minimum with $L = 0.5$H.

4.42 A balanced three-phase system has line-to-line voltages of 120 V. If three identical impedances, $Z = 52 + j36\ \Omega$, are connected in Y across the lines, find the line currents, the total power, and the load power factor. (b) If the same impedances are connected in delta across the lines, find the line currents, the total power, the power factor, and the current in each impedance.

4.43 Show that the number of loops equals the number of branches minus the number of nodes plus one for the network of Fig. 4.37(a).

4.44 (a) How many independent loop-voltage equations can be written for the circuit of (b) in Fig. 4.37? (b) How many independent nodal equations can be written for the circuit? (c) If $V = 10$ V, $R_1 = 1000\ \Omega$, $R_2 = 200\ \Omega$, $R_3 = 300\ \Omega$, $R_4 = 400\ \Omega$, and $R_5 = 500\ \Omega$, determine the current in R_4.

4.45 Write the loop equations for Fig. 4.37(b) in matrix form. Also write the nodal equations in matrix form.

4.46 The networks of Fig. 4.37(c) and (d) are connected in cascade as 2 to 3. Write the impedance matrix for the cascaded connection between 1 and 4.

4.47 A three-phase system has 220 V rms between lines. A resistive load totaling 8 kW is delta-connected and in parallel with a three-phase motor taking 20 kW at a lagging power factor of 0.75. Find the line currents.

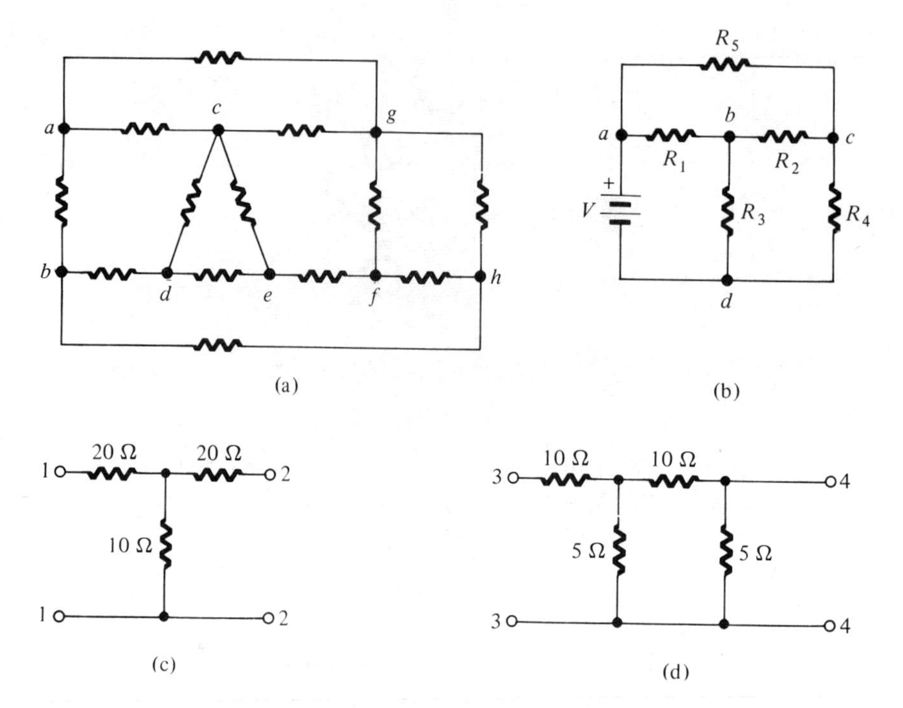

FIGURE 4.37

4.48 A voltmeter with a full-scale reading of 15 V is needed. The basic instrument movement of Problem 4.20 is available. (a) What multiplier series resistance should be used? (b) How much power is consumed in the complete voltmeter at full indication?

4.49 The standard voltage for domestic power distribution in the United States is now 117 V at 60 Hz (sinusoidal). (a) What is the rms voltage? (b) What is the peak voltage? (c) What is the angular frequency? (d) What is the period?

4.50 You have a PMMC instrument showing 300 V at full scale. The resistance at the terminals is 30,000 Ω. Upon opening the case, you find a series resistance of 29,950 Ω. (a) What is the sensitivity of this instrument? (b) By removing the series resistor and shunting the basic movement, you can make an ammeter to read 1 A full scale. What shunt resistance should be used?

REFERENCES

1. W. C. Johnson, *Mathematical and Physical Principles of Engineering Analysis*. McGraw-Hill Book Company, New York, 1944.
2. E. Guillemin, *The Mathematics of Circuit Analysis*. John Wiley & Sons, Inc., New York, 1949.
3. J. S. Frame, "Matrix Functions and Applications." *IEEE Spectrum*, 1, No. 3, 208; No. 4, 102; No. 5, 100; No. 6, 123; No. 7, 103 (1964).
4. M. E. Van Valkenburg, *Network Analysis*, 2d ed. Prentice-Hall, Inc., Englewood Cliffs, N.J., 1964.
5. W. E. Lewis and D. G. Pryce, *The Application of Matrix Theory to Electrical Engineering*. Chapman & Hall Ltd., London, 1965.

nonlinear elements; rectifier circuits

five

We have been working with linear bilateral circuit elements; we are now ready to consider nonlinear elements with which the principle of superposition is not valid. The voltage-current curve of a nonlinear circuit element provides the basic relation, but we have only a cumbersome graphical procedure or numerical methods with computers for circuit solution. For analytic use we are led to approximations which consider the nonlinear device as quasi-linear. That is, we approximate the nonlinear characteristic curve by several straight lines; this method is known as *piecewise linearization*. The quasi-linear device model is then employed as a circuit element, under restricted conditions of voltage or current.

Such methods will be applied here with the rectifier diode, one of our most common nonlinear elements. The diode is also a nearly unilateral device which passes current more easily in one direction than in the other. When used with filter circuits to give smooth unvarying output current from an ac supply, diode circuits provide examples for the study of transients as well as the use of phasors in ac circuits.

5.1 The semiconductor diode

A *diode* is a one-port nearly unilateral circuit device which acts as a polarity-controlled switch, with low resistance for one polarity and very high resistance for the opposite polarity. This is the property of a *rectifier*, used for conversion of alternating current to unidirectional direct current.

The semiconductor diode of Fig. 5.1(a) obeys the *diode equation*,

$$i = i_s(\varepsilon^{eV/kT} - 1) \tag{5.1}$$

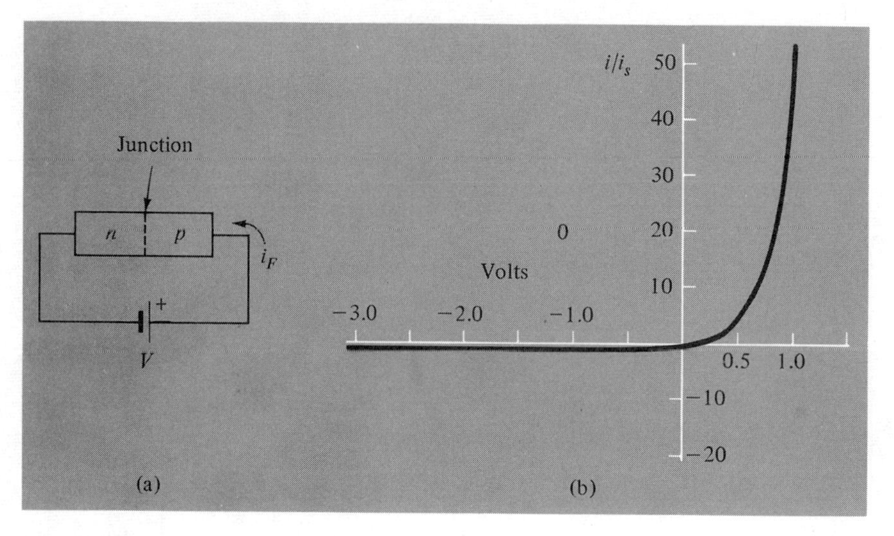

FIGURE 5.1
(a) Junction np diode; (b) volt-ampere characteristic.

where

$$e = \text{electronic charge magnitude} = 1.60 \times 10^{-19}\ \text{C}.$$

$$k = \text{Boltzmann's constant} = 1.38 \times 10^{-23}\ \text{J/°K}.$$

$$T = \text{temperature, °K}.$$

Usually made of one of the semiconductors, germanium or silicon, the diode has a p terminal, or anode, and an n terminal, or cathode. With forward potential ($+V$) applied to the p terminal of (a) and with V large with respect to kT/e ($kT/e = 0.026$ V at 300°K), the exponential term is large with respect to unity and

$$i = i_s \varepsilon^{eV/kT} \tag{5.2}$$

The forward voltage-current curve is of exponential form, as shown in Fig. 5.1(b). The forward voltage rarely exceeds 1 V in magnitude and is partly due to the resistance of the semiconductor body materials.

With $-V$ applied to the p terminal of the diode, the exponential term becomes negligible with respect to unity and $i = -i_s$. This is a very small reverse leakage current. Theory shows this current as constant for all negative voltages; actually at some high reverse voltage a *breakdown* occurs, although the diode is not damaged. The current rises to some high value but the voltage remains nearly constant. When used for voltage stabilization, the reverse-biased diode is known as a *Zener diode*.

5.2 The incremental forward resistance

Equation 5.2 shows the diode as a nonlinear element having an exponential voltage-current relation in the forward direction. For applied voltage V in Fig. 5.2(a), the current is

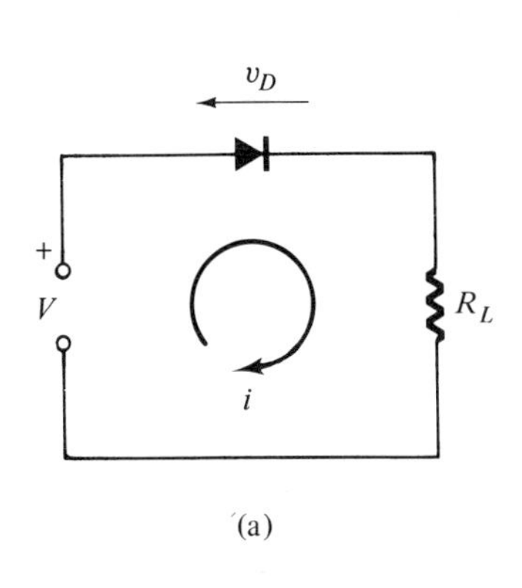

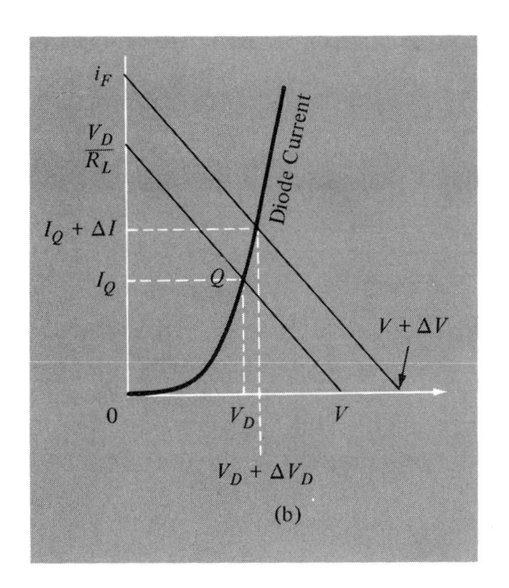

FIGURE 5.2
(a) Diode circuit; (b) graphical diode analysis.

$$i = f(v_D) \tag{5.3}$$

in general. Around the circuit

$$v_D = V - iR_L \qquad (i > 0) \tag{5.4}$$

This is the equation of a straight line, of the form $x = b + my$. The line has intercepts at V and at $i = V/R_L$ and is known as the *load line*, representing relations between current in the circuit and voltage across the load R_L, for positive currents. Equations 5.3 and 5.4 involve the same current i and can be solved simultaneously by a graphical method.

In Fig. 5.2(b) the load line is drawn on the diode characteristic for a given supply voltage V. The intersection of the two lines determines the current as I_Q. If the supply voltage is increased to $V + \Delta V$, a new load line must be drawn from the new voltage intercept but with the same slope, since the slope is the negative reciprocal of the load resistance R_L. The current is then $i = I_Q + \Delta I$ and $v_D = V_D + \Delta V_D$. Substitution in Eq. 5.4 gives

$$V_D + \Delta V_D = V + \Delta V - (I_Q + \Delta I)R_L$$

By Eq. 5.4 we have

$$V_D = V - I_Q R_L$$

at the intersection, and by subtraction we have the remainder

$$\Delta V_D = \Delta V - \Delta I R_L \tag{5.5}$$

The resistance of the circuit to a change is found by dividing by ΔI

$$\frac{\Delta V}{\Delta I} = R_L + \frac{\Delta V_D}{\Delta I} = R_L + r_D$$

where

$$r_D = \frac{\Delta V_D}{\Delta I} \tag{5.6}$$

For a change in input the diode appears to add a resistance $r_D = \Delta V_D/\Delta I$ to the circuit. This diode resistance r_D is the variational or *ac resistance*, which can be measured in terms of the reciprocal slope of the diode voltage-current curve at the operating point Q.

For variations in the operating point Q well above the lower sharp curvature, the diode curve can be reasonably approximated by the piecewise linear characteristic of Fig. 5.3(a). The diode current is zero for forward

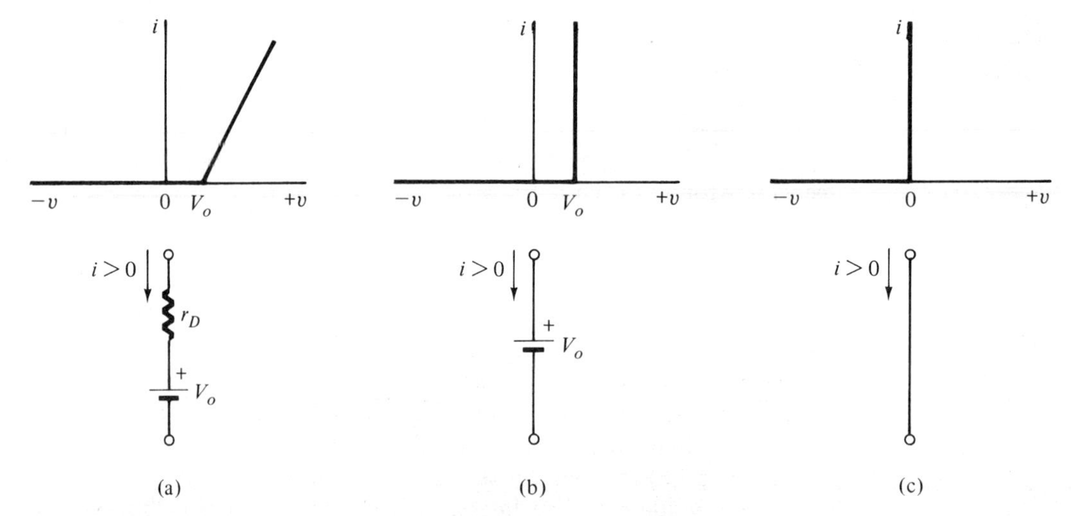

FIGURE 5.3
(a) Piecewise linear diode curve and forward current equivalent circuit; (b) neglecting the diode resistance; (c) ideal diode in forward direction.

voltages less than V_o, and the diode appears to have a constant slope or variational resistance for higher voltages. The circuit as given is approximate to the performance of the diode for all positive or forward currents. For voltages less than V_o the diode circuit appears open or with zero current. Voltage V_o is of the order of 0.2 V for a germanium diode and 0.5 V for a silicon unit.

As a further simplification of the diode equivalent circuit we have (b) in Fig. 5.3. We neglect the diode resistance in the forward direction as small with respect to any external circuit resistance but still include V_o. In the forward direction the diode is simulated by a battery V_o and appears as an open circuit in the reverse direction. At (c) in Fig. 5.3 the characteristic is that of the *ideal diode*, of zero forward voltage drop and zero resistance. This model is useful if the applied circuit potential is large with respect to V_o, as is usually the case in rectifier circuits.

5.3 The diode as a wave clipper

With the diode represented by its ideal characteristic it acts as a polarity-controlled switch, which is useful in study of circuits in which the wave form is changed by *clipping*.

In (a) in Fig. 5.4, we have a shunt diode which will clip off all positive voltages greater than V_1. For $v_i < V_1$ the diode is held open by V_1 as in (b), and for $v_i > V_1$ the diode is in its conducting region as in (c). The break point

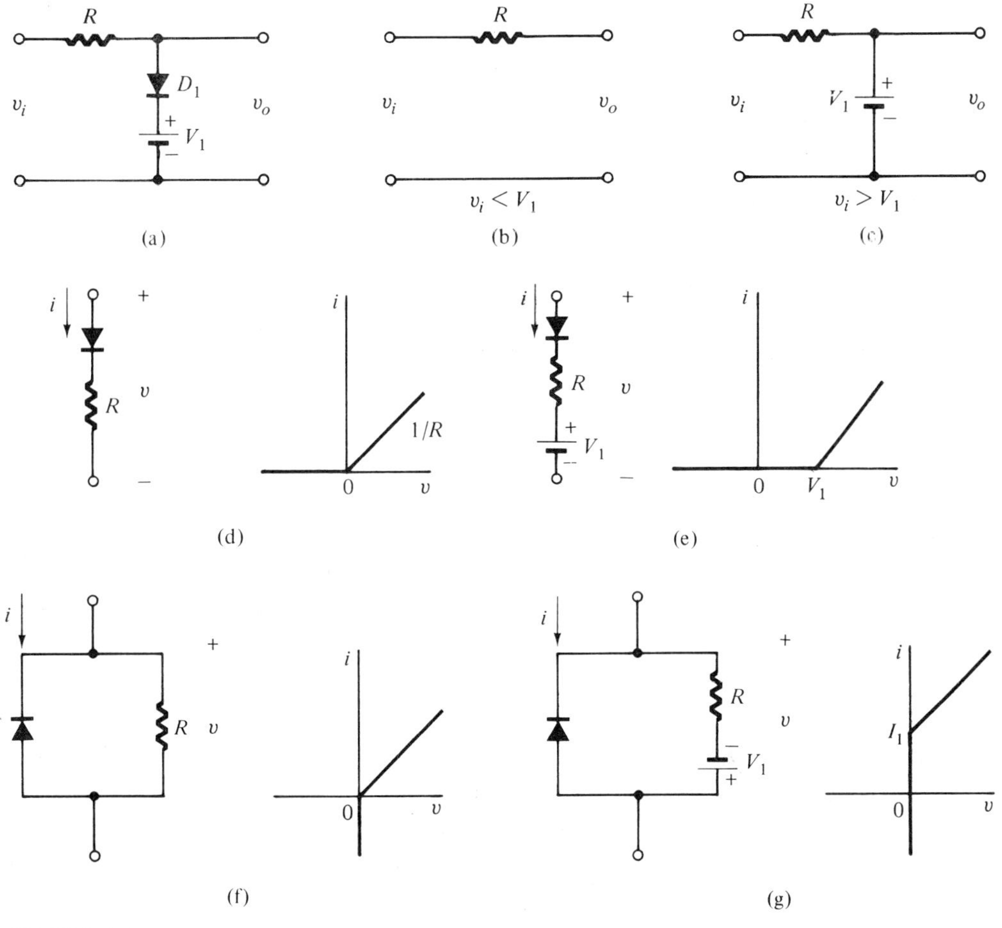

FIGURE 5.4

(a), (b), (c) Diode shunt clipper and state circuits; (d), (e), (f), (g) piecewise linear circuits and performance curves.

between the two states occurs at $v_i = V_1$. The slope of the characteristic in the forward direction may be changed by adding a resistance R in series with the diode, giving a forward direction slope of $1/R$ from the break point, as in Fig. 5.4(d). The break point may be shifted on the voltage axis by connecting a voltage source in series with the diode as in Fig. 5.4(e). Parallel connection of diode and resistance causes the diode to maintain zero voltage across the output for all negative voltages, with a slope of $1/R$ for positive voltages, as in Fig. 5.4(f). In Fig. 5.4(g), the break point is shifted up or down the current axis by addition of a voltage in series with R, giving a break point current of $I_1 = V_1/R$.

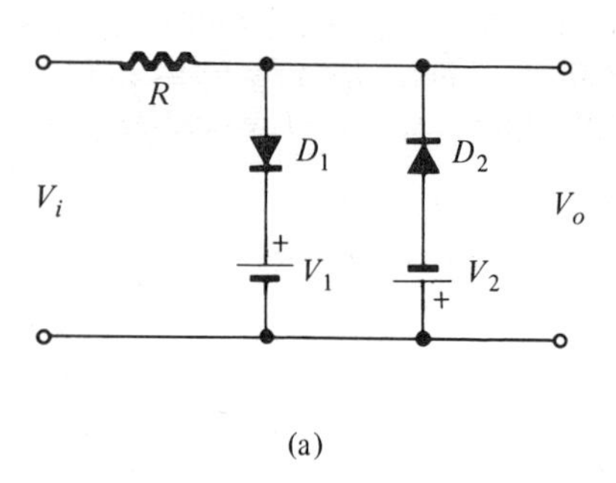

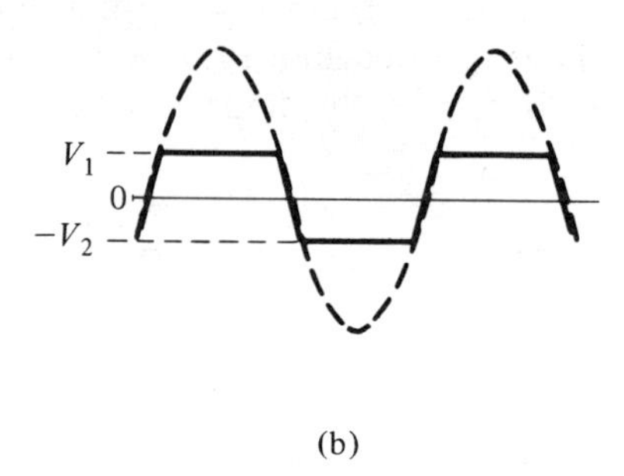

FIGURE 5.5
Double shunt clipper.

A double shunt clipper [Fig. 5.5(a)] can shape an applied sine wave as in Fig. 5.5(b), cutting at amplitudes fixed by V_1 and $-V_2$. This is one means of generation of waves of approximate rectangular form.

5.4 The diode as a logic switch

Digital computers employ pulse trains to represent binary numbers, with 1011 and 0111 being shown in Fig. 5.6(a). The respective negative and positive pulses represent 0 and 1 in the binary number system, the most significant digit appearing at the left. To add binary numbers, it is necessary to have a circuit which will yield an output when a pulse is simultaneously present for signals *A and B*, as for the least significant digits at the right, or when pulses are present in signals *A or B*, as for any of the digits.

The circuits are said to respond to the *logical operations* AND and OR. Typical *diode logic switches* for these operations are shown in Fig. 5.6(b) and (c), designed for the *positive logic* signals of (a); this designation indicates that the 1-level signal is more positive than the 0-level signal.

Consider the AND circuit of (b) which is designed for +10 V for a 1 signal and −10 V for a 0 signal. The summing resistor R is connected to +10 V to ensure the desired signal levels. The diodes are assumed ideal, switching on and off in accordance with the polarity of the signal. A 0-level

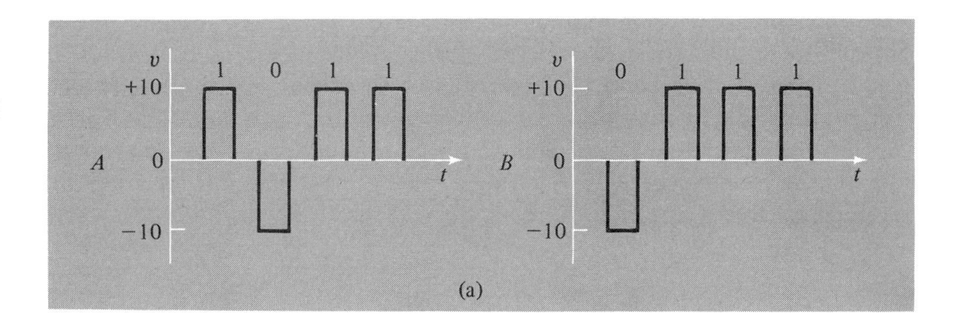

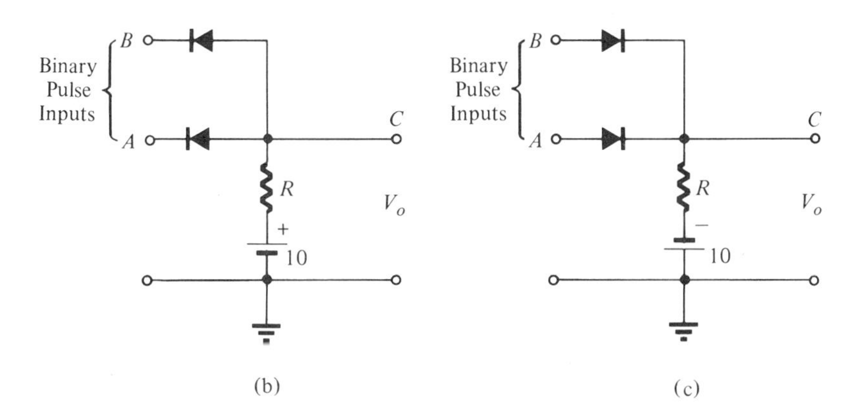

FIGURE 5.6

(a) Binary signals, 1011 and 0111; (b) AND circuit in positive logic; (c) OR circuit.

signal at A connects -10 V to that terminal and the diode is switched on; the output at C is then -10 V, or 0 level. A 0-level signal at B switches on its diode, but the output remains at -10 V. A 1-level signal at B places $+10$ V on that diode and the back-biased diode becomes an open circuit; however, the 0-level signal on A still controls the output to the 0 level. Simultaneous 1-level signals on A and B open both diodes and the output at C goes to $+10$ V, or the 1-level output.

The above operations can be summarized in a *table of combinations* for the AND operation as

A	B	C
0	0	0
0	1	0
1	0	0
1	1	1

which shows a 1-level output only when 1-level signals exist simultaneously on *A and B*; this is the operation of a logical AND circuit.

The OR circuit at (c) in Fig. 5.6 gives an output at C when a 1-level signal is present at *A or B* since a +10-V input turns on a diode and places +10 V on the output. Both diodes may be simultaneously turned on, but the output remains at +10 V, giving a table of combinations for a logical OR circuit as

A	B	C
0	0	0
0	1	1
1	0	1
1	1	1

This diode application illustrates the polarity-switching property as well as the circuit isolation obtainable by the use of diodes.

5.5 The diode in power rectification

The diode is employed in the circuit of Fig. 5.7(a) as a unilateral switch, a *half-wave rectifier*. The circuit current can be expressed as

$$i_b = \frac{V_m \sin \omega t - v_D}{R_L} \qquad (i > 0) \tag{5.7}$$

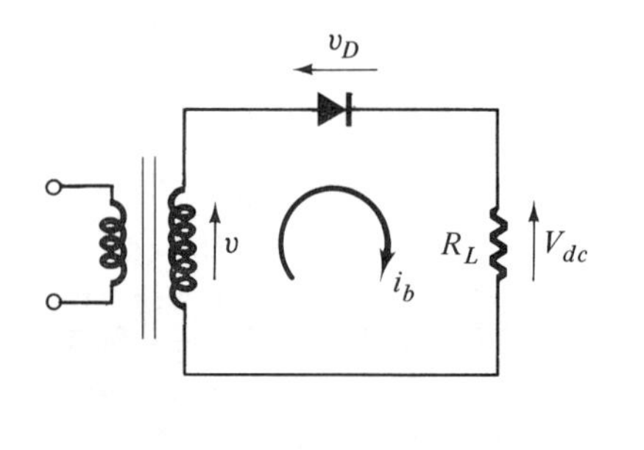

(a)

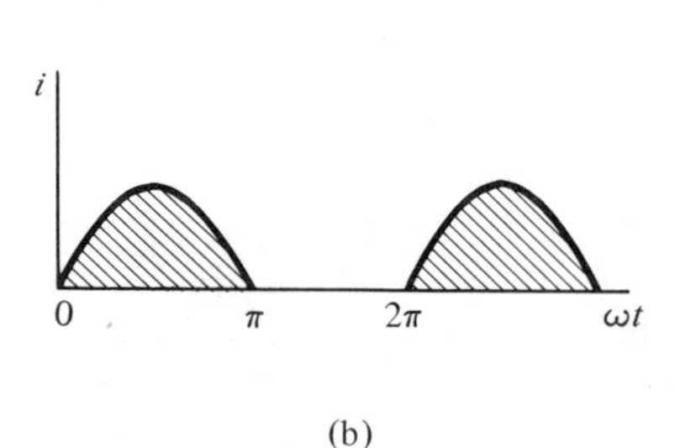

(b)

FIGURE 5.7
(a) Half-wave rectifier circuit; (b) current in load.

This current can only be positive; negative currents are not possible in the ideal diode. Use of the ideal diode characteristic reduces v_D to zero; the current wave form appears in Fig. 5.7(b).

The *full-wave rectifier circuit* of Fig. 5.8 is more efficient; with positive polarity alternately present on each diode anode, the diodes supply current to the load R_L on alternate half cycles, as in Fig. 5.8(b). Such a voltage wave

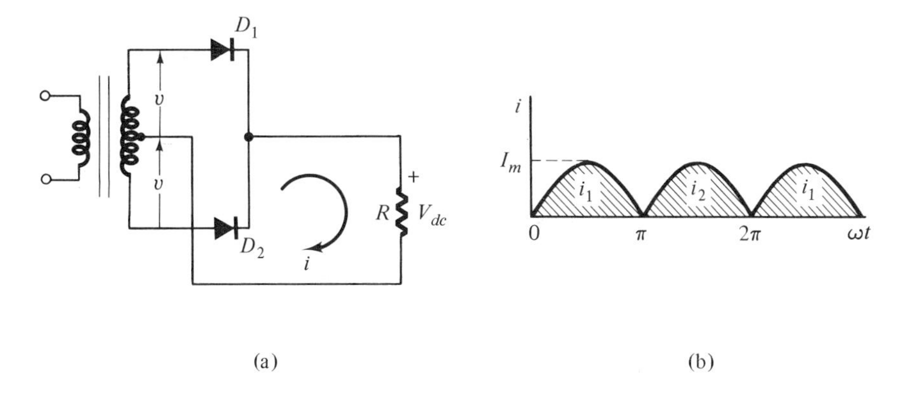

FIGURE 5.8
(a) Full-wave rectifier circuit; (b) current in load.

form has been expressed by a Fourier expansion in Chapter 4 as

$$v = \frac{2V_m}{\pi} - \frac{4V_m}{\pi}\left(\frac{1}{3}\cos 2\omega t + \frac{1}{15}\cos 4\omega t + \frac{1}{35}\cos 6\omega t + \cdots\right) \tag{5.8}$$

The dc or average term is the desired output of the rectifier and is

$$V_{\mathrm{dc}} = \frac{2V_m}{\pi} = \frac{2I_m R}{\pi} \tag{5.9}$$

A second form of full-wave rectifier circuit appears in Fig. 5.9(a) as the *bridge rectifier circuit.* In this arrangement a transformer of one half the total voltage and about 70 percent greater current rating can supply the same dc voltage and power as given by the center-tapped circuit; two extra diodes are required, however. Diodes D_1 and D_3 conduct in series when the upper transformer terminal is positive, and diodes D_2 and D_4 conduct when the lower terminal is positive. The current pulses in the load are in the same direction, and the output current is again that of Fig. 5.8(b).

When rectifiers supply electronic amplifiers, the output must be a smooth, unvarying direct current, free of extraneous ripple from the alternating harmonic terms of Eq. 5.8. The *ripple factor* measures the smoothness of the output current wave as

$$\gamma = \frac{\text{rms value of the ac components}}{\text{average value of the load voltage}}$$

Analysis of the wave forms leads to the general expression

$$\gamma = \sqrt{\frac{I_{\mathrm{rms}}^2}{I_{\mathrm{dc}}^2} - 1} \tag{5.10}$$

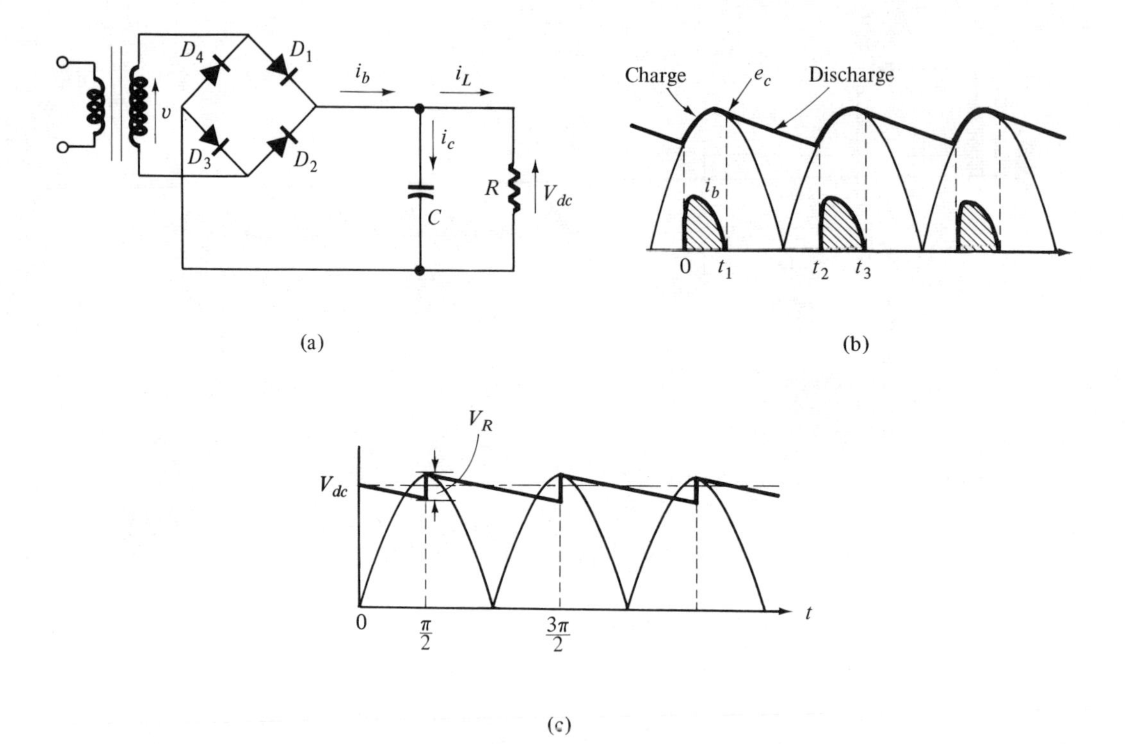

FIGURE 5.9
Capacitor filter.

For the output wave form of Fig. 5.8(b), $I_{\text{rms}} = I_m/\sqrt{2}$, and

$$\gamma = \sqrt{\frac{I_m^2/2}{4I_m^2/\pi^2} - 1} = 0.485$$

which is too large a ripple to be tolerated for supplying electronic amplifiers.

5.6 The capacitor filter

Filter circuits are used to smooth the output voltage of rectifier circuits. The capacitor filter of Fig. 5.9(a) changes the load voltage wave form to that of the heavy line in Fig. 5.9(b). The capacitor charges alternately through the conducting diodes to the peak of the voltage wave and maintains the load voltage near that value by reason of a large time constant RC.

At $t = 0$ the diode starts to charge C and supplies a current $i_b = i_C + i_L$, the crosshatched wave form in (b). After the ac wave peaks, the voltage v from the transformer falls below the capacitor voltage v_C. The potential across diodes D_1 and D_3 reverses at $t = t_1$ and this causes the diodes to open and disconnect the supply from the load. The capacitor then supplies current to the load at voltage $V_L(s)$, and we can write the circuit equation in the s domain as

$$V_L(s) = RI_L(s) = \frac{I_C(s)}{sC} + \frac{V_L(0+)}{s} \tag{5.11}$$

However, $I_L(s) = -I_C(s)$ during this discharge interval, so

$$RI_L(s) = V_L(0+)\left(\frac{1}{s + 1/RC}\right) \tag{5.12}$$

At $t = t_1$ when the capacitor starts to supply the load current, $V_L(0+) = V_m \sin \omega t_1$, and reference to a table of transforms provides the solution for $i_L(t)$, after which

$$v_L = Ri_L = V_m \sin \omega t_1 \varepsilon^{-\omega t/\omega CR} \tag{5.13}$$

after changing the variable to ωt to permit time to be expressed in the wave period.

Equation 5.13 predicts an exponential fall in the load voltage for the interval of diode nonconduction t_1 to t_2. The parameter ωCR appears and is important in filter design; $\omega CR = 2\pi CR/T$ and is a function of the ratio of the filter-load time constant CR to the period T of the supply wave.

A small ripple is desired and ωCR should be large. Exact analysis is difficult because of a lack of knowledge of the diode conduction angle. However, with ωCR large the wave form may be approximated by (c) in Fig. 5.9, and an analysis is then possible.

In effect we assume the period of diode conduction to be negligibly short for large ωCR. In the discharge period, which is then $\cong \pi$, the capacitor voltage falls V_R. The average rate of fall over the period $(t_2 - t_1) \cong 1/(2f)$, where f is the supply frequency, can be written

$$\frac{V_R}{T} = \frac{V_R}{1/2f} = \frac{d}{dt}\frac{q}{C}$$
$$V_R = \frac{1}{2fC}\frac{dq}{dt} = \frac{1}{2fC} I_{dc} = \frac{\pi I_{dc}}{\omega C} \tag{5.14}$$

since $i_R = dq/dt \cong I_{dc}$ in the interval.

By the usual root-mean-square method the effective value of the triangular ripple wave is found to be

$$V_{R\text{-rms}} = \frac{V_R}{2\sqrt{3}} = \frac{\pi I_{dc}}{2\sqrt{3}\omega C} \tag{5.15}$$

Using $V_{dc} = I_{dc}R$, the ripple factor γ can be written from its definition as

$$\gamma = \frac{V_{R-\text{rms}}}{V_{dc}} = \frac{\pi}{2\sqrt{3}\omega CR} \tag{5.16}$$

for the full-wave rectifier with shunt-capacitor filter. The ripple, being inverse to R, increases with increase in load current. The result is plotted against ωCR in Fig. 5.10. Neglect of the conduction angle leads to little error for $\omega CR > 20$.

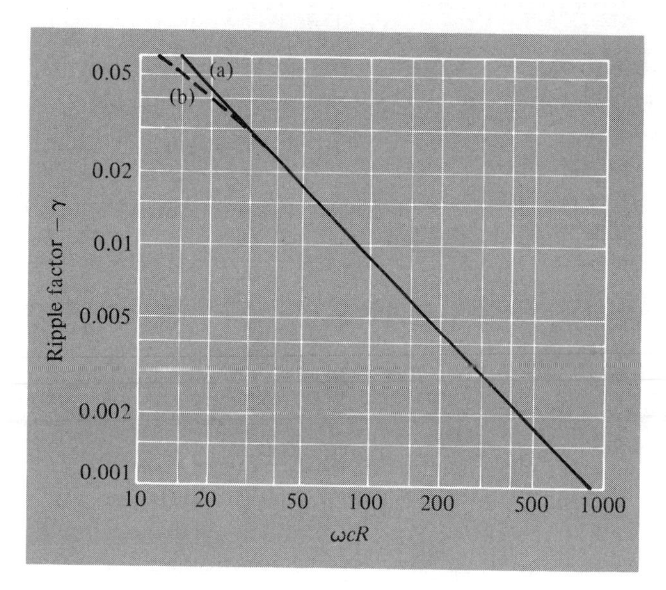

FIGURE 5.10
Full-wave circuit: ripple factor vs. ωCR (a) by Eq. 5.16; (b) conduction angle introduced.

The average or dc load voltage can be written as

$$V_{dc} = V_m - \frac{V_R}{2} = V_m - \frac{\pi I_{dc}R}{2\omega CR}$$

from which

$$\frac{V_{dc}}{V_m} = \frac{1}{1 + (\pi/2\omega CR)} \tag{5.17}$$

This represents the ratio of the dc load voltage to the peak of the applied ac voltage from the transformer and is plotted in Fig. 5.11 as a function of ωCR. Again the result is conservative when compared with a solution which includes the conduction angle.

Constancy of the dc voltage with variation in dc load current is important in electronic circuits, and selection of ωCR above 30 will place

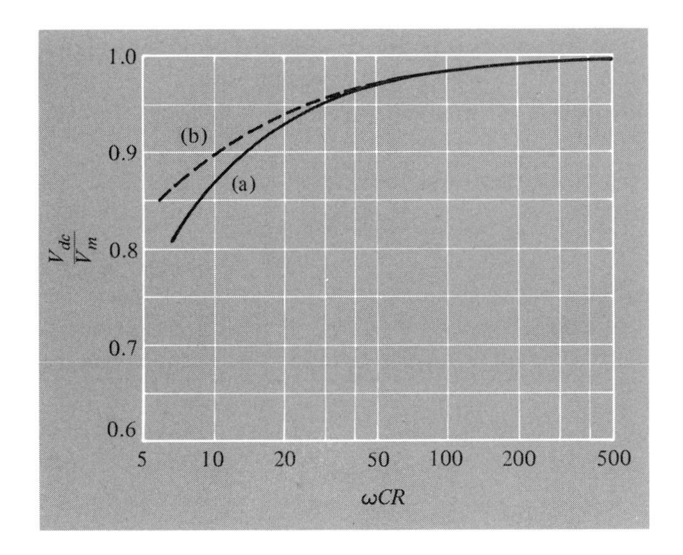

FIGURE 5.11

V_{dc} vs. ωCR: (a) by Eq. 5.17; (b) conduction angle included.

the operation on the upper plateau of the voltage output curve, ensuring constancy of voltage within a few percent.

The rms voltage of the transformer in Fig. 5.9, or from diode to center tap in Fig. 5.8, can be found as

$$V = \frac{V_m}{\sqrt{2}} = \frac{V_{dc}}{\sqrt{2}}\left(1 + \frac{\pi}{2\omega CR}\right) \tag{5.18}$$

5.7 The π filter

For applications requiring ripple factors much less than 1 percent, the cascaded filtering action of the π network of Fig. 5.12 is useful. Capacitor

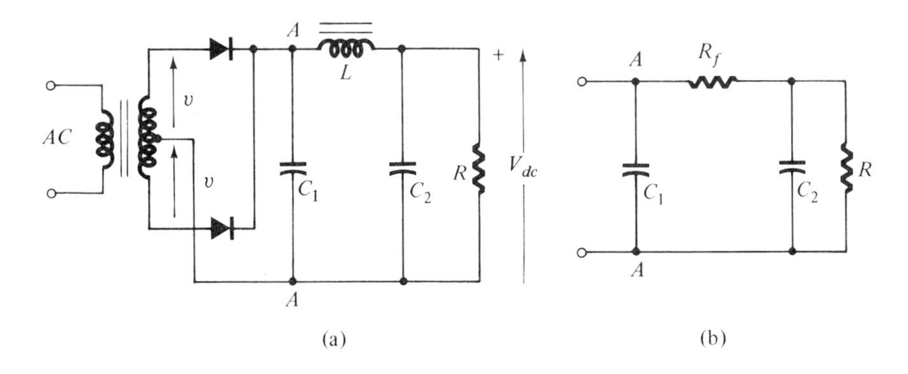

FIGURE 5.12

Full-wave rectifier and π filter.

C_1 acts in a transient manner and charges to the peak of the applied voltage, as does the shunt capacitor of the CR filter of the previous section. The ripple voltage V_R then appears at A,A and is given by Eq. 5.15 as

$$V_{AA} = V_{R-\text{rms}} = \frac{\pi I_{\text{dc}}}{2\sqrt{3}\omega C_1}$$

Note that the angular frequency of this voltage is 2ω, where ω is that of the ac supply. We make the reactance of C_2 small with respect to the load R, and by voltage division the portion of V_{AA} appearing across C_2 and the load is

$$V_2 = \frac{-j/2\omega C_2}{j2\omega L - j/2\omega C_2} V_{AA} = \frac{\pi I_{\text{dc}}}{2\sqrt{3}\omega C_1(4\omega^2 LC_2 - 1)}$$

We make $4\omega^2 LC_2 \gg 1$ and so can write the ripple factor as

$$\gamma = \frac{V_2}{V_{\text{dc}}} \cong \frac{(\pi I_{\text{dc}})/(8\sqrt{3}\omega^3 C_1 C_2 L)}{I_{\text{dc}} R} \cong \frac{0.216}{R}\left(\frac{X_{C_1} X_{C_2}}{X_L}\right) \tag{5.19}$$

where the reactances are computed at the supply frequency. The ripple increases with load current in this filter.

Usual filter components might be: $C_1 = 50\,\mu\text{F}$, $C_2 = 100\,\mu\text{F}$, $L = 10\,\text{H}$, and then $\gamma \cong 0.08/R$. With $R = 200$ ohms the ripple will approximate 0.04 percent.

With a resistance R_c in the inductor L, and using the methods of the previous section, the output voltage is found to be

$$V_{\text{dc}} = V_m - I_{\text{dc}}\left(R_c + \frac{\pi}{2\omega C_1}\right) \tag{5.20}$$

so that the output voltage varies with C_1. For small I_{dc} the output dc voltage nears the value of the peak of the transformer supply voltage V_m.

5.8 Rectifying ac voltmeters

Because of the greater current sensitivity of PMMC instruments in comparison with ac-measuring electrodynamometers, a PMMC movement is often combined with rectifier diodes to form an ac voltmeter. The bridge rectifier circuit of Fig. 5.13(a) is used, leading to ac voltmeters with sensitivities as high as 2000 Ω/V.

The diodes supply the movement with full-wave rectified current, and the deflection is due to the average of these sine half-waves because of the PMMC characteristic. That is,

$$I_{av} = \frac{2I_m}{\pi_{ac}} = 0.636 I_m \tag{5.21}$$

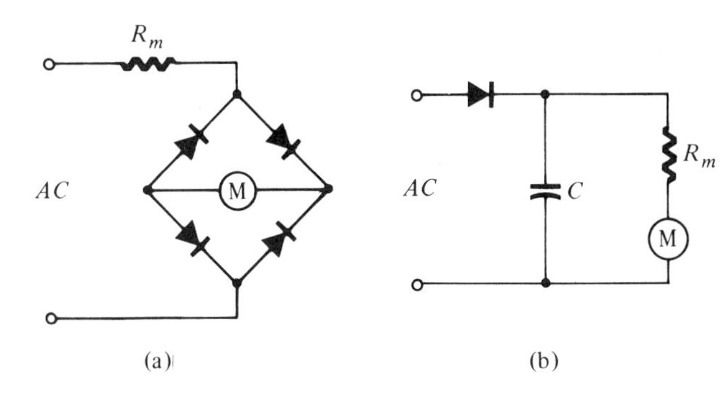

FIGURE 5.13
(a) Average-reading voltmeter; (b) peak-reading voltmeter.

where I_m is the peak of the sine wave. The instrument scale is calibrated in rms values, and the scalc introduces a factor of 0.707/0.636 = 1.11. If the input wave is not a sinusoid, the instrument readings may be in error.

A series multiplier resistor R_m increases the applied voltage range and also suppresses the effect on the calibration of diode resistance variations, which may be a result of temperature changes.

Another form of rectifier voltmeter appears in Fig. 5.13(b), where a half-wave diode is used with a capacitive filter. The voltage applied to R_m and the instrument is proportional to the positive peak of the applied wave. The scale is calibrated to read in rms values by inclusion of 0.707 in the calibration. Again, errors are possible if the wave form is not sinusoidal.

5.9 Summary

The semiconductor diode is a nonlinear circuit device which transmits current more readily with one polarity than with the oppositc. Resistance is determined as the slope of the volt-ampere curve, and with the varying slope of the diode, we have a nonlinear resistance to changes in current or voltage.

The semiconductor diode is approximated by a battery and a resistance in the forward direction and an open circuit in the reverse direction. For most applications this circuit can be further simplified to the ideal diode with zero forward voltage drop and zero reverse current. The diode then becomes an ideal polarity-controlled switch.

A major application of such circuit elements is in the rectification of alternating currents to direct current. The average or dc component is the desired output, but other ac components are also present. Filter circuits are used to separate the dc and ac components. The filters are excited by recurrent transient pulses, and reasonable approximations to the circuit performance of the filters lead to an understanding of design factors for dc supplies.

PROBLEMS

5.1 For a silicon diode with $i_s = 25\ \mu$A at 300°K, find the forward voltage drop v_D at a forward current of 1 A.

5.2 Calculate the forward incremental resistance of a diode having $i_s = 10\ \mu$A at 300°K at a forward voltage of 0.35 V. Plot the forward and reverse voltage characteristic curve to 0.5 V.

5.3 A diode having $i = 0.30\ v^{3/2}$ is used in the circuit of (a) in Fig. 5.14. (a) Determine the current analytically. (b) Find the current by graphical methods.

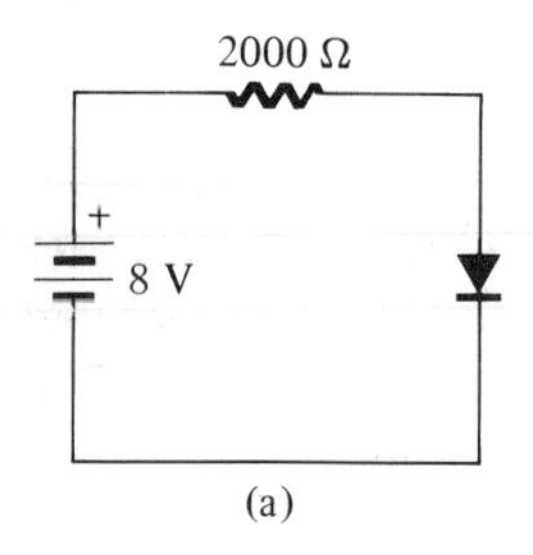

(a)

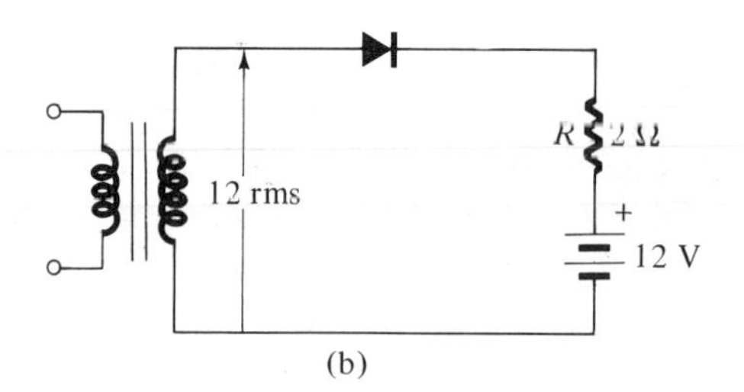

(b)

FIGURE 5.14

5.4 An ideal diode is in series with a 12-V rms source and a 2-Ω resistor to charge a 12-V storage battery. By writing Eq. 5.7, find the angles of conduction and turnoff for the diode. Determine the average charging current of the battery in Fig. 5.14(b).

5.5 With a resistance load, compare the rms current rating of the transformer winding for a bridge rectifier with that of one half the winding for a center-tap full-wave circuit.

5.6 A half-wave rectifier uses an applied voltage of 250 V rms, and the load is 1800 Ω. Find (a) the dc current in the load and (b) the I_{rms} of the total load current and the *total* power lost in the load. (c) What is the dc power delivered to the load? Explain the difference in power in (b) and (c).

5.7 A 60-Hz transformer having 40 V on each side of the center tap supplies a full-wave rectifier circuit; the diodes are ideal. The circuit load is 100 Ω with a shunt filter of 1000 μF. Find (a) the dc load voltage and (b) the ripple factor.

5.8 Prove that the rms value of the triangular ripple wave of Fig. 5.9(c) is given by $V_R/2\sqrt{3}$.

5.9 (a) Determine the ripple factor for the full-wave rectifier circuit at 60 Hz with a π filter having $C_1 = C_2 = 100\ \mu\text{F}$ and $L = 10$ H, with a load current of 500 mA at 60 V dc. (b) Compare with the same capacitors in parallel in a capacitor filter without inductor.

5.10 In a full-wave rectifier circuit, with $V_m = 100$ V and $R = 100\ \Omega$, what value of C is needed in a shunt-capacitor filter if the voltage is not to drop more than 10 percent between capacitor charging pulses?

5.11 Two 50-μF capacitors and a 10-H inductor of 5-Ω resistance are available. With a full-wave rectifier circuit supplying the π filter of these elements, connected to a load taking 0.5 A at 30 V dc, find the ripple and the transformer supply voltage for the bridge rectifier and determine the change in load voltage (dc) from zero load current to full load current.

5.12 Two diode clippers are assembled in the circuit of Fig. 5.15(a). Draw a voltage output curve for the circuit for positive and negative input voltages.

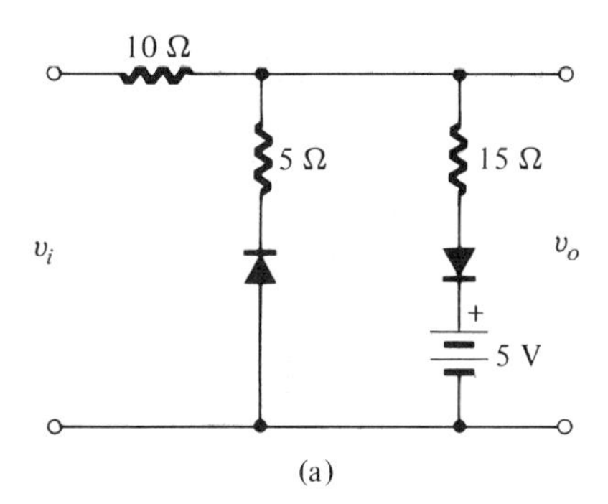

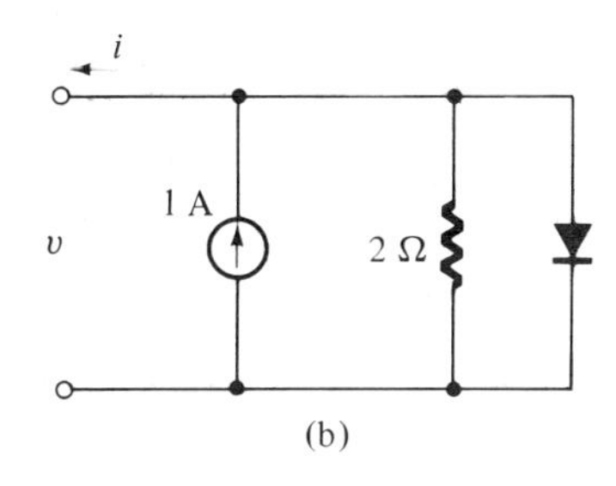

FIGURE 5.15

5.13 Draw the volt-ampere curve for the load for the circuit of Fig. 5.15(b) for positive and negative input voltages.

5.14 The double-diode shunt clipper of Fig. 5.5, with ideal diodes and $V_1 = V_2 = 50$ V, is connected to thc output of a 120 V_{rms} source, having 600 ohms series resistance. (a) Sketch the expected wave form across the diodes; and (b) what is the average power dissipated in the source resistance?

REFERENCES

1. D. L. Waidelich, "Analysis of Full-Wave Rectifier and Capacitive Input Filter." *Electronics*, 20, 120 (Sept. 1947).
2. J. D. Ryder, *Electronic Fundamentals and Applications*, 4th ed. Prentice-Hall, Inc., Englewood Cliffs, N.J., 1970.

network models and controlled sources

Simple circuits, which give the same terminal response as more complicated networks, are said to be *equivalent* to the complicated networks. We use the simple networks as models; when phasor circuit equations are written to describe the responses of the equivalent circuits, we consider the equations as representing *mathematical models*. These models can be utilized to analytically study the performance of the original circuits.

This chapter begins with the determination of such mathematical and physical models or equivalent circuits; the methods are extended to the modeling of active circuits as are found with transistor amplifiers. An ultimate goal is the prediction of network response, as well as the design of networks to give a desired response.

6.1 The two-port impedance model; z parameters

Let us assume that the box of Fig. 6.1(a) contains a passive linear electrical network of arbitrary internal connections. For access, conductors may be connected to any desired nodes in the box and brought out as terminals. Any pair of terminals is a *port*, and at (b) in Fig. 6.1 we have a *two-port* network; arbitrary *n*-port networks are possible.

A driving source is connected at one of the ports and a load at the other. The network may serve for transmission or transformation of power or for transmitting signals selectively with frequency.

The currents and voltages at the ports of the box are available for measurement. If one network can be substituted for another without change in the currents and voltages at the ports, the two networks are externally indistinguishable and *are said to be equivalent* at a given frequency. Nothing need be known of the internal network configuration.

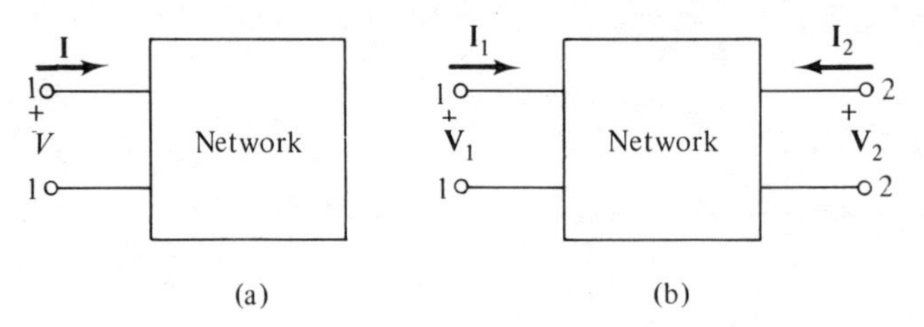

FIGURE 6.1

(a) One-port network; (b) two-port network.

The currents and voltages at the ports are conventionally defined as in Fig. 6.1. Inward currents are made positive at each port, and the analysis is then independent of designation of the ports for input and output.

At the terminals there are four variables, v_1, v_2, i_1, and i_2, taken for the moment as arbitrary time functions. Any pair of these quantities may be designated as independent variables, as

$$\begin{array}{ll} i_1, i_2 & v_2, i_2 \\ v_1, v_2 & v_1, i_1 \\ v_2, i_1 & v_1, i_2 \end{array}$$

allowing the writing of two equations which may then be solved for the other two independent variables. Three of the six possible variable pairs have been found useful in electronic circuit analysis: i_1, i_2; v_1, v_2; and v_2, i_1. The choice of v_2, i_2 is of importance in power transmission.

The choosing of i_1 and i_2 as the independent variable pair implies that

$$v_1 = f_1(i_1, i_2)$$

$$v_2 = f_2(i_1, i_2)$$

We wish to find the changes in current and voltage at the ports as a result of changes in source and load. Mathematically the response to changes can be found by introducing small variations, and by writing the total derivative, we have

$$\Delta v_1 = \frac{\partial v_1}{\partial i_1}\Delta i_1 + \frac{\partial v_1}{\partial i_2}\Delta i_2 \tag{6.1}$$

$$\Delta v_2 = \frac{\partial v_2}{\partial i_1}\Delta i_1 + \frac{\partial v_2}{\partial i_2}\Delta i_2 \tag{6.2}$$

With the small changes made in sinusoidal form, the partial derivatives can be recognized as impedances, and the equations can be written for effective values:

$$\mathbf{V}_1 = \mathbf{z}_i\mathbf{I}_1 + \mathbf{z}_r\mathbf{I}_2 \tag{6.3}$$

$$\mathbf{V}_2 = \mathbf{z}_f\mathbf{I}_1 + \mathbf{z}_o\mathbf{I}_2 \tag{6.4}$$

The **z** coefficients employ a subscript system which implies position in the impedance matrix.*

Measurements made at the ports with arbitrary terminations provide a means of correlating the **z** parameters of the above equations with the actual network inside the box. Open-circuit and short-circuit terminations are convenient to use, although the choice is arbitrary. With port 2 open-circuited, the output current $\mathbf{I}_2 = 0$, and using this condition in Eqs. 6.3 and 6.4, we have

$$\mathbf{z}_i = \frac{\mathbf{V}_1}{\mathbf{I}_1} \quad \text{with } \mathbf{I}_2 = 0 \tag{6.5}$$

$$\mathbf{z}_f = \frac{\mathbf{V}_2}{\mathbf{I}_1} \quad \text{with } \mathbf{I}_2 = 0 \tag{6.6}$$

It appears that $\mathbf{z}_i$ represents the *input impedance* with the output circuit open and that $\mathbf{z}_f$ is the *forward transfer impedance* between output voltage and input current with the output circuit open. A transfer impedance implies a relation between current at one port and voltage at a second port.

Similarly, we can leave port 1 open with $\mathbf{I}_1 = 0$ and measure the voltage and current at port 2. Then from Eqs. 6.3 and 6.4,

$$\mathbf{z}_r = \frac{\mathbf{V}_1}{\mathbf{I}_2} \quad \text{with } \mathbf{I}_1 = 0 \tag{6.7}$$

$$\mathbf{z}_o = \frac{\mathbf{V}_2}{\mathbf{I}_2} \quad \text{with } \mathbf{I}_1 = 0 \tag{6.8}$$

Impedance $\mathbf{z}_r$ is the open-circuit *reverse transfer impedance* and $\mathbf{z}_o$ is the open-circuit port 2 *output impedance*. Since $\mathbf{z}_i$, $\mathbf{z}_r$, $\mathbf{z}_f$, and $\mathbf{z}_o$ are all measured with open-circuit terminations, these parameters are known as the *open-circuit impedance parameters* of the two-port network.

At sufficiently low frequencies the **z** parameters become real and are then known as the r parameters.

6.2 y- and h-parameter models

If we had chosen v_1 and v_2 as the independent variables, with i_1 and i_2 as independent, a similar process would lead to

$$\mathbf{I}_1 = \mathbf{y}_i\mathbf{V}_1 + \mathbf{y}_r\mathbf{V}_2 \tag{6.9}$$

$$\mathbf{I}_2 = \mathbf{y}_f\mathbf{V}_1 + \mathbf{y}_o\mathbf{V}_2 \tag{6.10}$$

*Numerical subscripts such as $\mathbf{z}_{11}$, $\mathbf{z}_{12}$, $\mathbf{z}_{21}$, and $\mathbf{z}_{22}$ are also used, but i, r, f, and o are prefered by the IEEE in electronic circuit analysis and will be used here. Their use will also avoid triple subscripts later.

as the steady-state equations for the network in the box. Following the previous discussion but with short-circuit terminations at the ports and using zero values for the terminal voltages, we obtain

$$\mathbf{y}_i = \frac{\mathbf{I}_1}{\mathbf{V}_1} \qquad \text{with } \mathbf{V}_2 = 0 \tag{6.11}$$

$$\mathbf{y}_f = \frac{\mathbf{I}_2}{\mathbf{V}_1} \qquad \text{with } \mathbf{V}_2 = 0 \tag{6.12}$$

$$\mathbf{y}_r = \frac{\mathbf{I}_1}{\mathbf{V}_2} \qquad \text{with } \mathbf{V}_1 = 0 \tag{6.13}$$

$$\mathbf{y}_o = \frac{\mathbf{I}_2}{\mathbf{V}_2} \qquad \text{with } \mathbf{V}_1 = 0 \tag{6.14}$$

The *input admittance* at port 1 is $\mathbf{y}_i$, and $\mathbf{y}_o$ is the *output admittance* at port 2—both with short-circuit terminations. Also, $\mathbf{y}_f$ is the short-circuit *forward transfer admittance*, and $\mathbf{y}_r$ is the short-circuit *reverse transfer admittance.* The **y** parameters are known as the *short-circuit admittance parameters* for the two-port network.

At sufficiently low frequencies the **y** parameters become real and they are then known as the *g parameters.*

The third choice of independent variables is v_2 and i_1 and leads to the circuit equations

$$\mathbf{V}_1 = \mathbf{h}_i\mathbf{I}_1 + \mathbf{h}_r\mathbf{V}_2 \tag{6.15}$$

$$\mathbf{I}_2 = \mathbf{h}_f\mathbf{I}_1 + \mathbf{h}_o\mathbf{V}_2 \tag{6.16}$$

The **h** coefficients are known as the *hybrid parameters* of the two-port network. Following the previous analysis, and this time employing both open- and short-circuit terminations, justifying the hybrid name, we have

$$\mathbf{h}_i = \frac{\mathbf{V}_1}{\mathbf{I}_1} \qquad \text{with } \mathbf{V}_2 = 0 \tag{6.17}$$

$$\mathbf{h}_f = \frac{\mathbf{I}_2}{\mathbf{I}_1} \qquad \text{with } \mathbf{V}_2 = 0 \tag{6.18}$$

$$\mathbf{h}_r = \frac{\mathbf{V}_1}{\mathbf{V}_2} \qquad \text{with } \mathbf{I}_1 = 0 \tag{6.19}$$

$$\mathbf{h}_o = \frac{\mathbf{I}_2}{\mathbf{V}_2} \qquad \text{with } \mathbf{I}_1 = 0 \tag{6.20}$$

Dimensionally, $\mathbf{h}_i$ is the *input impedance* with short-circuit termination, $\mathbf{h}_o$ is the open-circuit *output admittance*, and $\mathbf{h}_r$ and $\mathbf{h}_f$ are dimensionless ratios, known as the open-circuit *reverse voltage gain* and the short-circuit *forward current gain*, respectively. At low frequencies the **h** parameters take on real values, and bold face designation will not often be used.

The three sets of network equations, with the parameters determined by terminal measurements on the network, serve as three mathematical models for the performance of the network box. The models also represent equivalent networks, since networks with equal parameters would be equivalent in performance at the ports.

Table 6.1 presents the relationships among the three sets of parameters.

TABLE 6.1 **RELATIONSHIPS AMONG THE TWO-PORT PARAMETERS**

	z		y		h	
z	$\mathbf{z}_i$	$\mathbf{z}_r$	$\frac{\mathbf{y}_o}{\Delta_y}$	$\frac{-\mathbf{y}_r}{\Delta_y}$	$\frac{\Delta_h}{\mathbf{h}_o}$	$\frac{\mathbf{h}_r}{\mathbf{h}_o}$
	$\mathbf{z}_f$	$\mathbf{z}_o$	$\frac{-\mathbf{y}_f}{\Delta_y}$	$\frac{\mathbf{y}_i}{\Delta_y}$	$\frac{-\mathbf{h}_f}{\mathbf{h}_o}$	$\frac{1}{\mathbf{h}_o}$
y	$\frac{\mathbf{z}_o}{\Delta_z}$	$\frac{-\mathbf{z}_r}{\Delta_z}$	$\mathbf{y}_i$	$\mathbf{y}_r$	$\frac{1}{\mathbf{h}_i}$	$\frac{-\mathbf{h}_r}{\mathbf{h}_i}$
	$\frac{-\mathbf{z}_f}{\Delta_z}$	$\frac{\mathbf{z}_i}{\Delta_z}$	$\mathbf{y}_f$	$\mathbf{y}_o$	$\frac{\mathbf{h}_f}{\mathbf{h}_i}$	$\frac{\Delta_h}{\mathbf{h}_i}$
h	$\frac{\Delta_z}{\mathbf{z}_o}$	$\frac{\mathbf{z}_r}{\mathbf{z}_o}$	$\frac{1}{\mathbf{y}_i}$	$\frac{-\mathbf{y}_r}{\mathbf{y}_i}$	$\mathbf{h}_i$	$\mathbf{h}_r$
	$\frac{-\mathbf{z}_f}{\mathbf{z}_o}$	$\frac{1}{\mathbf{z}_o}$	$\frac{\mathbf{y}_f}{\mathbf{y}_i}$	$\frac{\Delta_y}{\mathbf{y}_i}$	$\mathbf{h}_f$	$\mathbf{h}_o$

Δ = determinant of the parameter matrix.

6.3 Physical models for passive networks

If a network is to be equivalent to an unknown network in the box, the model must correctly establish eight quantities. These are the magnitudes and phase angles of the port variables $\mathbf{V}_1$, $\mathbf{V}_2$, $\mathbf{I}_1$, and $\mathbf{I}_2$. However, if a load $\mathbf{Z}_L$ is connected to port 2, then

$$\mathbf{V}_2 = -\mathbf{I}_2\mathbf{Z}_L \tag{6.21}$$

One magnitude and one phase angle are established by the load impedance, leaving three magnitudes and three phase angles to be determined. Three adjustable impedances are required for a network which will determine these quantities. Only two arrangements of three impedances are possible, and these appear at (a) and (b) in Fig. 6.2. The T network is at (a), redrawn as the Y configuration in (c). The π network is at (b), identical to the *delta* network in (d). The names of these networks follow from the drawn form of the circuit. The T and π nomenclature has become established in electrical

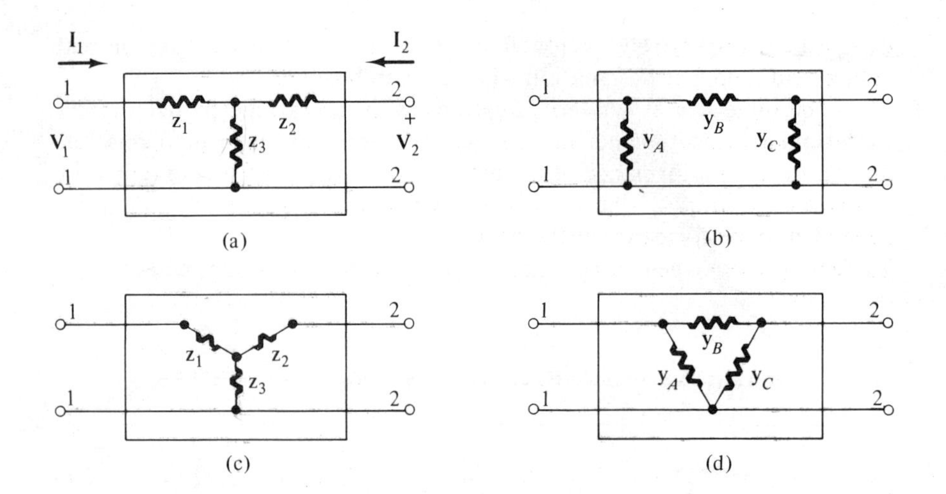

FIGURE 6.2

T-network equivalent; (b) π-network equivalent; (c) *Y* form; (d) delta form.

communications, while *Y* and delta are names used for the same equivalents when applied in electrical power systems.

With open-circuit terminations at the ports, the **z**-parameter definitions can be written by observation. That is,

$$\begin{aligned} \mathbf{z}_i &= \mathbf{Z}_1 + \mathbf{Z}_3 \\ \mathbf{z}_r &= \mathbf{Z}_3 \\ \mathbf{z}_f &= \mathbf{Z}_3 \\ \mathbf{z}_o &= \mathbf{Z}_2 + \mathbf{Z}_3 \end{aligned} \tag{6.22}$$

Solving for the impedances of the *T* network, we find that the two-port equivalent requires

$$\mathbf{Z}_1 = \mathbf{z}_i - \mathbf{z}_f \tag{6.23}$$

$$\mathbf{Z}_2 = \mathbf{z}_o - \mathbf{z}_f \tag{6.24}$$

$$\mathbf{Z}_3 = \mathbf{z}_f = \mathbf{z}_r \tag{6.25}$$

With short-circuit terminations and the **y**-parameter definitions, we can obtain design equations for the π network of Fig. 6.2(b). That is,

$$\begin{aligned} \mathbf{y}_i &= \mathbf{Y}_A + \mathbf{Y}_B \\ \mathbf{y}_r &= -\mathbf{Y}_B \\ \mathbf{y}_f &= -\mathbf{Y}_B \\ \mathbf{y}_o &= \mathbf{Y}_C + \mathbf{Y}_B \end{aligned} \tag{6.26}$$

The values for the arms of the equivalent π model are

$$\mathbf{Y}_A = \mathbf{y}_i + \mathbf{y}_f \tag{6.27}$$

$$\mathbf{Y}_B = -\mathbf{y}_f = -\mathbf{y}_r \tag{6.28}$$

$$\mathbf{Y}_C = \mathbf{y}_o + \mathbf{y}_f \tag{6.29}$$

Thus *any linear, bilateral, passive electrical network can be represented, at a single frequency, by a T (or π) network.* As derived in terms of **Z** or **Y**, the equivalent circuits are general, but when converted to R, L, and C values the circuits become equivalent only at the frequency of conversion.

Impossibly large or small values of L or C may occasionally be needed to fulfill the mathematical requirements; such values do not affect the validity of the equivalence, although the physical realizability may be limited.

The equivalent networks may be structured as in Fig. 6.3.

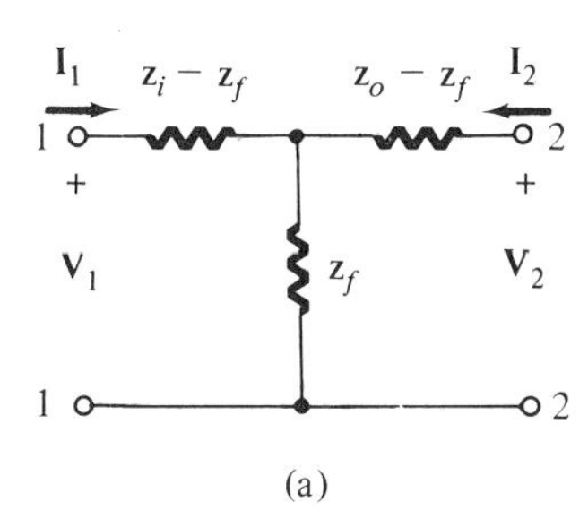

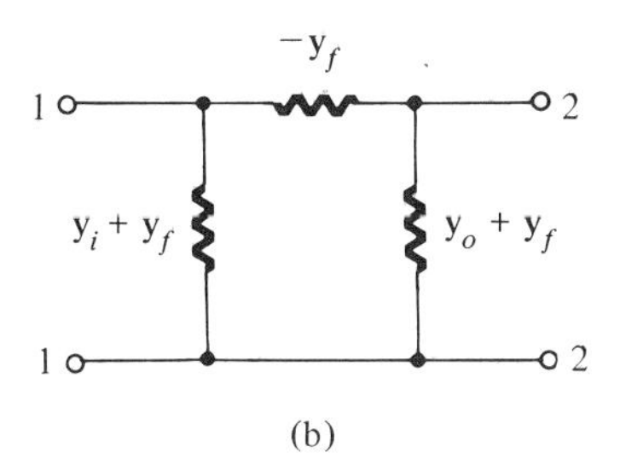

FIGURE 6.3
(a) Equivalent T for the box of Fig. 6.1(b); (b) equivalent π.

6.4 Conversions between T and π networks

Since either T or π networks may be made equivalent to the same two-port network, it is possible to convert from a T to a π or vice versa. For the π network the admittances may be converted to impedances by use of $\mathbf{Z} = 1/\mathbf{Y}$, and the two networks then have terminal measurements as follows:

T NETWORK	π NETWORK
$\mathbf{z}_i = \mathbf{Z}_1 + \mathbf{Z}_3$	$\mathbf{z}_i = \dfrac{\mathbf{Z}_A(\mathbf{Z}_B + \mathbf{Z}_C)}{\mathbf{Z}_A + \mathbf{Z}_B + \mathbf{Z}_C}$
$\mathbf{z}_r = \mathbf{z}_f = \mathbf{Z}_3$	$\mathbf{z}_r = \mathbf{z}_f = \dfrac{\mathbf{Z}_A\mathbf{Z}_C}{\mathbf{Z}_A + \mathbf{Z}_B + \mathbf{Z}_C}$
$\mathbf{z}_o = \mathbf{Z}_2 + \mathbf{Z}_3$	$\mathbf{z}_o = \dfrac{\mathbf{Z}_C(\mathbf{Z}_A + \mathbf{Z}_B)}{\mathbf{Z}_A + \mathbf{Z}_B + \mathbf{Z}_C}$

If the T and π networks are equivalent, then these measurements must be equal. Equating the parameters leads to values for a T-network equivalent to a π as

$$\mathbf{Z}_1 = \frac{\mathbf{Z}_A\mathbf{Z}_B}{\mathbf{Z}_A + \mathbf{Z}_B + \mathbf{Z}_C} \tag{6.30}$$

$$\mathbf{Z}_2 = \frac{\mathbf{Z}_B\mathbf{Z}_C}{\mathbf{Z}_A + \mathbf{Z}_B + \mathbf{Z}_C} \tag{6.31}$$

$$\mathbf{Z}_3 = \frac{\mathbf{Z}_A\mathbf{Z}_C}{\mathbf{Z}_A + \mathbf{Z}_B + \mathbf{Z}_C} \tag{6.32}$$

To change a T network to an equivalent π requires

$$\mathbf{Z}_A = \frac{\mathbf{Z}_1\mathbf{Z}_2 + \mathbf{Z}_2\mathbf{Z}_3 + \mathbf{Z}_1\mathbf{Z}_3}{\mathbf{Z}_2} \tag{6.33}$$

$$\mathbf{Z}_B = \frac{\mathbf{Z}_1\mathbf{Z}_2 + \mathbf{Z}_2\mathbf{Z}_3 + \mathbf{Z}_1\mathbf{Z}_3}{\mathbf{Z}_3} \tag{6.34}$$

$$\mathbf{Z}_C = \frac{\mathbf{Z}_1\mathbf{Z}_2 + \mathbf{Z}_2\mathbf{Z}_3 + \mathbf{Z}_1\mathbf{Z}_3}{\mathbf{Z}_1} \tag{6.35}$$

6.5 The cascade parameters

The fourth choice of independent variables, v_2 and i_2, leads to the circuit equations for sinusoidal quantities:

$$\mathbf{V}_1 = \mathbf{A}\mathbf{V}_2 - \mathbf{B}\mathbf{I}_2 \tag{6.36}$$

$$\mathbf{I}_1 = \mathbf{C}\mathbf{V}_2 - \mathbf{D}\mathbf{I}_2 \tag{6.37}$$

where **A**, **B**, **C**, and **D** are known as the *cascade parameters*, or the *general circuit constants.*

The former designation appears pertinent if we note that in the cascading of networks $\mathbf{V}_2$ and $-\mathbf{I}_2$ are the output quantities of a first network and are also $\mathbf{V}_3$ and $\mathbf{I}_3$ as the input quantities of a second network. This process of substitution can be carried out at each network junction, which in matrix form gives

$$\begin{bmatrix} \mathbf{V}_1 \\ \mathbf{I}_1 \end{bmatrix} = \begin{bmatrix} \mathbf{A}_1 & -\mathbf{B}_1 \\ \mathbf{C}_1 & -\mathbf{D}_1 \end{bmatrix} \times \begin{bmatrix} \mathbf{A}_2 & -\mathbf{B}_2 \\ \mathbf{C}_2 & -\mathbf{D}_2 \end{bmatrix} \times \cdots \times \begin{bmatrix} \mathbf{V}_x \\ \mathbf{I}_x \end{bmatrix} \tag{6.38}$$

where $\mathbf{V}_x$ and $\mathbf{I}_x$ are the output variables of the last network. The matrix of this system is then the product of the individual matrices.

The parameters are defined as

$$\mathbf{A} = \frac{\mathbf{V}_1}{\mathbf{V}_2} \quad (\mathbf{I}_2 = 0) \qquad \mathbf{C} = \frac{\mathbf{I}_1}{\mathbf{V}_2} \quad (\mathbf{I}_2 = 0)$$

$$\mathbf{B} = -\frac{\mathbf{V}_1}{\mathbf{I}_2} \quad (\mathbf{V}_2 = 0) \qquad \mathbf{D} = -\frac{\mathbf{I}_1}{\mathbf{I}_2} \quad (\mathbf{V}_2 = 0)$$

Applying these definitions to the equivalent T network of Fig. 6.3(a) and using the defined open- and short-circuit terminations, we can determine equivalences in terms of the **z** parameters. For instance, under open circuit at port 2, $\mathbf{A} = \mathbf{V}_1/\mathbf{V}_2$. But from the network we can obtain

$$\frac{\mathbf{V}_1}{\mathbf{I}_1} = \mathbf{z}_i$$

Also under open circuit at port 2,

$$\mathbf{V}_2 = \mathbf{I}_1\mathbf{z}_f$$

and

$$\frac{\mathbf{V}_2}{\mathbf{I}_1} = \mathbf{z}_f$$

Then by the definition for **A**,

$$\mathbf{A} = \frac{\mathbf{V}_1}{\mathbf{V}_2}\,(\text{open circuit}) = \frac{\mathbf{V}_1/\mathbf{I}_1}{\mathbf{V}_2/\mathbf{I}_1} = \frac{\mathbf{z}_i}{\mathbf{z}_f} \tag{6.39}$$

Using a short-circuit termination at port 2 and with the current division factor applied to $\mathbf{I}_1$,

$$-\mathbf{I}_2 = \mathbf{I}_1\frac{\mathbf{z}_f}{\mathbf{z}_o}$$

Also,

$$\mathbf{V}_1 = \mathbf{I}_1\left(\mathbf{z}_i - \frac{\mathbf{z}_f^2}{\mathbf{z}_o}\right)$$

$$-\mathbf{I}_2 = \mathbf{V}_1\left(\frac{\mathbf{z}_f}{\mathbf{z}_i\mathbf{z}_o - \mathbf{z}_f^2}\right)$$

from which

$$\mathbf{B} = -\frac{\mathbf{V}_1}{\mathbf{I}_2}\,(\text{short circuit}) = \frac{\mathbf{z}_i\mathbf{z}_o - \mathbf{z}_f^2}{\mathbf{z}_f} \tag{6.40}$$

To write a value for $\mathbf{C}$, we use

$$\mathbf{V}_2 = \mathbf{I}_1 \mathbf{z}_f$$

with an open circuit at 2, and

$$\mathbf{C} = \frac{\mathbf{I}_1}{\mathbf{V}_2} \text{ (open circuit) } = \frac{1}{\mathbf{z}_f} \tag{6.41}$$

For D, with a short circuit at port 2, we have

$$-\mathbf{I}_2 = \mathbf{I}_1 \frac{\mathbf{z}_f}{\mathbf{z}_o}$$

by use of current division. Then

$$\mathbf{D} = -\frac{\mathbf{I}_1}{\mathbf{I}_2} \text{ (short circuit) } = \frac{\mathbf{z}_o}{\mathbf{z}_f} \tag{6.42}$$

A relation which exists among the cascade parameters is

$$\mathbf{AD} - \mathbf{BC} = 1 \tag{6.43}$$

which can be proved by using the **z**-parameter equivalences as

$$\frac{\mathbf{z}_i}{\mathbf{z}_f}\frac{\mathbf{z}_o}{\mathbf{z}_f} - \frac{\mathbf{z}_i}{\mathbf{z}_f}\frac{\mathbf{z}_o}{\mathbf{z}_f} + 1 = 1$$

6.6 Transfer functions

We have used the across-through variable ratio to define an input impedance at a port; this is really a ratio of response to excitation at the port. It is also convenient to describe the performance of a network (or any linear system) by its overall response-to-excitation ratio. Such functions, written as the ratio of exponential response to an exponential excitation of a linear system, are called *transfer functions.*

Since either voltage or current may be considered as response or excitation, we define four general transfer functions of a network as

$$Z_T(s) = \frac{V_2(s)}{I_1(s)} \quad \text{(transfer impedance)} \tag{6.44}$$

$$Y_T(s) = \frac{I_2(s)}{V_1(s)} \quad \text{(transfer admittance)} \tag{6.45}$$

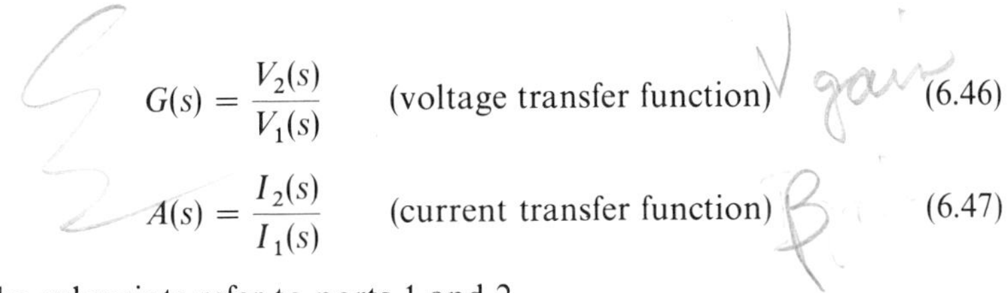

$$G(s) = \frac{V_2(s)}{V_1(s)} \qquad \text{(voltage transfer function)} \tag{6.46}$$

$$A(s) = \frac{I_2(s)}{I_1(s)} \qquad \text{(current transfer function)} \tag{6.47}$$

where the subscripts refer to ports 1 and 2.

We may write the above for the network of Fig. 6.4 and specialize it to the sinusoidal steady state by using $s = j\omega$. In Fig. 6.4 the T network

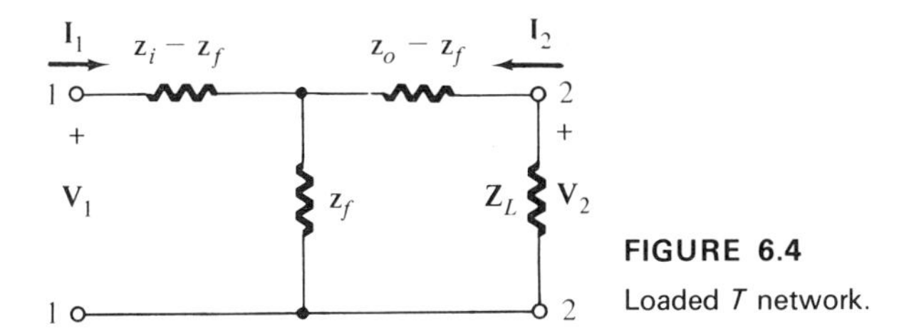

FIGURE 6.4
Loaded T network.

represents any linear passive network, and a load $\mathbf{Z}_L$ appears at port 2. It is useful to first obtain the input impedance at port 1:

$$\mathbf{Z}_{11} = \frac{\mathbf{V}_1}{\mathbf{I}_1} = \mathbf{z}_i - \mathbf{z}_f + \frac{(\mathbf{z}_o - \mathbf{z}_f + \mathbf{Z}_L)\mathbf{z}_f}{\mathbf{z}_o + \mathbf{Z}_L}$$

$$= \mathbf{z}_i - \frac{\mathbf{z}_f^2}{\mathbf{z}_o + \mathbf{Z}_L} \tag{6.48}$$

The current transfer function may be written directly by use of the current division factor:

$$\mathbf{A} = \frac{\mathbf{I}_2}{\mathbf{I}_1} = \frac{-\mathbf{z}_f}{\mathbf{z}_o + \mathbf{Z}_L} \tag{6.49}$$

The voltage transfer function follows through use of $\mathbf{A}$ and $\mathbf{Z}_{11}$, giving

$$\mathbf{G} = \frac{\mathbf{V}_2}{\mathbf{V}_1} = \frac{-\mathbf{I}_2\mathbf{Z}_L}{\mathbf{I}_1\mathbf{Z}_{11}} = -\mathbf{A}\frac{\mathbf{Z}_L}{\mathbf{Z}_{11}}$$

$$= \frac{\mathbf{z}_f\mathbf{Z}_L}{\mathbf{z}_i(\mathbf{z}_o + \mathbf{Z}_L) - \mathbf{z}_f^2} \tag{6.50}$$

The transfer impedance is

$$\mathbf{Z}_T = \frac{\mathbf{V}_2}{\mathbf{I}_1} = \frac{-\mathbf{I}_2\mathbf{Z}_L}{\mathbf{I}_1} = -\mathbf{A}\mathbf{Z}_L$$

$$= \frac{\mathbf{z}_f\mathbf{Z}_L}{(\mathbf{z}_o + \mathbf{Z}_L)} \tag{6.51}$$

and the transfer admittance is

$$\mathbf{Y}_T = \frac{\mathbf{I}_2}{\mathbf{V}_1} = \frac{\mathbf{I}_2}{\mathbf{I}_1\mathbf{Z}_{11}} = \frac{\mathbf{A}}{\mathbf{Z}_{11}}$$

$$= \frac{-\mathbf{z}_f}{\mathbf{z}_i(\mathbf{z}_o + \mathbf{Z}_L) - \mathbf{z}_f^2} \tag{6.52}$$

Note that $\mathbf{Y}_T$ is not the reciprocal of $\mathbf{Z}_T$.

6.7 Equivalent *T* network for inductive coupling

As another example of a physical representation of an equivalent network, consider the inductive coupling of two circuits, discussed in Section 1.12. An equivalent *T* network can now be derived.

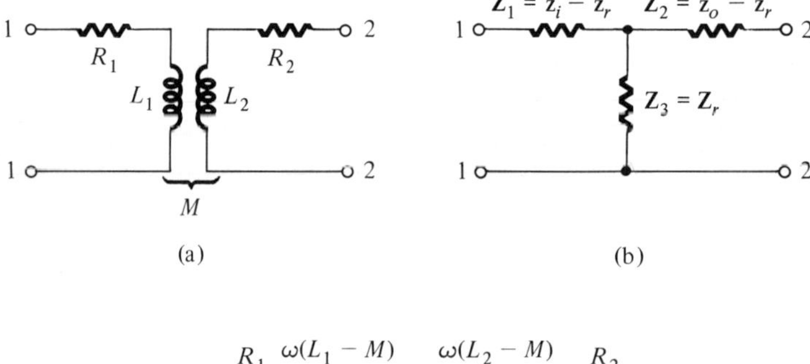

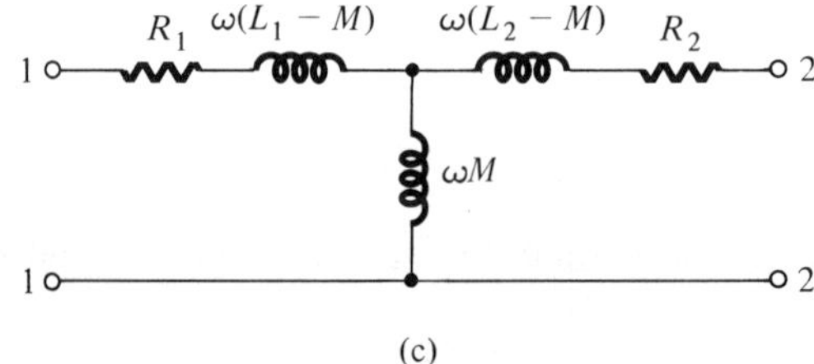

FIGURE 6.5

(a) Magnetically coupled circuit; (b) equivalent *T* network; (c) inductive circuit equivalent.

By applying an open circuit to the ports of the inductively coupled coils of Fig. 6.5(a), we use the **z**-parameter definitions and obtain

$$\mathbf{z}_i = R_1 + j\omega L_1$$

$$\mathbf{z}_o = R_2 + j\omega L_2$$

$$\mathbf{z}_f = \mathbf{z}_r = j\omega M$$

In Fig. 6.5(b) we have a general T network defined in terms of the $\mathbf{z}$ parameters. By direct comparison it can be seen that

$$\mathbf{Z}_1 = \mathbf{z}_i - \mathbf{z}_r = R_1 + j\omega(L_1 - M) \tag{6.53}$$

$$\mathbf{Z}_2 = \mathbf{z}_o - \mathbf{z}_r = R_2 + j\omega(L_2 - M) \tag{6.54}$$

$$\mathbf{Z}_3 = \mathbf{z}_f = \mathbf{z}_r = j\omega M \tag{6.55}$$

and the resultant T network is drawn at (c). It is equivalent to the inductively coupled circuit at the design angular frequency ω and will be applied in the analysis of coupled circuits.

6.8 Active circuit models

Assume Fig. 6.6(a) to be an active network, containing one or more sources and passive elements, with a port available at 1,1 to which is connected an impedance $\mathbf{Z}_L$. We intend to show that an active network as at (a) can be replaced with the simple equivalent model at (d).

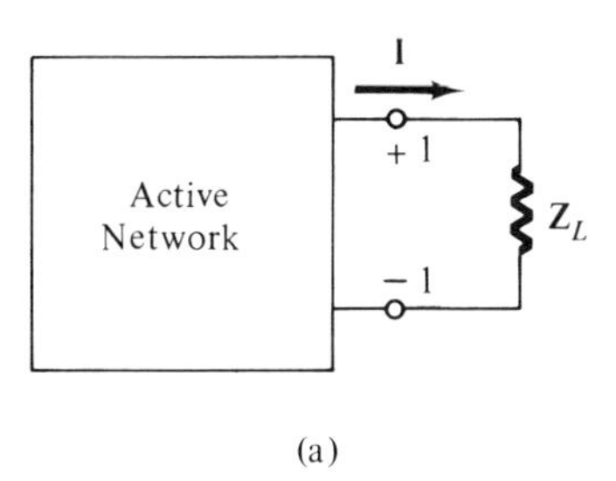

(a)

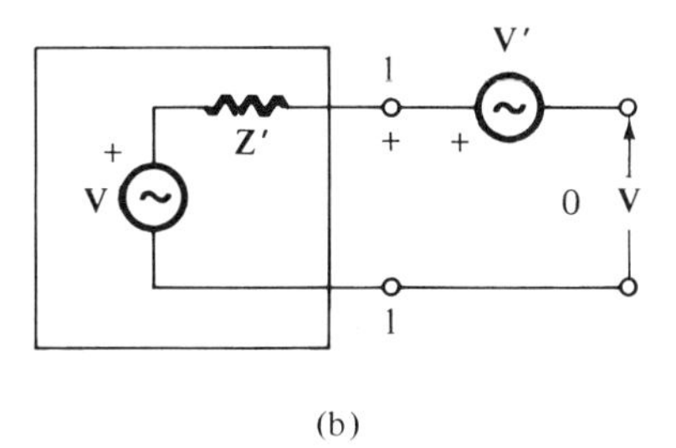

(b)

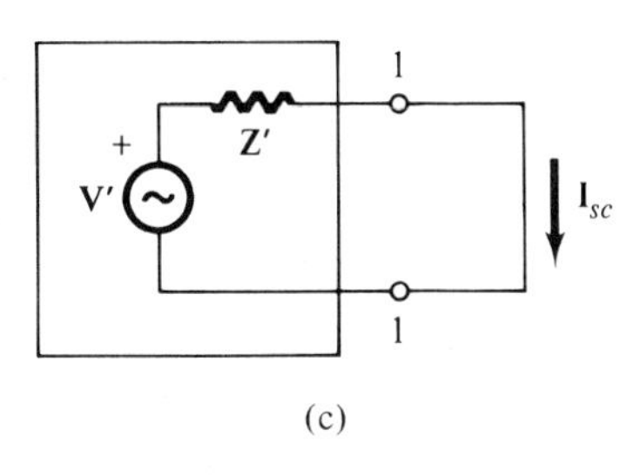

(c)

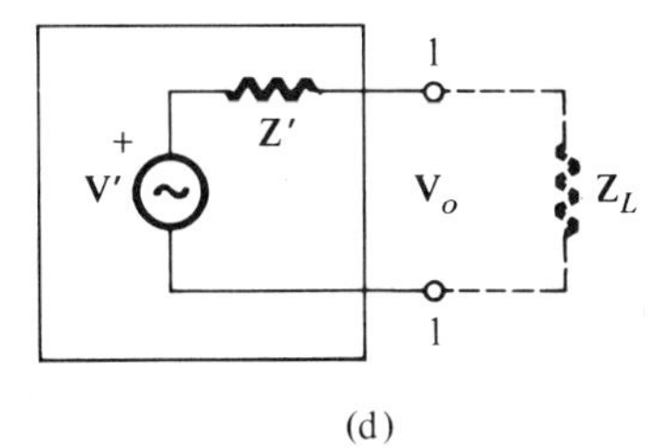

(d)

FIGURE 6.6

(a) Active network; (b) port open-circuited; (c) port short-circuited; (d) Thévenin's equivalent network.

Our concept of an active source is that of an ideal source acting through an impedance $\mathbf{Z}'$, as in (b). As before, we shall use open- and short-circuit terminations to determine the box parameters. Insert an ideal generator $\mathbf{V}'$ in series with the load branch as at (b) and adjust this source until the current $\mathbf{I}$ in the load is equal to zero at all time. With zero current present the load can be removed without changing the current, and it is evident that for 0 V at the load terminals, as in (b), voltage $\mathbf{V}'$ is equal to the *open-circuit voltage* of the box at the port and that $\mathbf{V} = \mathbf{V}'$ but with opposite polarity.

Having established the voltage of the internal generator as the open-circuit voltage at the port, we now short-circuit the port as in (c) and can write

$$\mathbf{I}_{sc} = \frac{\mathbf{V}'}{\mathbf{Z}'}$$

from which

$$\mathbf{Z}' = \frac{\mathbf{V}'}{\mathbf{I}_{sc}} = \frac{\text{open-circuit voltage at port 1}}{\text{short-circuit current at port 1}} \tag{6.56}$$

Thus we have arrived at the parameters $\mathbf{V}'$ and $\mathbf{Z}'$ of an equivalent circuit for an active network at a port of the network as at (d). This equivalence is called *Thévenin's theorem*, and the result is a *voltage-source equivalent circuit*. The theorem states that *at a port it is possible to replace any linear network containing energy sources with an equivalent network having an ideal voltage source* $\mathbf{V}'$ *in series with an impedance* $\mathbf{Z}'$.

A second form of equivalent circuit for an active network is known as the Norton circuit. With a load $\mathbf{Z}_L$ on the port at (d) in Fig. 6.6, we can write

$$\mathbf{Z}'\mathbf{I} + \mathbf{Z}_L\mathbf{I} = \mathbf{V}'$$

But $\mathbf{I} = \mathbf{V}_o/\mathbf{Z}_L$ and also dividing by $\mathbf{Z}'$

$$\frac{\mathbf{V}_o}{\mathbf{Z}_L} + \frac{\mathbf{V}_o}{\mathbf{Z}'} = \frac{\mathbf{V}'}{\mathbf{Z}'} = \mathbf{I}_{sc} \tag{6.57}$$

This equation represents a current summation at a node, with a current source $\mathbf{I}_{sc} = \mathbf{V}'/\mathbf{Z}'$ supplying an impedance $\mathbf{Z}'$ and an impedance $\mathbf{Z}_L$ in parallel. More simply,

$$\mathbf{Y}_L\mathbf{V}_o + \mathbf{Y}'\mathbf{V}_o = \mathbf{Y}'\mathbf{V}' \tag{6.58}$$

Corresponding circuits appear in Fig. 6.7, and these are known as *Norton's equivalent* or as a *current-source equivalent circuit*. The Norton circuit follows mathematically from the Thévenin circuit, and the two circuits are interchangeable, with $\mathbf{V}'$ and $\mathbf{Z}' = 1/\mathbf{Y}'$ having the same values in both circuits.

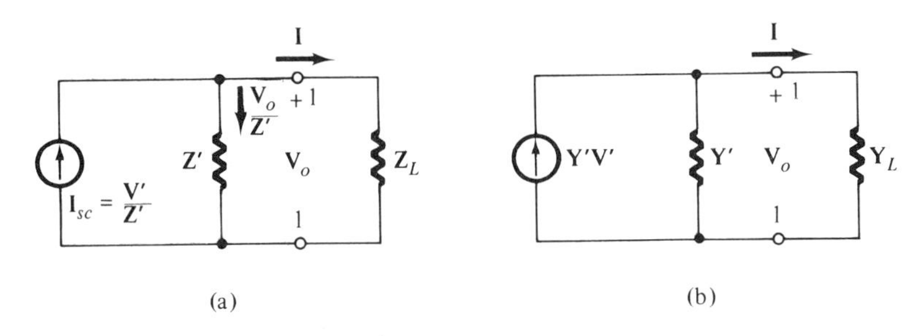

FIGURE 6.7

Norton current-source equivalent circuit.

As a practical matter, the impedance $\mathbf{z}'$ can be obtained by measurement or inspection of the circuit. For $\mathbf{z}'$ we find the input impedance at the output port, with all internal independent voltage sources removed by short-circuiting, but with their impedances present. Independent current sources are removed by open-circuiting, leaving their shunted impedances in the circuit. As we will find, controlled dependent sources are maintained with their magnitudes as established by the designated circuit variables.

Batteries, rectifier power supplies, and rotating generators are well approximated as voltage sources. Practical current sources are not generally available, although we simulate such forms with a voltage generator and a very high series impedance. The output current is then $\mathbf{V}'/(\mathbf{Z}' + \mathbf{Z}_L)$; with $|\mathbf{Z}'| \gg |\mathbf{Z}_L|$ the current is nearly independent of load value. This independence of current output from load is the basic property of a current source.

EXAMPLE. Find both Thévenin and Norton equivalent circuits for the active network of Fig. 6.8(a).

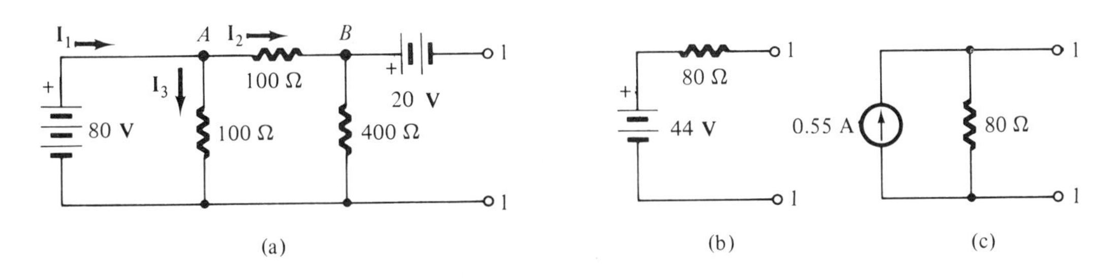

FIGURE 6.8

(a) Example; (b) Thévenin equivalent; (c) Norton equivalent.

To determine $\mathbf{V}'$, use an open circuit at terminals 1,1. The node equations at A are

$$\mathbf{I}_2 + \mathbf{I}_3 - \mathbf{I}_1 = 0$$

Also, $V_A = 80$ V; then

$$\mathbf{I}_3 = \frac{80}{100} = 0.80 \text{ A}$$

$$\mathbf{I}_2 = \frac{80}{500} = 0.16 \text{ A}$$

The voltage $\mathbf{V}_B$ is then

$$\mathbf{V}_B = 400\, I_2 = 400 \times 0.16 = 64 \text{ V}$$

The voltage at the 1,1 terminals is

$$\mathbf{V}' = \mathbf{V}_B - 20 = 64 - 20 = 44 \text{ V}$$

To determine $\mathbf{Z}'$, short-circuit the 80-V battery and 20-V battery (independent sources) to eliminate their effect. This also short-circuits one resistor of 100 Ω. The resultant circuit has 100 Ω in parallel with 400 Ω, and so

$$\mathbf{Z}' = \frac{100 \times 400}{100 + 400} = 80\ \Omega$$

The Thévenin equivalent circuit is drawn at (b) in Fig. 6.8.

In the Norton circuit, $\mathbf{I} = \mathbf{V}'/\mathbf{Z}' = 44/80 = 0.55$ A. The Norton circuit appears at (c) in Fig. 6.8.

6.9 Transfer of maximum power to a load

In some *transducers* used to convert energy to electrical signals, the complexity and cost is great and the output electrical power is small. It is economically important to deliver the maximum possible power to the associated electrical load, and the impedance conditions for transfer of maximum power to a load can be readily found. Examples of such devices include strain gages, phonograph pickups, microphones, and amplifiers.

As already demonstrated, a linear network containing energy sources can be represented at the 1,1 port by the voltage-source circuit of Fig. 6.9(a). If the internal impedance is resistive or $\mathbf{Z}' = R'$ and the load is R_L, the current in the load is

$$\mathbf{I} = \frac{\mathbf{V}'}{R' + R_L}$$

The average power delivered to the load is

$$P = I^2 R_L = \frac{(V')^2 R_L}{(R' + R_L)^2} \tag{6.59}$$

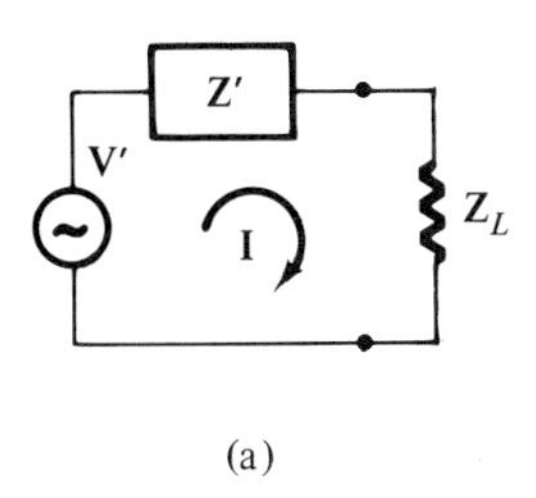

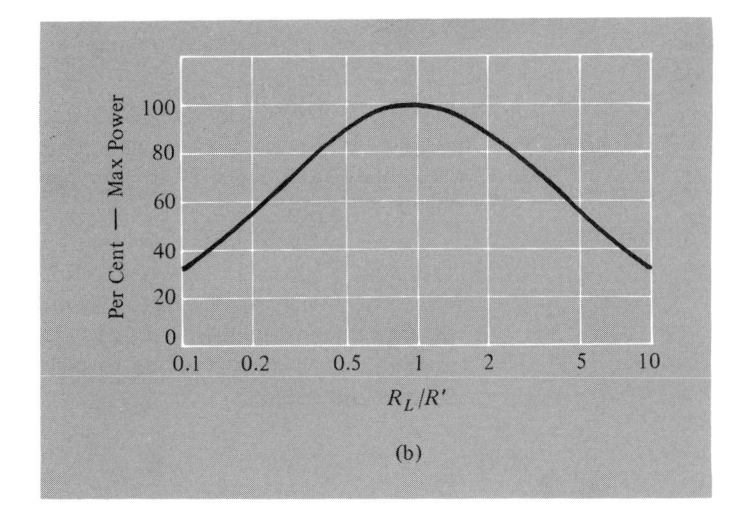

FIGURE 6.9

(a) Voltage-source active circuit; (b) variation of load power vs. R_L/R'.

Maximizing P with respect to R_L gives

$$\frac{dP}{dR_L} = \frac{(V')^2[(R' + R_L)^2 - 2R_L(R' + R_L)]}{(R' + R_L)^4} = 0$$

from which

$$R'^2 - R_L^2 = 0$$
$$R_L = R' \tag{6.60}$$

The load R_L must be equal or *matched* to the source resistance R' for maximum transfer of power to the load. This is demonstrated in (b) in Fig. 6.9.

If both generator and load include reactances in their internal impedances,

$$\mathbf{Z}' = R' + jX'$$
$$\mathbf{Z}_L = R_L + jX_L$$

then the average power delivered to the load is

$$P = \frac{(V')^2 R_L}{(R' + R_L)^2 + (X' + X_L)^2}$$

If we choose $X_L = -X'$, then the above reduces to Eq. 6.60, and we find maximum power transferred with $R' = R_L$.

The conditions for a load which will take maximum power from a source are

$$R_L = R'$$
$$X_L = -X' \tag{6.61}$$

or *the load should be the conjugate of the source impedance.*

With $R_L = R'$, the internal I^2R' loss is equal to the power delivered to the load, I^2R_L, and the power efficiency is only 50 percent. This is satisfactory for small sources where the power level is low and no overheating is encountered. Such an efficiency is not economic in electrical power systems and they operate with higher load values, since high power efficiency and not maximum power output is the operating criterion.

EXAMPLE. An active resistive one-port is short-circuited at port 1 and the output current is measured as 0.224 A dc. With a load of 1000 Ω of resistance the port voltage is found to be 79.5 V dc. Find the maximum possible power output from the port.

We know that

$$I_{sc} = \frac{V'}{R'} = 0.224 \text{ A}$$

and so

$$V' - 0.224R' = 0 \tag{1}$$

With the load of 1000 Ω, the output current is $V_o/1000 = 79.5/1000 = 0.0795$ A. Then from the circuit

$$V' - 0.0795R' = 79.5 \tag{2}$$

Having two equations and two unknowns, we can solve for the parameters of the Thévenin circuit.

From (1),

$$R' = \frac{V'}{0.224}$$

and insertion in (2) yields

$$V' - \frac{0.0795V'}{0.224} = 79.5$$

from which

$$V' = 123 \text{ V}$$

Then

$$R' = \frac{V'}{0.224} = \frac{123}{0.224} = 550\ \Omega$$

Maximum power output will be obtained if we use a matching load, or $R_L = R' = 550\ \Omega$. With this load, the load current is

$$I = \frac{V'}{R' + R_L} = \frac{123}{1100} = 0.112 \text{ A}$$

The *load* power is

$$P = I^2 R_L = 0.112^2 \times 550 = 6.88 \text{ W}$$

The power lost in the generator resistance is an equal amount since $R' = R_L$, and the total power in generator and load is then $6.88 \times 2 = 13.76$ W. Only 6.88 W represents power *output* from the generator, however.

This example illustrates an alternative method for determining Z' when the active network cannot be safely short-circuited. That is, we need two load values to determine the two unknowns, V' and Z', and we have previously chosen open circuit and short circuit as convenient loads. However, when short circuit currents might be damaging any other load may be chosen, and in the example we chose open circuit and 1000 ohms to gain sufficient experimental data to solve the problem.

6.10 Duality

The Thévenin circuit yields

$$\mathbf{V}_{oc} - \mathbf{Z}'\mathbf{I}_L = \mathbf{Z}_L\mathbf{I}_L \tag{6.62}$$

and from the Norton circuit we have

$$I_{sc} - \mathbf{Y}'\mathbf{V}_{oc} = \mathbf{Y}_L\mathbf{V}_{oc} \tag{6.63}$$

The parallelism between these equations, of the same mathematical form but with one written for voltages and the other for currents, demonstrates the relation of *duality*. The Thévenin and Norton circuits are known as *circuit duals*.

Consider Fig. 6.10 as an example, where a series RLC circuit is driven by a voltage source. Then

$$Ri + L\frac{di}{dt} + \frac{1}{C}\int i\,dt = v$$

Duality predicts a similar equation as

$$Gv + C\frac{dv}{dt} + \frac{1}{L}\int v\,dt = i$$

and this relation actually describes the dual circuit at (b) in Fig. 6.10. The relation between the T and π circuits is also that of duals.

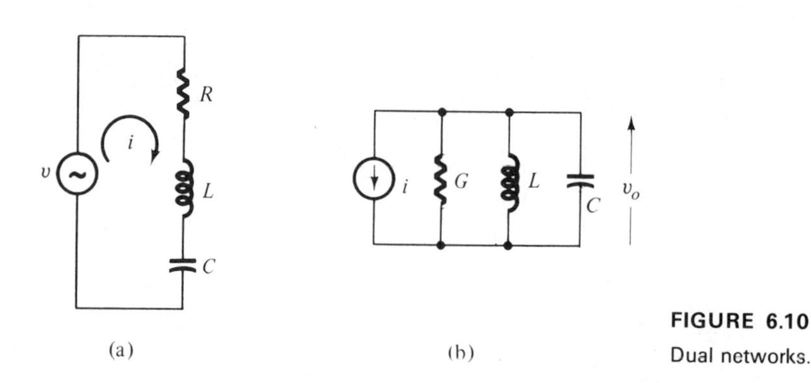

FIGURE 6.10
Dual networks.

The principle of duality comes from the parallelism of Kirchhoff's two laws and further demonstrates that choice of voltage or current as the independent variable or forcing function is quite arbitrary.

6.11 The controlled source

We now consider another circuit element, the *amplifier* of (a) in Fig. 6.11. Its performance is defined by the current-gain transfer function $\mathbf{A} = \mathbf{I}_2/\mathbf{I}_1$ or the voltage-gain transfer function $\mathbf{G} = \mathbf{V}_2/\mathbf{V}_1$. The amplifier can be included in networks with other circuit elements, and as an active circuit it will supply energy.

The transfer function may be written for the s domain or as a function of ω, as dictated by the conditions of application. Functions **A** and **G** are related as

$$\mathbf{G} = \frac{\mathbf{V}_2}{\mathbf{V}_1} = \frac{\mathbf{I}_2\mathbf{Z}_L}{\mathbf{I}_1\mathbf{Z}_{in}} = \mathbf{A}\frac{\mathbf{Z}_L}{\mathbf{Z}_{in}} \tag{6.64}$$

and we use the symbol K to imply a gain function of either form.

The amplifier introduces the concept of the *controlled source*, which is widely employed in electronic and control systems. It is a 2-port dependent source, with the magnitude and phase angle proportional to one of the current or voltage variables of the network. It also provides signal gain for the network and introduces a unilateral coupling between two physically separate ports.

Figure 6.11 illustrates four forms of *idealized* controlled sources. In (b) and (c) are current sources controlled by a current or voltage, respectively; the control parameters are h_f, a dimensionless *current ratio*, and g_m, a *transconductance*. In (d) and (e) are voltage sources controlled by a current or a voltage, with the control parameters being r_m, a *transresistance*, and μ, a dimensionless *voltage gain*. With h_f, g_m, r_m, and μ as constants, the sources are linear between control signal and output variable.

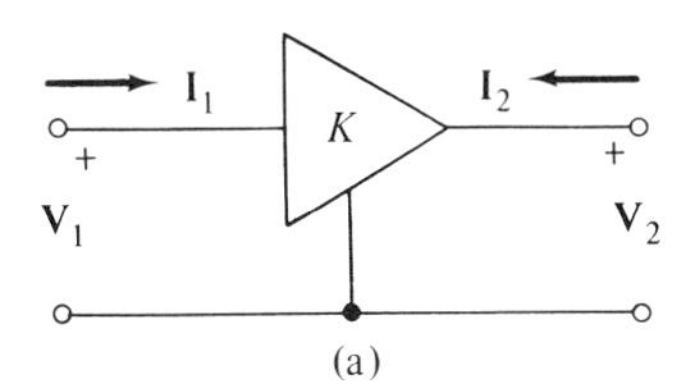

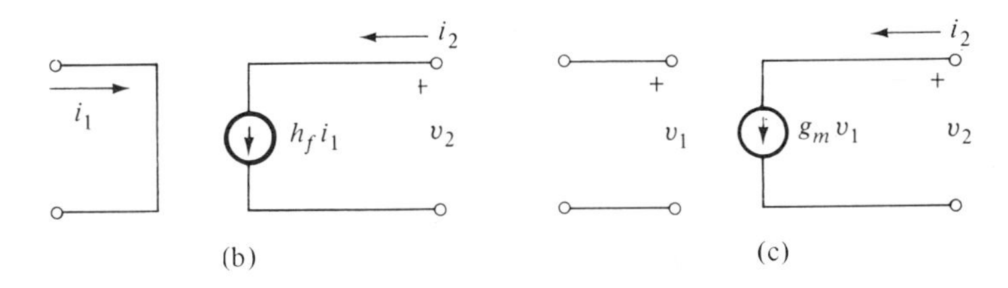

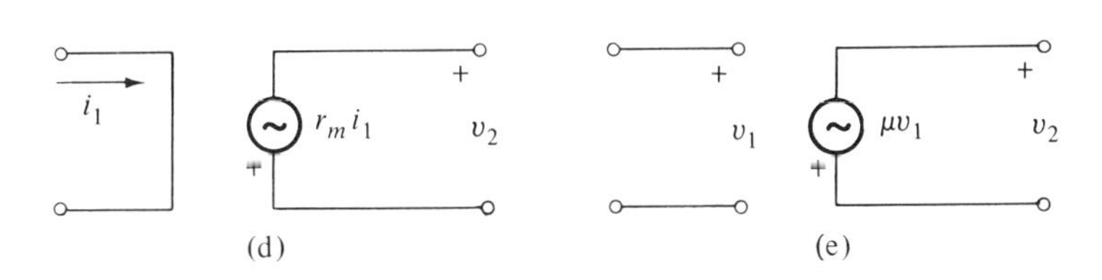

FIGURE 6.11

(a) Amplifier; (b) current source, current controlled; (c) current source, voltage controlled; (d) voltage source, current controlled; (e) voltage source, voltage controlled.

The input resistance is zero for the ideal current-controlled sources of (b) and (d). In (c) and (e) the input resistance is infinite for the ideal voltage-controlled sources. The output circuits can be considered as Thévenin or Norton equivalents, in which internal shunt impedances are infinite and internal series impedances are zero.

Output characteristics for controlled current and voltage sources of an ideal nature are shown in Fig. 6.12. Representing linear circuit elements, these curves extend through all quadrants. For current sources, the output family in (a) demonstrates that output current i_2 is independent of load voltage v_2, or the shunt output admittance is zero. For unit curve spacing of the i_1 curves, the vertical curve spacing represents the h_f value. Similarly, voltage sources have the characteristics of (b), with output voltage v_2 shown independent of load current i_2, or having zero output series impedance. For unit-spaced v_1 curves, the horizontal curve spacing represents the value of μ.

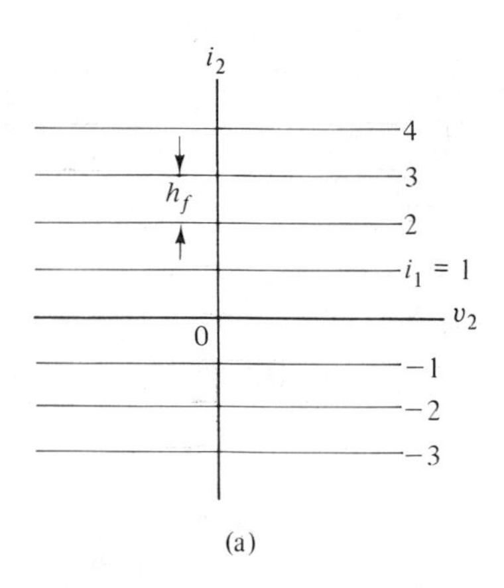

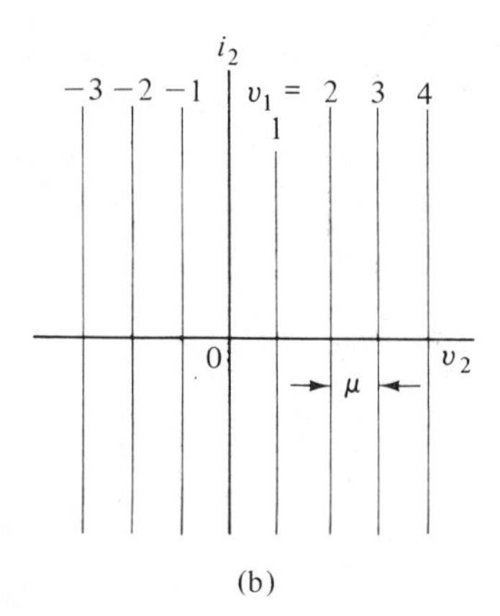

FIGURE 6.12

(a) Output characteristics, ideal current source; (b) same, for an ideal voltage source.

6.12 Models for two-port active circuits

We have discussed one-port active network models, but it is possible to have two-port active circuits as well; the controlled sources constitute examples of the class. Equivalent circuits may be derived for the two-port active network by again using the methods of Section 6.1.

Choosing $\mathbf{I}_1$ and $\mathbf{I}_2$ as the independent variables, along with the open-circuit $\mathbf{z}$ parameters, the general two-port relations can be written

$$\begin{bmatrix} \mathbf{V}_1 \\ \mathbf{V}_2 \end{bmatrix} = \begin{bmatrix} \mathbf{z}_i & \mathbf{z}_r \\ \mathbf{z}_f & \mathbf{z}_o \end{bmatrix} \begin{bmatrix} \mathbf{I}_1 \\ \mathbf{I}_2 \end{bmatrix} \tag{6.65}$$

These equations are represented by the equivalent circuit at (a) in Fig. 6.13. The circuit is active, and in general $\mathbf{z}_r \neq \mathbf{z}_f$. The circuit is complicated by the presence of two transfer voltage generators, dependent on the two terminal currents.

However, addition and subtraction of $\mathbf{z}_r\mathbf{I}_1$ to the second equation allows us to write

$$\begin{bmatrix} \mathbf{V}_1 \\ \mathbf{V}_2 - (\mathbf{z}_f - \mathbf{z}_r)\mathbf{I}_1 \end{bmatrix} = \begin{bmatrix} \mathbf{z}_i & \mathbf{z}_r \\ \mathbf{z}_r & \mathbf{z}_o \end{bmatrix} \begin{bmatrix} \mathbf{I}_1 \\ \mathbf{I}_2 \end{bmatrix} \tag{6.66}$$

and these equations can be represented by the circuit at (b) in Fig. 6.13. The network matrix is that of a passive T circuit, with $\mathbf{z}_r = \mathbf{z}_f$ between ports 1 and 3. The generator $(\mathbf{z}_f - \mathbf{z}_r)\mathbf{I}_1$ provides the active control source, with voltage proportional to $\mathbf{I}_1$, and *transimpedance* $\mathbf{z}_f - \mathbf{z}_r$. This circuit was widely used in transistor amplifier analysis at one time.

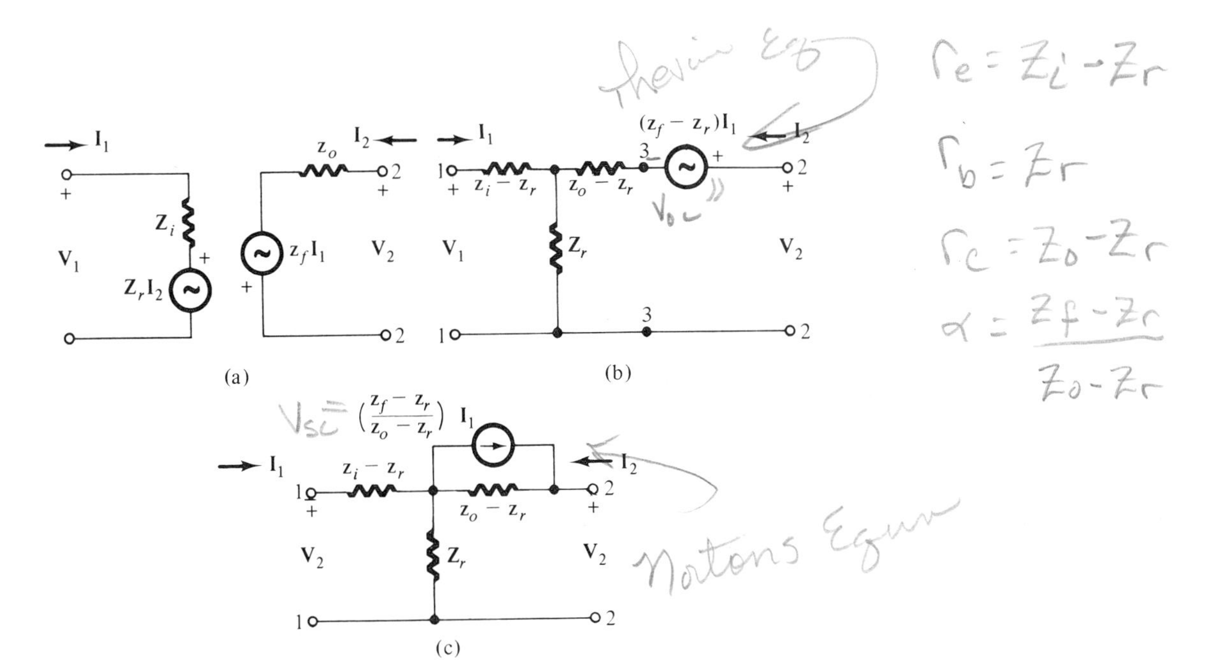

FIGURE 6.13

(a) Active **z**-parameter circuit (b) one-generator **z**-parameter equivalent; (c) one-generator current-source equivalent.

A second circuit form, in Fig. 6.13(c), results from change of the transfer generator to a controlled current source. This source is bridged across $\mathbf{z}_o - \mathbf{z}_r$ and supplies a current

$$\left(\frac{\mathbf{z}_f - \mathbf{z}_r}{\mathbf{z}_o - \mathbf{z}_r}\right)\mathbf{I}_1 \tag{6.67}$$

controlled by $\mathbf{I}_1$. It therefore introduces across $\mathbf{z}_o - \mathbf{z}_r$ a potential $(\mathbf{z}_f - \mathbf{z}_r)\mathbf{I}_1$, as did the controlled voltage generator of (b).

The short-circuit **y** parameters yield the equations

$$\begin{bmatrix} \mathbf{I}_1 \\ \mathbf{I}_2 \end{bmatrix} = \begin{bmatrix} \mathbf{y}_i & \mathbf{y}_r \\ \mathbf{y}_f & \mathbf{y}_o \end{bmatrix} \begin{bmatrix} \mathbf{V}_1 \\ \mathbf{V}_2 \end{bmatrix} \tag{6.68}$$

with $\mathbf{y}_r \neq \mathbf{y}_f$ because of the active circuit conditions. The two controlled generators appear in Fig. 6.14(a) as a circuit model. If we manipulate the

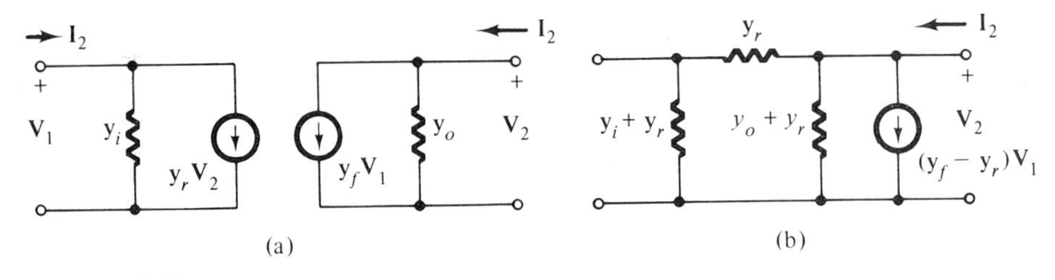

FIGURE 6.14

(a) Active **y**-parameter network; (b) one-generator active equivalent.

second equation by adding and subtracting $\mathbf{y}_r\mathbf{V}_1$, we can write

$$\begin{bmatrix} \mathbf{I}_1 \\ \mathbf{I}_2 - (\mathbf{y}_f - \mathbf{y}_r)\mathbf{V}_1 \end{bmatrix} = \begin{bmatrix} \mathbf{y}_i & \mathbf{y}_r \\ \mathbf{y}_r & \mathbf{y}_o \end{bmatrix} \begin{bmatrix} \mathbf{V}_1 \\ \mathbf{V}_2 \end{bmatrix} \tag{6.69}$$

and these equations are represented by the circuit at (b) in Fig. 6.14. We have a passive π network ($\mathbf{y}_r = \mathbf{y}_f$) and a controlled current source, controlled by $\mathbf{V}_1$. The circuit is employed for study of the high-frequency performance of the bipolar transistor.

The hybrid parameters, with equations

$$\begin{bmatrix} \mathbf{V}_1 \\ \mathbf{I}_2 \end{bmatrix} = \begin{bmatrix} \mathbf{h}_i & \mathbf{h}_r \\ \mathbf{h}_f & \mathbf{h}_o \end{bmatrix} \begin{bmatrix} \mathbf{I}_1 \\ \mathbf{V}_2 \end{bmatrix} \tag{6.70}$$

can be used to describe the circuit of Fig. 6.15. The first equation is a voltage equation, the second is a current summation. It follows that one source is a voltage generator: the other is a current source. At present this circuit is widely used as an equivalent for the bipolar transistor and is often simplified

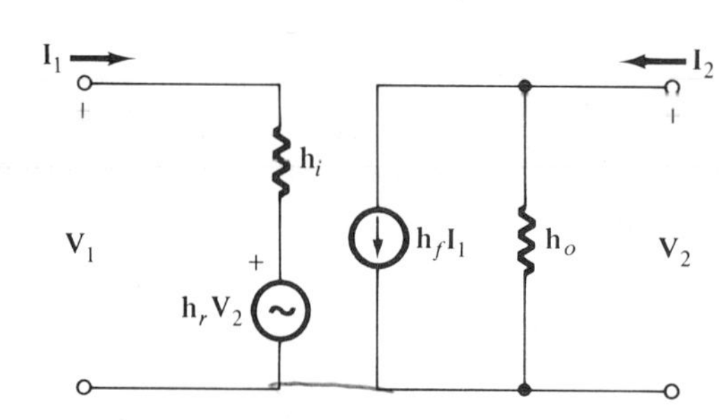

FIGURE 6.15
h-Parameter active equivalent circuit.

by noting that sine $\mathbf{h}_r$ is small, its effect may be neglected and considered to be zero. The circuit then simplifies to a current-controlled current generator, with finite input and output impedances.

Controlled sources are dependent generators and are so treated in network measurements. When measuring terminal impedances we short-circuit independent voltage generators and open independent current generators; for dependent generators the output is given as called for by the controlling current or voltage under the resultant conditions.

6.13 Gain; the decibel

For any network the ratio of the output voltage to the input voltage has been defined as a transfer function, the *voltage gain*:

$$\mathbf{G} = \frac{\mathbf{V}_o}{\mathbf{V}_i} \tag{6.71}$$

With sinusoidal signals this is a complex quantity; the phase angle is of importance.

Another useful transfer ratio is the *current gain*,

$$\mathbf{A} = \frac{\mathbf{I}_o}{\mathbf{I}_i} \tag{6.72}$$

which, with sinusoidal forcing, will also be complex.

With resistive input R_i and a resistive load R_L, the ratio of output power to the input power is known as the *power gain* of the network:

$$\text{P.G.} = \frac{V_o I_o}{V_i I_i} \tag{6.73}$$

Expressing the power gain as a power of 10,

$$\text{P.G.} = 10^{\alpha}$$

and if we have a cascade of amplifiers, the overall power gain will be

$$\text{P.G.}_{\text{ov}} = 10^{\alpha} \times 10^{\beta} \times 10^{\gamma} \times \cdots \tag{6.74}$$

Taking the logarithm to the base 10,

$$\log \text{P.G.}_{\text{ov}} = \alpha + \beta + \gamma + \cdots$$

Use of a logarithmic unit allows cascaded gains to be added or subtracted, a simpler process than the multiplication of Eq. 6.74.

Consequently we have the *decibel* (dB), named for Alexander Graham Bell and defined as

$$\text{Number of decibels} = 10 \log \frac{P_2}{P_1} \tag{6.75}$$

as a *unit of power ratio*.

A positive number of decibels indicates that P_2 is larger than P_1 and that there is a power gain in the network; a negative number of decibels indicates a power reduction or loss. With $P_2/P_1 = 2$, the logarithm is 0.301 and using this in Eq. 6.75 we find that 3.01 dB represents a doubling of the power. A figure of -3.01 dB represents half-power, and the half-power points of the resonance curve are also called the -3 dB band limit frequencies.

The power ratio may also be written as

$$\text{dB} = 10 \log \frac{V_2^2/R_2}{V_1^2/R_1}$$

and *if* $R_2 = R_1$, then

$$\text{dB} = 20 \log \frac{V_2}{V_1} \tag{6.76}$$

Although *technically improper*, it has become customary to define voltage gains (or current gains) by Eq. 6.76 whether $R_2 = R_1$ or not. It is suggested that when voltage gains are so expressed in decibels the symbol be changed to dBV to so indicate.

The decibel is not a measure of absolute power but is used in that manner when a *reference level* for P_1 is adopted and stated. In electrical work this reference level is usually taken as 1 mW. Thus an amplifier with 7 W output is

$$10 \log \frac{7}{0.001} = 38.45 \text{ dB}$$

above reference level, sometimes designated dBm, when 1 mW is used as the reference.

The dB, as a logarithmic unit, has a good psychological basis when related to human hearing, the ultimate end point of much amplifier output. It was hoped that 1 dB would approximate the minimum audible power change, but actually the ear requires a change approximating 2.5 dB for a minimum noticeable stimulus.

6.14 Summary

Any passive, linear, two-port network can be replaced with a simple T or π equivalent circuit insofar as terminal performance is concerned. One-port active networks can be replaced with one-port voltage-source or current-source equivalent generators. The equivalent circuit is repeatedly used in the "black-box" method of analysis by replacement of a functional unit with its equivalent circuit. The Thévenin and Norton circuits, and the T and π circuits, were each shown to be circuit duals, contributing to the flexibility of choice in the selection of equivalent circuits.

Maximum possible power is transferred from a source to a load having an impedance equal to the conjugate of the source impedance at the port; this constitutes *power matching*.

We use the two-port active equivalent circuit to represent amplifiers and relate transistor terminal measurements to the elements of the equivalent network. Thus we physically realize the circuit models predicted by the mathematical models.

Controlled sources were introduced as a concept in circuit elements. They are unilateral dependent generators, with output linearly controlled by a current or voltage at a port in the network.

In this chapter we have attempted to lay a foundation for the generalized study of network response, amplifier response, and combinations of the two as in the operational amplifier and the active filter. The network will usually be considered in its equivalent T or π form and the amplifier as a *gain box*. As in any treatment utilizing equivalent circuits, we are concerned only with terminal performance—not with the complexities which usually are inherent with internal details of circuitry.

PROBLEMS

6.1 Determine the **z** parameters for the network of Fig. 6.16(a).

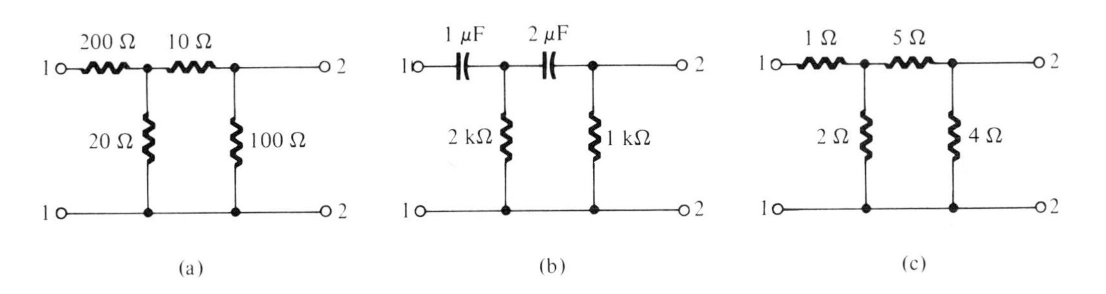

FIGURE 6.16

6.2 Find the **y** parameters for the circuit of Fig. 6.16(a) and draw an equivalent circuit.

6.3 Find the values of the **h** parameters for the circuit of Fig. 6.16(c).

6.4 Find the **z** parameters and draw an equivalent T network for the circuit of Fig. 6.16(b) when operated at 100 Hz.

6.5 Find the values of the **h** parameters of the circuit of Fig. 6.16(b) when operated at 10 Hz.

6.6 Determine the **y** parameters for the circuit of Fig. 6.16(b) when operated at $\omega = 100$ r/s.

6.7 With a load of 600 Ω at port 2, determine the port 1 input impedance for the circuit of Fig. 6.16(b) with $f = 100$ Hz. Also find the transfer impedance $\mathbf{Z}_T$.

6.8 Find the Thévenin equivalent at port 1 for the circuit of Fig. 6.17(a). Convert your Thévenin circuit to a Norton current-source circuit.

6.9 Find a current-source equivalent at the port for Fig. 6.17(b).

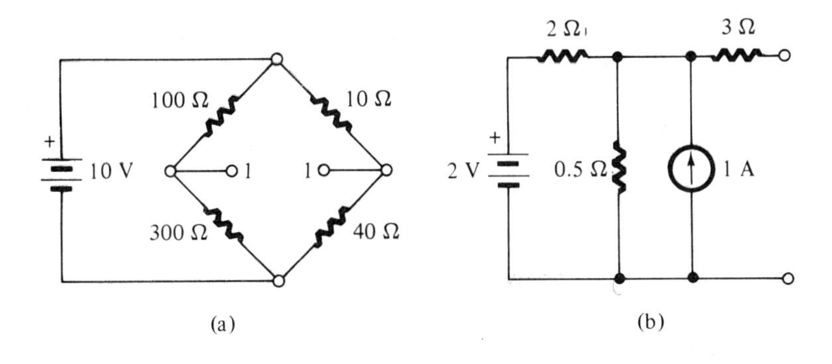

FIGURE 6.17

6.10 (a) Find the current $\mathbf{I}_L$ in the circuit of Fig. 6.18(a). (b) Find the voltage-source equivalent to the left of terminals 1,1. (c) Prove that you have the same $\mathbf{I}_L$ value when using your voltage-source circuit.

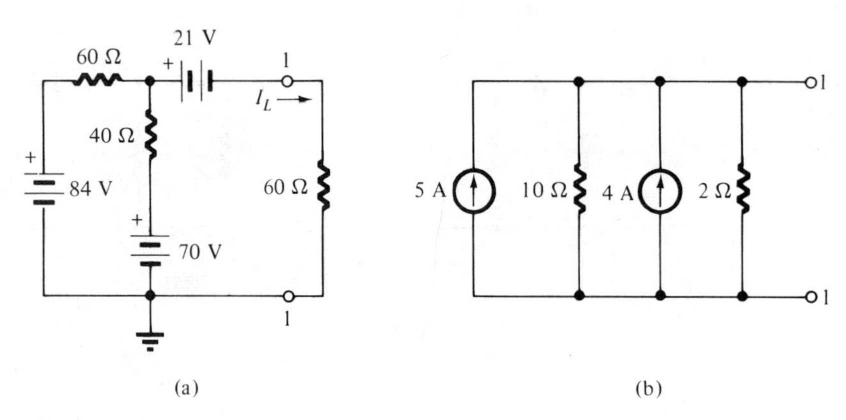

FIGURE 6.18

6.11 Develop a voltage-source equivalent at the port for the circuit of Fig. 6.18(b).

6.12 Find the value of maximum power obtainable at the port from the network of Fig. 6.17(b).

6.13 Determine the suitable load for power matching and the maximum power output from the network of Fig. 6.18(b).

6.14 In the circuit of Fig. 6.19(a), calculate the power-matched load at A, A and determine the maximum possible power output with a frequency of 100 Hz from the generator.

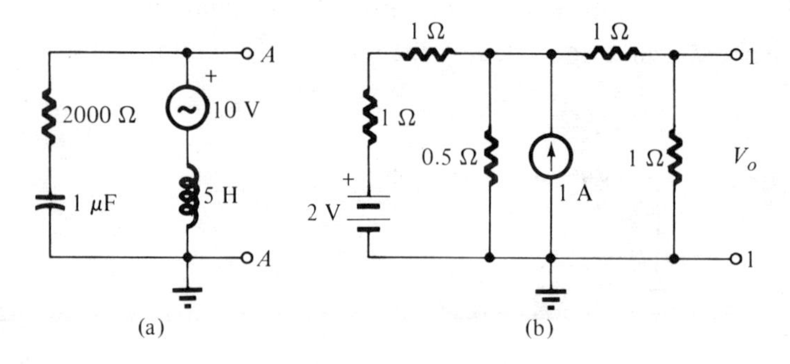

FIGURE 6.19

6.15 Find the maximum power output possible at the 1,1 terminals of the network in Fig. 6.19(b); state the load used.

6.16 Find a current-source equivalent circuit for the network of Fig. 6.19(b) at the 1,1 terminals.

6.17 Find the voltage-source equivalent circuit to the left of 1,1 for the network of Fig. 6.18(a). (a) What load will give maximum power output? (b) What is the maximum possible power output?

6.18 If the 100-Ω resistor is reduced to 30 Ω, what is the Thévenin equivalent for the circuit of Fig. 6.17(a) at the 1,1 terminals?

6.19 Find the current-source equivalent circuit for the network of Fig. 6.19(a) at 60 Hz.

6.20 Draw an equivalent circuit for Fig. 6.19(b) using one voltage-source generator.

6.21 Find the current-source equivalent for the circuit of Fig. 6.18(b); also draw the voltage-source equivalent and show the values of the circuit parameters.

6.22 In the circuit of Fig. 6.20(a), find voltage $\mathbf{V}_{Ao}$ and the current supplied by the 10-V source if the circuit to the right of the 1,1 terminals is replaced by its Thévenin equivalent.

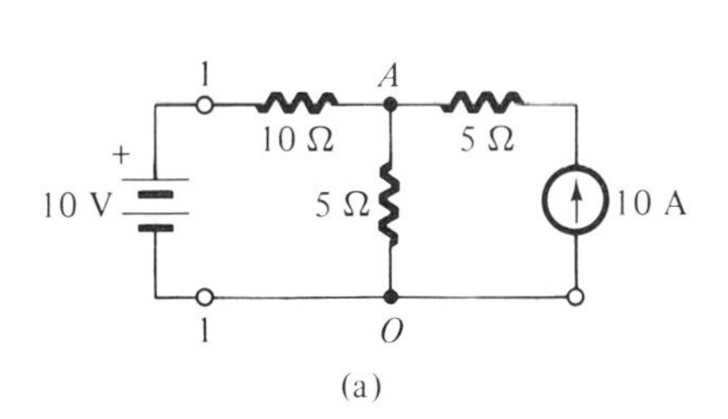

(a)

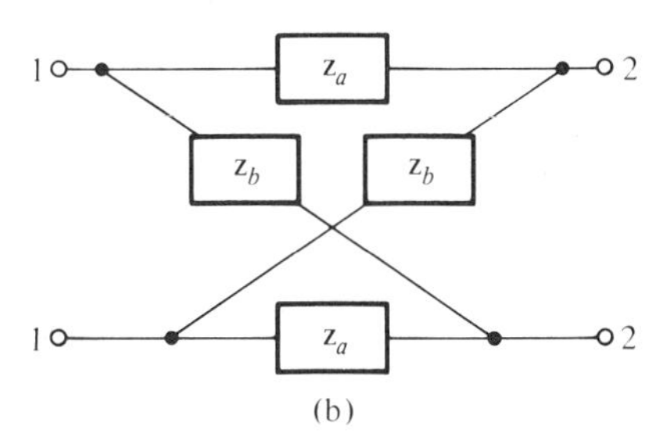

(b)

FIGURE 6.20

6.23 Draw a dual network for the circuit of Fig. 6.18(b).

6.24 A network including a generator has an open-circuit ac voltage of 105 V and on short circuit produces a current of 5.05 A. When a 10-Ω resistance is used as a load, the current is 4.05 A. Find an equivalent voltage-source circuit for the network. How could you determine the sign of the included reactance?

6.25 The following measurements are made on a box containing impedances: At 1,1 with an open circuit at 2,2, $\mathbf{Z}_{in} = 40\underline{/0°}\ \Omega$; at 1,1 with a short circuit at 2,2, $\mathbf{Z}_{in} = 20.3\underline{/30°}$; at 2,2 with a short circuit at 1,1, $\mathbf{Z}_{in} = 27.2\underline{/-27°}\ \Omega$. Find the values for an equivalent T network and draw the circuit. What maximum power could be obtained if supplied by a generator of 25 V at the 1,1 terminals?

6.26 For the symmetrical lattice network of Fig. 6.20(b), show that

$$\mathbf{z}_i = \tfrac{1}{2}(\mathbf{Z}_a + \mathbf{Z}_b) \qquad \text{and} \qquad \mathbf{z}_r = \tfrac{1}{2}(\mathbf{Z}_b - \mathbf{Z}_a)$$

6.27 A generator of voltage V is connected to the 1 port of Fig. 6.20(b). Find the Thévenin equivalent at the 2 port.

6.28 In Fig. 6.18(a), change the 40-Ω resistor to a value which will establish its current as 0.2 A.

6.29 The input in Fig. 6.21(a) is $V_i = 1$ mV. With the reference level at 1 mW, determine the output power in the 10K load in watts and decibels above reference.

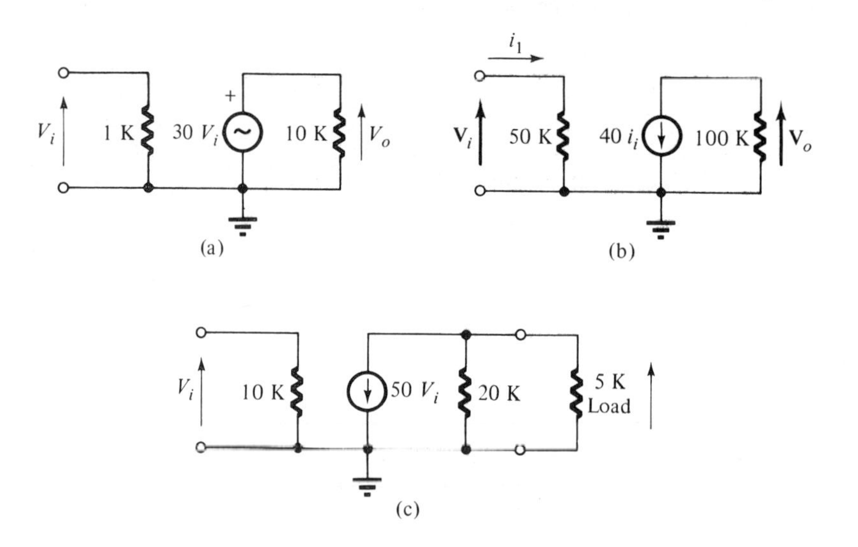

FIGURE 6.21

6.30 Find the voltage gain V_o/V_i for the circuit of Fig. 6.21(b); state the result in dBV.

6.31 With $V_i = 0.05$ V, find the dBV gain to the 5K load of Fig. 6.21(c). What load will give maximum power output?

6.32 A radio receiver has an input impedance which is 70 Ω resistive. The antenna supplies 50 μV to this input for a given signal. If the output to the loudspeaker is to be 18 W, find (a) the receiver dB power gain; (b) the dB input power level, 1-mW reference; and (c) the reduction in dB gain, for the same output, when the input signal changes to 2000 μV.

6.33 A microphone can be represented as a voltage source of 50-Ω resistance and generator output at −54 dBV below a 1-mW reference. (a) If directly connected to an 8-Ω loudspeaker, find the power delivered in watts. If the microphone drives an amplifier of 50-dB gain, what is the power output to the loudspeaker?

6.34 A voltmeter having an integral resistance of 10,000 Ω on the 10-V scale and 50,000 Ω on the 50-V scale is connected to the output port of a circuit. When using the 10-V scale of the instrument, the reading is

7.5 V; when using the 50-V scale, the reading is 30 V. Determine the circuit values for a voltage-source equivalent for the one-port box.

6.35 A certain active network has an open-circuit voltage of 125 V ac, and on short circuit produces a current of 5.59 A rms. When a 10-Ω resistive load is used, the output current is 4.41 A rms. Find the equivalent voltage source circuit for the active network, giving reactance and resistance values.

REFERENCE

1. W. A. Lynch and J. G. Truxal, *Signals and Systems in Electrical Engineering*. McGraw-Hill Book Company, New York, 1962.

frequency response of networks

seven

The reactances of inductance and capacitance vary oppositely with frequency. When the magnitudes are properly chosen, their effects cancel and an LC circuit becomes resistive; the circuit is then said to be *resonant*. The variation of circuit impedance as a function of frequency is used to select one band of frequencies out of a wider band. Our concern will center on the design of such networks as will be useful in providing frequency selectivity in communication systems.

7.1 Series resonance

A series circuit of R, L, and C appears in Fig. 7.1(a). The reactance of the inductor varies linearly with frequency as ωL and that of the capacitor inversely as $-1/\omega C$; these curves are shown in Fig. 7.1(b). The circuit impedance is

$$\mathbf{Z} = R_s + j\left(\omega L - \frac{1}{\omega C}\right) \tag{7.1}$$

and the current is

$$\mathbf{I} = \frac{\mathbf{V}}{R_s + j[\omega L - (1/\omega C)]} \tag{7.2}$$

The net reactance curve appears dashed in (b), and when the reactance is zero at f_o, the current is maximum in the series circuit and

$$\mathbf{I} = \frac{\mathbf{V}}{R_s} \tag{7.3}$$

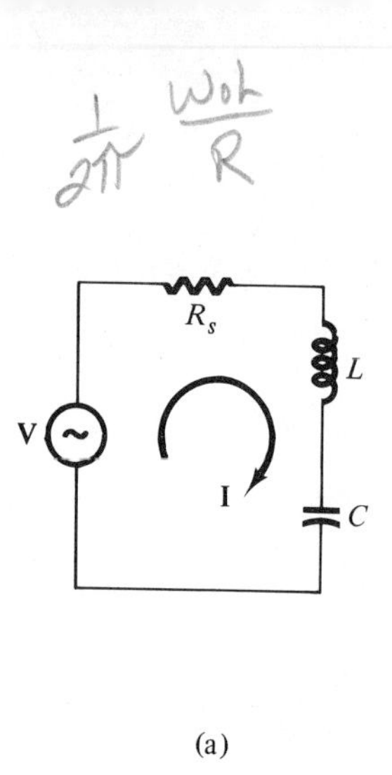

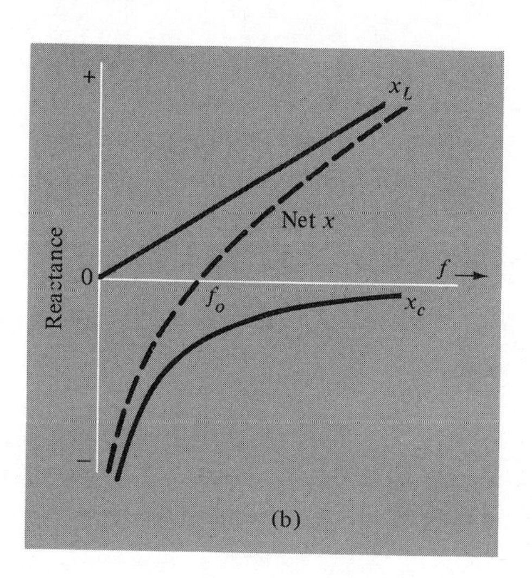

FIGURE 7.1
Series *RLC* circuit.

With the j term at zero and the current and voltage in phase, we have the *condition of resonance*. That is,

$$\omega_o L - \frac{1}{\omega_o C} = 0$$

from which

$$\omega_o^2 LC = 1 \quad \text{or} \quad \omega_o = \frac{1}{\sqrt{LC}} \tag{7.4}$$

which states *the frequency of resonance* for the series *RLC* circuit. The resistance R_s is the total for the circuit, including R_c of the inductor.

Using $\omega_o^2 LC = 1$ and substituting for C in Eq. 7.2, we have

$$I = \frac{\mathbf{V}/R_s}{1 + j\left(\dfrac{\omega_o L}{R_s}\right)[(\omega/\omega_o) - (\omega_o/\omega)]} \tag{7.5}$$

At frequencies where $\omega > \omega_o$, the j term is positive, but it yields a negative angle for **I** when the expression is rationalized. The current *lags the voltage* and the circuit appears as *inductive above resonance*. When $\omega < \omega_o$, the expression yields a positive angle for **I**, and **I** *leads the voltage*, so that the circuit *appears capacitive below resonance*. This also appears in Fig. 7.1(b).

The current and phase angle vary with frequency as in Fig. 7.2, and the effect of resonance is to emphasize the frequencies near f_o.

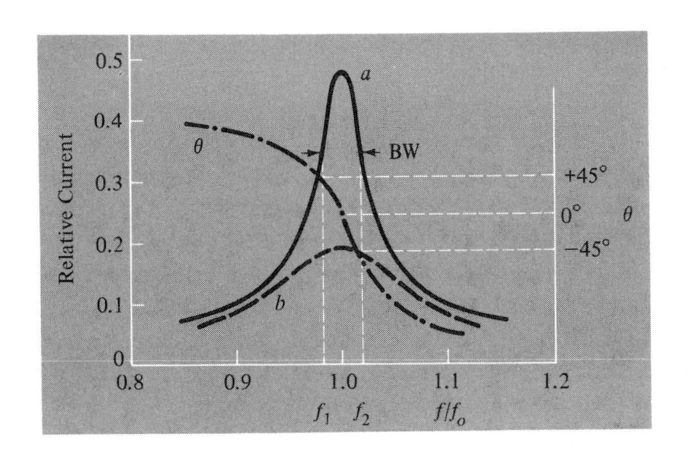

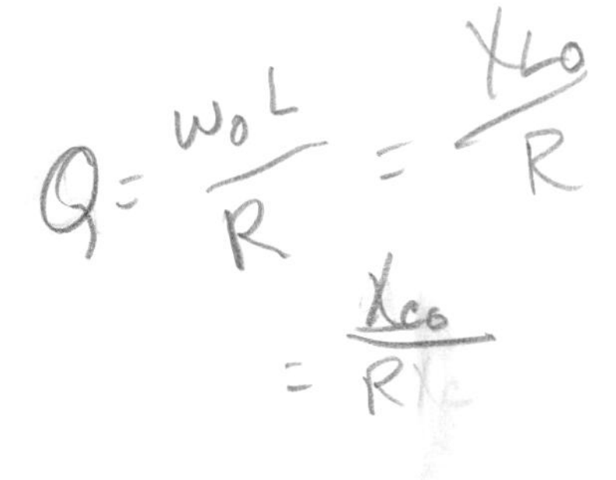

FIGURE 7.2

Resonance curve: (a) $Q = 30$; (b) $Q = 12$; θ is current angle, voltage reference.

7.2 Q, the circuit quality factor

The ability of a series RLC circuit to select a narrow band of frequencies is dependent on its energy storage at the resonant frequency as compared to its energy losses. The *quality factor* for a circuit at resonance is called Q, defined as

$$Q = 2\pi \frac{\text{maximum energy stored at resonance}}{\text{energy lost per cycle at resonance}} \tag{7.6}$$

Returning to the fundamental definitions, the energy stored in an inductor with a current $i = I_m \sin \omega t$ is

$$W_L = \frac{Li^2}{2} = \frac{LI_m^2}{2} \sin^2 \omega t$$

Since $v = (1/C) \int i \, dt$, the energy stored in the capacitance with the same current is

$$W_C = \frac{Cv^2}{2} = \frac{I_m^2}{2\omega^2 C} \cos^2 \omega t$$

In view of $\omega_o^2 LC = 1$ at resonance, the total energy stored is

$$W = \frac{LI_m^2}{2} (\sin^2 \omega t + \cos^2 \omega t) = \frac{LI_m^2}{2} \tag{7.7}$$

The energy flows from L to C twice per cycle, as indicated by the double frequency term inherent in $\sin^2 \omega t$ or $\cos^2 \omega t$.

The energy lost per cycle in a series circuit is $I_m^2 R_s/2f$, and so

$$Q = \frac{2\pi L I_m^2}{I_m^2 R_s/f} = \frac{\omega_o L}{R_s} = \frac{1}{R_s}\sqrt{\frac{L}{C}} \tag{7.8}$$

The latter form shows that the circuit Q is dependent on the choice of the L/C ratio by the designer.

Capacitor losses usually are small with respect to those of inductors, and the Q of a circuit depends primarily on the quality of the inductor and any resistor present. The quality of a coil is affected by wire size and coil diameter, the dielectric quality of the coil form, and the presence of magnetic material such as powdered iron or ferrite materials. The circuit Q will vary slowly with frequency and peak in some frequency range, as shown in Fig. 7.3.

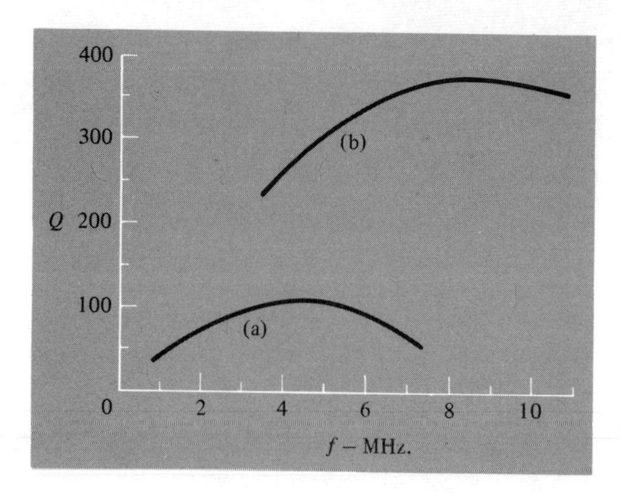

FIGURE 7.3
Variation of circuit Q vs. frequency; (a) 50 μH, no. 26 AWG wire, plastic form; (b) 8 μH, no. 14 AWG wire, no form.

7.3 Voltages in the series *RLC* circuit at resonance

In the series RLC circuit the current at resonance is $\mathbf{I} = \mathbf{V}/R_s$, and the voltage across the inductive reactance is

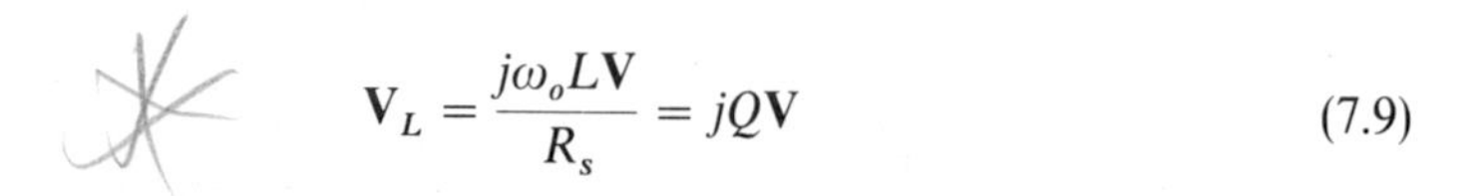

$$\mathbf{V}_L = \frac{j\omega_o L\mathbf{V}}{R_s} = jQ\mathbf{V} \tag{7.9}$$

Similarly, the voltage across the capacitive reactance is

$$\mathbf{V}_C = \frac{-j\mathbf{V}}{\omega_o C R_s} = -jQ\mathbf{V} \tag{7.10}$$

since $1/\omega_o C = \omega_o L$. The phasors for $\mathbf{V}_L$ and $\mathbf{V}_C$ are oppositely directed and cancel, and the voltage applied to the circuit needs only to be equal to the resistance drop at resonance.

The circuit may be considered as a voltage amplifier, having a gain Q to either $\mathbf{V}_L$ or $\mathbf{V}_C$. Since Q is usually greater than 10 and may have a value of several hundred, the voltages across L and C individually can be large and may be at dangerous levels.

7.4 The parallel-resonant circuit

The parallel form of a resonant circuit is shown in Fig. 7.4, where a current source **I** supplies a parallel combination of L, C, and resistance, R_p. The last is made up of the internal parallel resistance of the current source and any added circuit resistance. Voltage **V** is the circuit response.

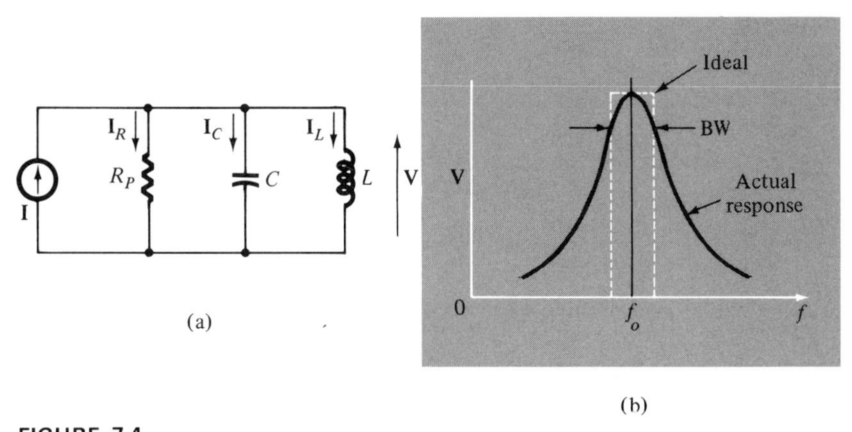

FIGURE 7.4
(a) Parallel-resonant circuit; (b) frequency response.

The currents in the circuit at (a) in Fig. 7.4 can be summed as

$$\mathbf{I} = \frac{\mathbf{V}}{R_p} + \frac{\mathbf{V}}{-j/\omega C} + \frac{\mathbf{V}}{j\omega L} = \mathbf{V}\left[\frac{1}{R_p} + j\left(\omega C - \frac{1}{\omega L}\right)\right] \tag{7.11}$$

For resonance we make the j term equal to zero, and this occurs at the resonant angular frequency ω_o, as

$$\omega_o = \frac{1}{\sqrt{LC}} \tag{7.12}$$

which is the same expression as was obtained for the series-resonant circuit.

The parallel circuit has a common voltage **V** across the elements, and the energy present in storage can be written in terms of that common value of **V**, following Section 7.2. The circuit Q can then be obtained from the maximum values of stored energy $C\mathbf{V}_m^2/2$, while the loss per cycle is $\mathbf{V}_m^2/(2R_p f)$. Then

$$Q = 2\pi \frac{C\mathbf{V}_m^2}{\mathbf{V}_m^2/(R_p f)} = \omega C R_p = \frac{R_p}{\omega_o L} = R_p\sqrt{\frac{C}{L}} \tag{7.13}$$

after using the resonance condition $\omega_o^2 LC = 1$. The value of Q can be controlled by adjustment of the shunting resistance R_p, or by choice of the ratio of C to L, and these are choices for the circuit designer when the circuit response bandwidth is to be adjusted, as will be shown.

The impedance of the parallel circuit is written from Eq. 7.11 as

$$\mathbf{Z} = \frac{1}{(1/R_p) + j[\omega C - (1/\omega L)]} \tag{7.14}$$

and with the j term made zero as is required for resonance

$$\mathbf{Z}_o = R_p \tag{7.15}$$

Using Eq. 7.13, we have

$$\mathbf{Z}_o = \frac{Q}{\omega_o C} = Q\omega_o L \tag{7.16}$$

as additional expressions for the resonant resistance of the parallel circuit. The value of $\mathbf{Z}_o$ is large and equal to R_p at resonance, and since $\mathbf{V} = \mathbf{IZ}$, the voltage falls with frequency departure from resonance on either side of the peak, as plotted in Fig. 7.4(b).

7.5 The resonant circuit with resistance in the inductive branch

No inductance is perfect and we may include the resistance of the inductance in the parallel circuit as Fig. 7.5(a). This circuit form can be made equivalent to the circuit at (b) if R_s and R' are properly related, and the previous resonance analysis can then be applied to (b).

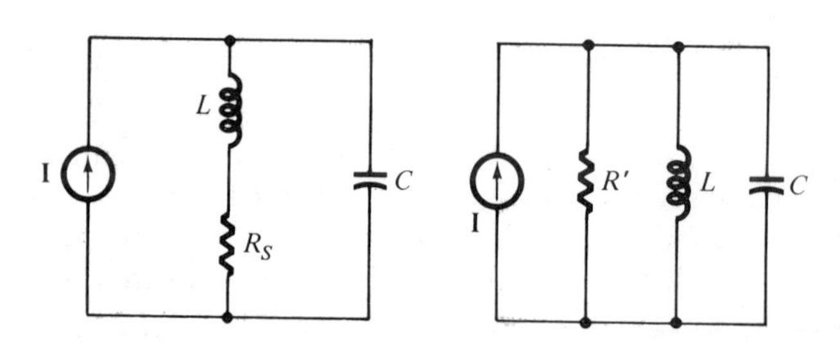

FIGURE 7.5
(a) With resistance R_s; (b) R_s transformed to R'.

If we proceed from (b) to (a), we can write the effect of R' and $j\omega L$ in parallel as

$$\mathbf{Z}_{\text{eq}} = \frac{j\omega L R'}{R' + j\omega L} = \frac{\omega^2 L^2 R' + j\omega L R'^2}{R'^2 + \omega^2 L^2} \tag{7.17}$$

It will usually be true that $R'^2 \gg \omega^2 L^2$, and so the denominator reduces, giving

$$\mathbf{Z}_{eq} = \frac{\omega^2 L^2}{R'} + j\omega L = R_s + j\omega L \tag{7.18}$$

This expression represents the series R, L branch at (a). The inductance does not appreciably change in the transformation from series to parallel form, but

$$R_s = \frac{\omega^2 L^2}{R'}$$

and

$$R' = \frac{\omega^2 L^2}{R_s} \tag{7.19}$$

This is a simple relation for passing from the series R form to the parallel R form of circuit or vice versa. If R' and R_s are so related, the circuits are essentially equivalent.*

Resistance R' may then be combined with any shunt resistance of the current source to yield an effective R_p, and the results of Section 7.4 may be applied for resonance study.

The circuits will have the same Q since

$$Q = \omega_o C R' \tag{7.20}$$

for the circuit of (b). By Eq. 7.19 we have

$$Q = \omega_o C \frac{\omega_o^2 L^2}{R_s}$$

but

$$\omega_o^2 LC = 1$$

at resonance, and so

$$Q = \frac{\omega_o L}{R_s} \tag{7.21}$$

which is the result for the circuit form at (a).

*The neglect of $\omega^2 L^2$ with respect to R'^2 in the denominator of Eq. 7.17 assumes that

$$\omega^2 L^2 \ll R'^2 = Q^2 \omega^2 L^2$$

which is true if Q is large with respect to unity.

7.6

Current gain in the parallel-resonant circuit

At resonance in Fig. 7.4(a) the voltage is $\mathbf{V}_o = \mathbf{I}R_p$, and the current in the capacitive branch is

$$\mathbf{I}_C = j\omega_o C\mathbf{V}_o = j\omega_o CR_p\mathbf{I} = jQ\mathbf{I} \tag{7.22}$$

Similar analysis for the current in the inductive branch gives

$$\mathbf{I}_L = \frac{\mathbf{V}_o}{j\omega_o L} = \frac{-jR_p\mathbf{I}}{\omega_o L} = -jQ\mathbf{I} \tag{7.23}$$

since $\omega_o L = 1/\omega_o C$ at resonance.

The branch currents are phase opposed and add to zero. However, each current is Q times the source current, which means that a large current circulates through L and C. Because of this circulating current and consequent energy in the L and C elements, these elements are often referred to as representing a *tank circuit*. The parallel circuit is evidently a current amplifier if the ratio of source to tank current is considered as a gain.

EXAMPLE. A generator of 0.1 A, 10^6 Hz, supplies a resonant parallel circuit which includes a 200-pF capacitor. The circuit is shunted by a resistance of 25,000Ω. Then $\omega = 2\pi \times 10^6$ and

$$\omega_o C = 2\pi \times 10^6 \times 200 \times 10^{-12} = 12.6 \times 10^{-4}$$

$$X_C = \frac{-1}{\omega_o C} = \frac{-1}{12.6 \times 10^{-4}} = -796\,\Omega$$

The resonant resistance of the circuit is $R_p = 25{,}000\,\Omega$, so

$$Q = \omega_o CR_p = 12.6 \times 10^{-4} \times 25 \times 10^3 = 31.5$$

The inductance needed for resonance is

$$\omega_o L = 796\,\Omega$$

$$L = \frac{796}{2\pi \times 10^6} = 127 \times 10^{-6} = 127\,\mu\text{H}$$

The tank circuit current is $QI = 31.6 \times 0.1 = 3.16$ A, compared to the generator current of 0.1 A.

7.7

Impedance variation near resonance

Equation 7.14 for the impedance of the parallel-resonant circuit involves a difference of the reactances, and these terms are nearly equal close to resonance. To improve the accuracy in calculation at frequencies near resonance and to provide a useful equation form, we define a *fractional deviation from the frequency of resonance* ω_o *as* δ:

$$\delta = \frac{\omega - \omega_o}{\omega_o} \tag{7.24}$$

Then

$$\frac{\omega}{\omega_o} = 1 + \delta \tag{7.25}$$

Equation 7.14 can be modified to

$$\mathbf{Z} = \frac{R_p}{1 + j\omega CR_p[1 - (1/\omega^2 LC)]} = \frac{R_p}{1 + jQ[1 - (\omega_o^2/\omega^2)]}$$

Using Eq. 7.25, the frequency ratio can be written as

$$\frac{\omega_o^2}{\omega^2} = \frac{1}{(1 + \delta)^2}$$

and expansion by the binomial theorem gives

$$\frac{1}{(1 + \delta)^2} = 1 - 2\delta + 3\delta^2 - 4\delta^3 + \cdots$$

Since δ is small, we can drop the higher-order terms and the impedance is obtained as

$$\mathbf{Z} = \frac{R_p}{1 + jQ(1 - 1 + 2\delta)} = \frac{R_p}{1 + j2Q\delta} \tag{7.26}$$

We can then write the ratio of the impedance at some frequency $(1 + \delta)f_o$ to that at resonance f_o as

$$\frac{\mathbf{Z}}{R_p} = \frac{1}{1 + j2Q\delta} = A\underline{/\theta} \tag{7.27}$$

The magnitude A and the phase angle θ are plotted in Fig. 7.6 as a *universal resonance curve* in terms of the parameter $Q\delta$. This is suited to any parallel-resonant circuit.

7.8 Bandwidth of the parallel-resonant circuit

An ideal response curve for a frequency-selective circuit would be a rectangle of the desired bandwidth and centered at the resonant frequency. Figure 7.6 indicates that the ideal is

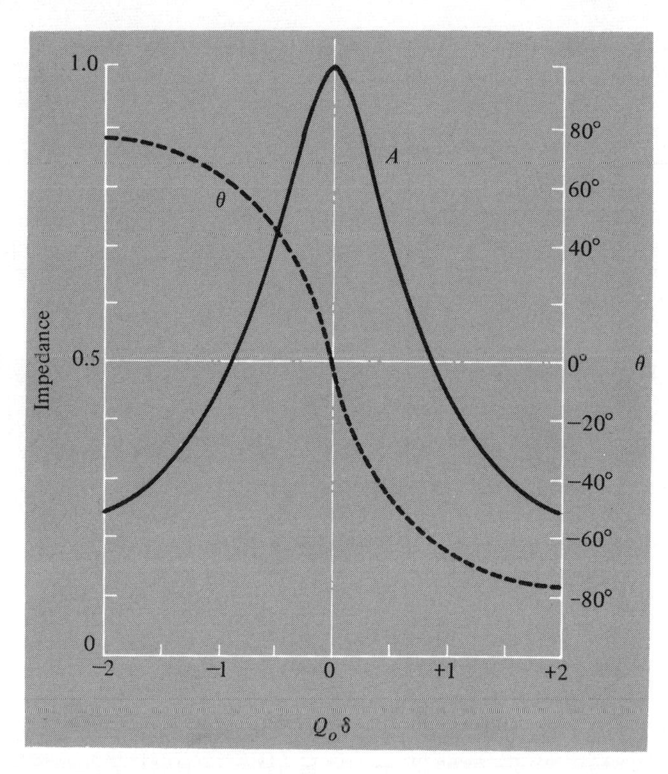

FIGURE 7.6

Universal resonance curve as a function of $Q\delta$.

only approximated with the response of a resonant circuit, but such circuits are nevertheless widely used for frequency selection because of their simplicity.

To compare the response of various circuits, we define the *bandwidth* of the circuit as the width in hertz between frequencies on each side of resonance, at which the power in the circuit is one half of the power at resonance. At resonance the power is $P_o = I^2 R_p$, and it follows, with constant current I supplied, that

$$P_{1/2} = \frac{I^2 R_p}{2} = \left(\frac{V_o}{\sqrt{2}}\right)^2 \frac{1}{R_p}$$

where V_o is the circuit voltage at resonance; the half-power point on the resonance curve of Fig. 7.4 is found at $V = V_o/\sqrt{2}$.

In the circuit the resistance R_p is in parallel with the net reactance $X = (\omega L - 1/\omega C)$ and the impedance at the half-power frequency is

$$\mathbf{Z}_{1/2} = \frac{jXR_p}{R_p + jX}$$

The impedance magnitude is

$$|\mathbf{Z}_{1/2}| = \frac{XR_p}{\sqrt{R_p^2 + X^2}}$$

Now with $\mathbf{V}_o = IR_p$, we have

$$|V_{1/2}| = \frac{|V_o|}{\sqrt{2}} = \frac{IR_p}{\sqrt{2}}$$

But also

$$|V_{1/2}| = I|Z_{1/2}| = \frac{XR_p}{\sqrt{R_p^2 + X^2}}I$$

and so

$$\frac{R_p}{\sqrt{2}} = \frac{XR_p}{\sqrt{R_p^2 + X^2}}$$

From this statement we arrive at

$$R_p^2 + X^2 = 2X^2$$

$$R_p = \pm X \tag{7.28}$$

At a half-power frequency the circuit resistance is equal to the circuit reactance, and the phase angle is $\pm 45°$. This is *a fundamental relation for the circuit impedance at a half-power frequency.*

We can use the condition of Eq. 7.28 in Eq. 7.27 and find

$$2Q\delta = 1$$

at the half-power frequencies. At the low-frequency half-power frequency we have $f_1 < f_o$ and $-\delta_1 = f_1/f_o - 1$, so that

$$-2Q\left(\frac{f_1}{f_o} - 1\right) = 1$$

At the upper half-power frequency $\delta_2 = f_2/f_o - 1$ and

$$2Q\left(\frac{f_2}{f_o} - 1\right) = 1$$

Adding these relations, we have

$$2Q\frac{(f_2 - f_1)}{f_o} = 2 \tag{7.29}$$

and the bandwidth between f_2 and f_1 is

$$\text{BW} = f_2 - f_1 = \frac{f_o}{Q} \tag{7.30}$$

The bandwidth is inversely proportional to circuit Q; thus if we vary R_p, we can control Q and the bandwidth.

It can also be determined from the above that

$$\omega_o = \sqrt{\omega_1 \omega_2} \tag{7.31}$$

so that the resonant frequency is the geometric mean of the two half-power frequencies.

Since R_p includes the effective coil resistance due to interaction of the coil fields with metals or dielectrics, we can obtain accurate Q values only with the coil mounted in the equipment. Thus we often measure the bandwidth in the equipment and determine the Q by use of Eq. 7.30.

With $2Q\delta = 1$ at the half-power band limits, we locate these points on Fig. 7.6 at $Q\delta = \frac{1}{2}$. Between these band limits the impedance phase angle changes in almost linear manner from $+45°$ to $-45°$. The average rate of change of phase angle is

$$\frac{d\theta}{d\omega} = \frac{-90°}{\text{BW}} = -90° \frac{Q}{\omega_o} \tag{7.32}$$

A higher Q leads to a greater rate of phase change, and this is important in the maintenance of constant frequency in oscillator circuits.

It can be shown that the bandwidth of the series-resonant circuit is also given by Eq. 7.30.

7.9 Narrow-band resonant circuit response

The parallel-resonant circuit, with impedance R_p at resonance, is often used as a load for an amplifier, combining frequency selectivity with gain. In Fig. 7.7, with the voltage-controlled current source simulating a transistor, the gain can be written as

$$\mathbf{G} = \frac{\mathbf{V}_2}{\mathbf{V}_1} = -g_m \frac{R_p}{1 + j2Q\delta} \tag{7.33}$$

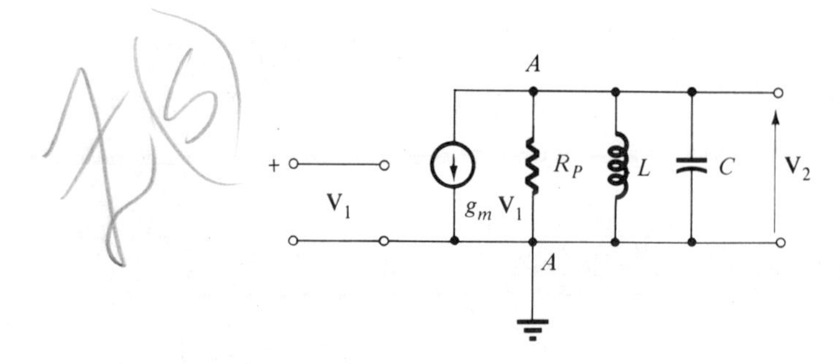

FIGURE 7.7
Parallel-resonant amplifier load.

which reduces to $-g_m R_p$ at resonance. Resistance R_p represents the paralleled value of the inductor resistance and the transistor output resistance.

The Q of the circuit is usually chosen to provide the needed bandwidth or pass band of frequencies. For *narrow-band response*, where the pass band may be only 1 or 2 percent of f_o (10 kz at 1 MHz in the broadcast band), the gain curve is like that of Fig. 7.6.

To demonstrate the analysis of resonant circuits from a study of their poles and zeros, we write the circuit impedance of Fig. 7.7 in the s form at the A, A port:

$$Z(s) = \frac{(L/C)R_p}{R_p(sL + 1/sC + L/CR_p)}$$

$$= \frac{1}{C}\left[\frac{s}{s^2 + (1/CR_p)s + (1/LC)}\right] \tag{7.34}$$

$$= \frac{1}{C}\frac{s}{(s + s_1)(s + s_2)} \tag{7.35}$$

The roots of the denominator locate poles of the impedance function and the numerator contributes a zero at the origin. The roots are

$$s_1, s_2 = -\frac{1}{2CR_p} \pm \sqrt{\left(\frac{1}{2CR_p}\right)^2 - \frac{1}{LC}}$$

$$= -\frac{B}{2} \pm j\omega_o\sqrt{1 - \frac{B^2}{4\omega_o^2}} \tag{7.36}$$

where $B = 1/CR_p$ and $\omega_o = 1/\sqrt{LC}$ as before.

Two important parameters are evident from the impedance relation of Eq. 7.34:

1. The resonant angular frequency is equal to the square root of the constant term of the quadratic denominator in $Z(s)$.
2. Since $B = \omega_o/Q$, the bandwidth in radians per second is equal to the coefficient of the linear term of the denominator.

Resonant circuit loads for narrow-band amplifiers operate with high $Q = \omega_o/B = \omega_o CR_p$ and are underdamped; thus B is small. Plotting a pole-zero diagram on the complex plane, with $B/2 \ll \omega_o$, the conjugate poles will be located approximately at

$$-\frac{B}{2} \pm j\omega_o \tag{7.37}$$

as in Fig. 7.8(a), with the zero at the origin. Since B is small, the poles are very close to the $j\omega$ axis. The magnitude of Eq. 7.36 gives the distance of the poles from the origin; this is found to be ω_o.

The amplitude and phase characteristic can be determined from Fig. 7.8(b). Using $s = j\omega$ for the steady state,

$$\mathbf{Z}(\omega) = \frac{1}{C} \frac{j\omega}{(j\omega + s_1)(j\omega + s_2)} \tag{7.38}$$

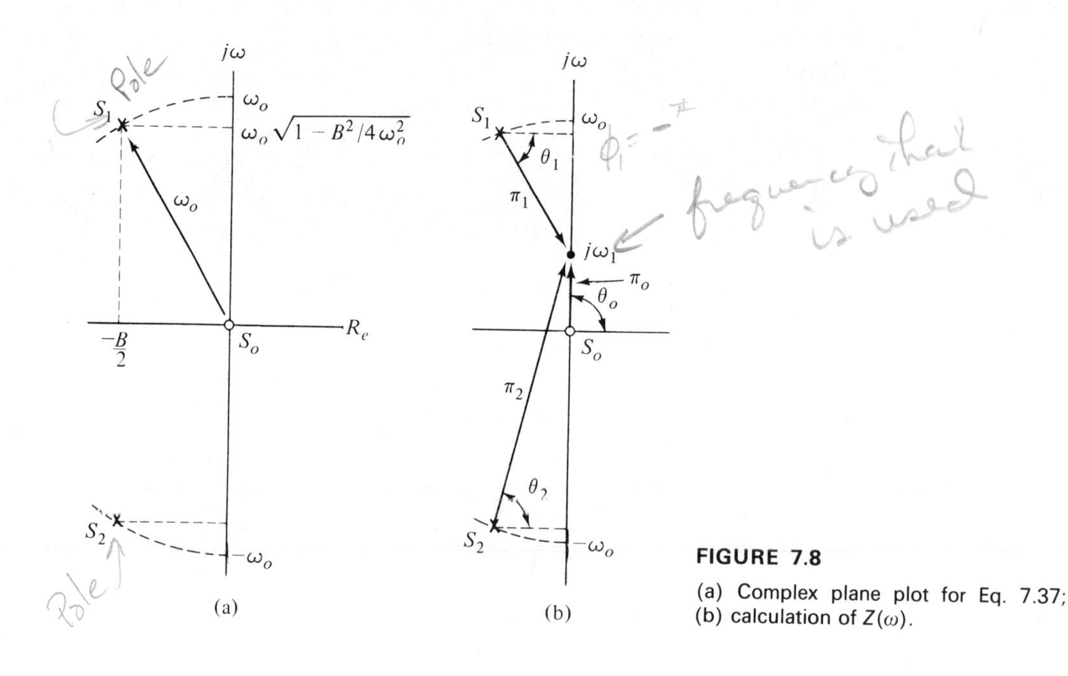

FIGURE 7.8

(a) Complex plane plot for Eq. 7.37; (b) calculation of $Z(\omega)$.

The factors $j\omega + 0$, $j\omega + s_1$, and $j\omega + s_2$ can be replaced with their polar equivalents, written as $\pi_o\angle\theta_o$, $\pi_1\angle\theta_1$, and $\pi_2\angle\theta_2$. The π magnitudes are the distances from each pole or zero to the $j\omega$ point on the axis of imaginaries. Then

Fourier Xform →

$$\mathbf{Z}(\omega) = \frac{1}{C} \frac{\pi_o}{\pi_1 \pi_2} \angle \theta_o - \theta_1 - \theta_2 \tag{7.39}$$

The impedance of the parallel circuit can be determined, for each value of ω chosen, by measurement of the π quantities with a scale and their angles with a protractor.

The pole locations are close to the $j\omega$ axis for high Q, and as $j\omega$ climbs vertically, π_2 quickly becomes nearly vertical and almost equal to $2\pi_o$. Angle θ_2 is nearly 90° and θ_o is 90°. Then

$$\mathbf{Z}(\omega) = \frac{1}{2C} \frac{1}{\pi_1} \angle -\theta_1 \tag{7.40}$$

where θ_1 is negative below resonance, as shown in Fig. 7.8(b), and $\mathbf{Z}$ has a positive angle and is inductive.

It is apparent that a pole, or zero, far removed from the active pole, as s_2 and s_o, has little effect on the variation of the function with frequency and contributes only a constant. As a result, it is possible to focus on the variation of π_1 from s_1 to $j\omega$ in Fig. 7.9(a).

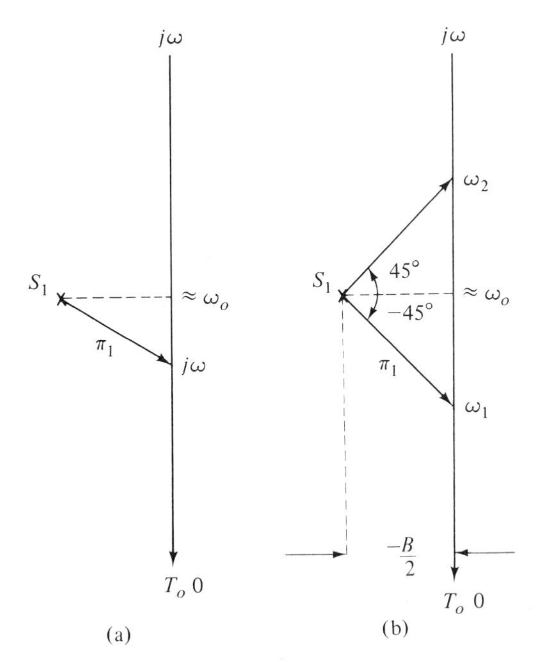

FIGURE 7.9
Determination of the bandwidth.

With pole s_1 close to the $j\omega$ axis, the variation of π_1 as it passes ω_o is very great and the resonant impedance curve plotted for high Q is sharp. At $j\omega = j\omega_o$ the length π_1 becomes minimum and the angle $\theta_1 = 0°$. This is the defined condition of resonance, with $\mathbf{Z}(\omega)$ a maximum. At resonance

$$\pi_1 \underline{/\theta_1} = \frac{B}{2} \underline{/0°} = \frac{1}{2CR_p} \underline{/0°}$$

Substitution into Eq. 7.40 gives

$$\mathbf{Z}(\omega_o) = \mathbf{Z}_o = R_p \tag{7.41}$$

as previously shown.

The bandwidth limits are reached at frequencies ω_1 and ω_2 at which π_1 has angles of $\pm 45°$, as drawn in Fig. 7.9(b). From the geometry

$$\omega_2 - \omega_1 = \omega_2 - \omega_o - (\omega_o - \omega_1)$$

$$\mathrm{BW} = \frac{B}{2} + \frac{B}{2} = B = \frac{1}{CR_p} \tag{7.42}$$

and the percent bandwidth is

$$\frac{\omega_2 - \omega_1}{\omega_o} = \frac{1}{\omega_o C R_p} = \frac{1}{Q} \tag{7.43}$$

as previously determined.

Thus Q, the bandwidth, and the resonant frequency can be found directly from the quadratic denominator, and the frequency-response curve can be drawn by measurements taken from the pole-zero diagram.

Maximum power transfer may not always be possible, because we cannot always match R_p to the source impedance. In the next chapter we shall undertake the analysis of several methods by which load and source impedances may be transformed to values suitable for maximum power transfer between circuits.

7.10 The meaning of the poles and zeros

We have just seen how the pole and zero locations of the impedance function can be used to derive the frequency response of the circuit, and with that experience we can look at further information derivable from the function poles and zeros in the s plane.

We could write the RLC circuit impedance of Eq. 7.34 as

$$Z(s) = \frac{A(s + s_1)(s + s_3)\cdots}{(s + s_2)(s + s_4)\cdots} \tag{7.44}$$

In Eq. 7.44, $A = 1/C$, $s_1 = s_3 = 0$ and s_2 and s_4 are denominator roots, reducing the form to that of Eq. 7.34, and this is the general polynomial form of a network function. A factor $(s + s_x)^n$ in the numerator of the circuit polynomial contributes a zero of order n at the point $s = -s_x$, where n is a finite integer power and s_x is a point in the complex plane. A pole of order n is located at $s = -s_x$ if the polynomial contains a factor $(s + s_x)^n$ in its denominator. In Eq. 7.44, $s_1, s_3, \ldots$ are roots of the numerator and therefore zeros, and $s_2, s_4, \ldots$ are roots of the denominator and create poles.

Driving this general impedance with a current source results in a voltage

$$V = \frac{IA(s + s_1)(s + s_3)\cdots}{(s + s_2)(s + s_4)\cdots}$$

An infinite response, which we recognize as a resonance, will occur if $s = -s_2, -s_4, \ldots$, i.e., when the complex frequency is at any one of the pole frequencies, these being the pole frequencies of the impedance function itself.

Thus the poles of the impedance function correspond to the natural frequencies of the circuit, with the current-source removed by open circuiting, as is usual. These frequencies are the parallel resonances.

Similarly, if we drive the general impedance with a voltage source V, we have

$$I = \frac{V(s + s_2)(s + s_4)\cdots}{A(s + s_1)(s + s_3)\cdots}$$

An infinite response, or a resonance, will occur if $s = -s_1, -s_3, \ldots,$ or at the value of complex frequency equal to any of the impedance zeros, with the input closed through the voltage generator.

Thus the zeros of the impedance function are the natural frequencies, with the circuit closed by short-circuiting the voltage source, as is usual. These frequencies are the series resonances.

We shall also work with transfer impedances; for example,

$$Z_{T_{12}} = \frac{V_2}{I_1} = \frac{H(s + s_1)(s + s_3)\cdots}{(s + s_2)(s + s_4)\cdots} \tag{7.45}$$

If we drive the circuit with a current I at the 1 port, we measure the response V_2 at the 2 port. However, an infinite response at one port will give an infinite response at any other port (see Section 8.1), and so the poles of the transfer impedance are the same as the poles of the impedance function; i.e., they are the open-circuit natural frequencies or parallel resonances.

For a zero of the transfer impedance, we simply have $V_2 = 0$, regardless of the input current. This means that at frequencies $s = s_1, s_3, \ldots$ there is no transmission through the network. This could happen if a series LC circuit were shunted across the port or a parallel LC circuit were in series; a balanced bridge (lattice) can also give a zero of transmission. We conclude that the zeros of a transfer impedance occur at frequencies where the network transmission is zero.

Now suppose that we excite the impedance of Eq. 7.44 with an impulse of current so that we obtain a set of natural responses, dependent on the network frequencies. That is,

$$v(s) = \frac{K(s + s_1)(s + s_3)\cdots}{(s + s_2)(s + s_4)\cdots} = \frac{a_2}{s + s_2} + \frac{a_4}{s + s_4} + \cdots$$

and so

$$v(t) = a_2\varepsilon^{-s_2 t} + a_4\varepsilon^{-s_4 t} + \cdots \tag{7.46}$$

If the poles are located on the real axis, the response is composed of a set of exponential terms. In order that the result remain finite in magnitude as time increases, we see that the roots must be negative or that the pole locations must be restricted to the left half of the s plane. This is a conclusion previously reached.

Purely reactive networks have no damping elements and their poles are imaginary and located on the $j\omega$ axis ($s_1 = j\omega_1$, etc.). The response contains sine waves only.

Networks with resistance and only one type of energy-storage element, as RC or RL networks, have poles and zeros only on the negative real axis; the response is composed of damped exponential transients.

Networks having RLC elements can have poles anywhere in the left half plane, but complex roots must occur in conjugate pairs. The response due to a conjugate pair of poles is a damped sine wave.

We can now conclude that the poles of a transfer impedance must be limited to the left half of the s plane since they are identical with the poles of the impedance function. However, the zeros of a transfer impedance are only zeros of transmission and they can occur anywhere on the s plane.

7.11 Magnitude-frequency and phase-frequency plots

The method of the *Bode plot* gives frequency response data in logarithmic form, covering a wide range of the variables on a given diagram. We normalize frequency against the resonant frequency as ω/ω_o, or against the band limit frequencies as ω/ω_1 or ω_2/ω, and plot the normalized frequency on a logarithmic scale as

$$u = \log \omega/\omega_o$$

A change of angular frequency by a factor of 10 results in a change of one unit or one decade in u.

Now consider a voltage-gain transfer function of the general form

$$G(s) = K\frac{s/\omega_1}{(s + \omega_2)(s + \omega_3)} \tag{7.47}$$

In the steady state, with $j\omega = s$, this equation may be written as

$$G(s) = \frac{K}{\omega_2\omega_3}\frac{j\omega/\omega_1}{(1 + j\omega/\omega_2)(1 + j\omega/\omega_3)} \tag{7.48}$$

which provides normalized frequencies for plotting purposes.

The method is based on the properties of the logarithm of a complex quantity:

$$\ln \mathbf{G}(\omega) = \ln |\mathbf{G}(\omega)| + j\theta \tag{7.49}$$

This result separates the magnitude performance from the phase shift and we can treat them separately in plotting response variation with frequency. It is usual to employ decibels for the gain, so we have

$$\text{dBV} = 20 \log |\mathbf{G}(\omega)|$$

for the magnitude response.

The method is most useful for first-order terms, where all the poles and zeros are on the negative-real axis. Then, if transformed to decibel gain, the steady-state response of Eq. 7.48 gives

$$\begin{aligned}\text{dBV} &= 20\log\frac{K}{\omega_2\omega_3} + 20\log\left|\frac{j\omega}{\omega_1}\right| - 20\log\left|1+\frac{j\omega}{\omega_2}\right| \\ &\quad - 20\log\left|1+\frac{j\omega}{\omega_3}\right| \\ &= 20\log\frac{K}{\omega_2\omega_3} + 20\log\frac{\omega}{\omega_1} - 10\log\left[1+\left(\frac{\omega}{\omega_2}\right)^2\right] \\ &\quad - 10\log\left[1+\left(\frac{\omega}{\omega_3}\right)^2\right] \end{aligned} \tag{7.50}$$

since

$$20\log\left|1+\frac{j\omega}{\omega_2}\right| = 20\log\sqrt{1+\left(\frac{\omega}{\omega_2}\right)^2} = 10\log\left[1+\left(\frac{\omega}{\omega_2}\right)^2\right]$$

The first term of Eq. 7.50 is a constant, establishes the absolute gain level, and plots as a horizontal straight line. The remaining terms plot as frequency variables, and the total gain variation is obtained as the sum of all.

Two typical functions of ω appear above, and two more are often encountered. We shall study

$$\text{(a)} \quad F_1(j\omega) = \frac{j\omega}{\omega_a} \tag{7.51}$$

$$\text{(b)} \quad F_2(j\omega) = \frac{\omega_a}{j\omega} \tag{7.52}$$

$$\text{(c)} \quad F_3(j\omega) = 1 + \frac{j\omega}{\omega_a} \tag{7.53}$$

$$\text{(d)} \quad F_4(j\omega) = \frac{1}{1 + j\omega/\omega_a} \tag{7.54}$$

Higher-order terms can often be reduced to these types by factoring.

Consider form (a): In decibels this is

$$\text{dBV} = 20\log\frac{\omega}{\omega_a} \tag{7.55}$$

For $\omega = 0$ the value is $-\infty$ dBV; at $\omega = \omega_a$ the gain is zero dBV, and for $\omega = \infty$ the dBV gain is infinite. The gain curve has a slope of $+20$ dBV/frequency decade and is linear. It passes through the zero dBV level at $\omega/\omega_a = 1$, and the function plots as line (1) in Fig. 7.10(a).

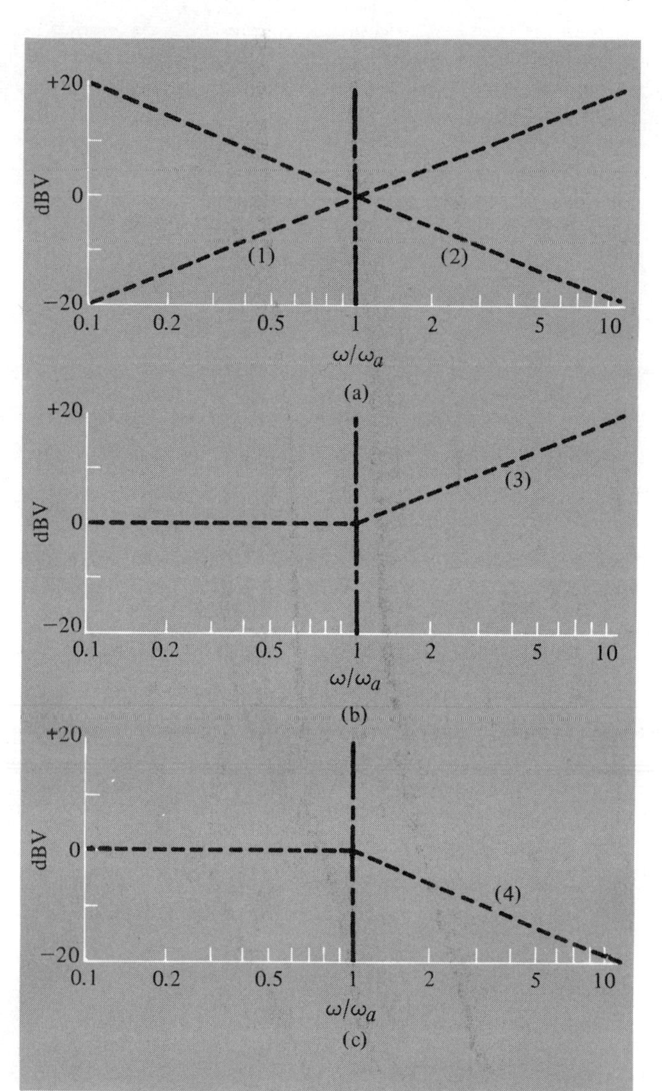

FIGURE 7.10
Asymptotes for logarithmic plots.

Repeating the process for form (b), we have

$$\text{dBV} = 20 \log \frac{\omega_a}{\omega} = -20 \log \frac{\omega}{\omega_a} \tag{7.56}$$

This relation represents another straight line from $+\infty$ at $\omega = 0$ to $-\infty$ at $\omega = \infty$, passing through $\omega/\omega_a = 1$ at zero gain, with a slope of -20 dBV/frequency decade. The curve appears as line (2).

A term of form (c) has a gain expression

$$\text{dBV} = 10 \log \left[1 + \left(\frac{\omega}{\omega_a} \right)^2 \right] \tag{7.57}$$

At *low frequencies*, where ω/ω_a is small with respect to unity, the decibel gain is at the zero level. At *high frequencies*, where ω/ω_a is large with respect to unity, the decibel gain becomes

$$\text{dBV} = 10\log\left(\frac{\omega}{\omega_a}\right)^2 = 20\log\frac{\omega}{\omega_a} \tag{7.58}$$

This expression represents a straight line, starting from zero dBV at $\omega/\omega_a = 1$ and rising at a slope of +20 dBV/frequency decade to infinite gain at $\omega = \infty$. This line is the asymptote which the actual curve approaches at large ω/ω_a values. The asymptote starts at the *corner frequency* $\omega = \omega_a$, which is the intersection of the low- and high-frequency gain asymptotes. Figure 7.10(b) shows this asymptotic form.

The form of (d) is the reciprocal of (c):

$$\text{dBV} = 20\log\frac{1}{1 + j\omega/\omega_a} = -10\log\left[1 + \left(\frac{\omega}{\omega_a}\right)^2\right] \tag{7.59}$$

With $\omega/\omega_a \ll 1$, the gain is zero dBV and at reference level; this is the low-frequency gain asymptote. When $\omega/\omega_a \gg 1$,

$$\text{dBV} = -20\log\frac{\omega}{\omega_a} \tag{7.60}$$

At $\omega/\omega_a = 1$ the gain is zero and this is a corner frequency. The high frequency asymptote starts there and falls with a slope of −20 dBV/frequency decade. The asymptotic curves appear in (c) in Fig. 7.10.

These asymptotes accurately represent the gains when the frequency is remote from the corner frequencies, and the total gain asymptotes can be

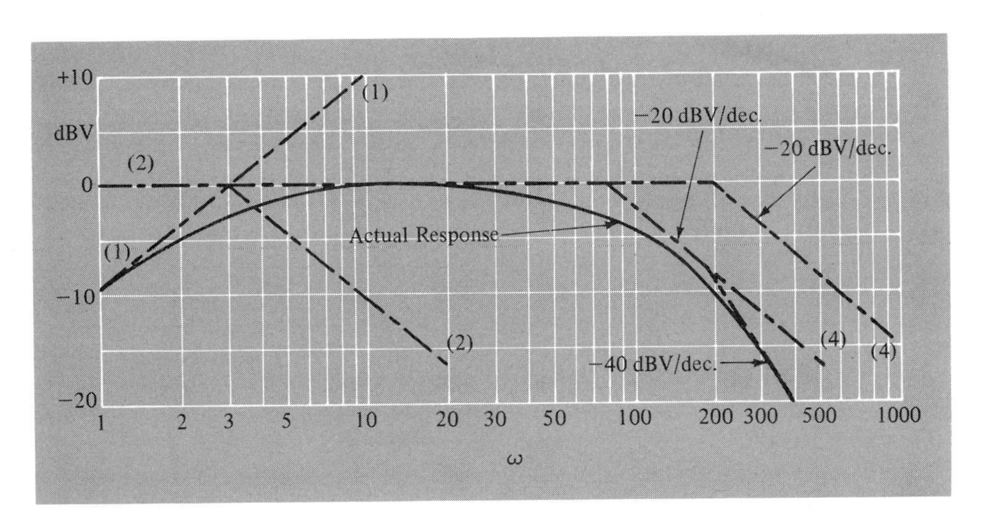

FIGURE 7.11
Use of the asymptotic plots.

obtained by adding the respective ordinate values. The procedure is illustrated by Fig. 7.11, where the gain function

$$G(\omega) = \frac{j\omega/3}{(1 + j\omega/3)(1 - j78/\omega)(1 + 200/\omega)}$$

is illustrated. Corner frequencies occur at $\omega = 3$, 78, and 200 r/s.

For the broad overview needed in the analysis of many systems these asymptotic curves are sufficiently accurate. To be more correct we note that at a corner frequency where $\omega/\omega_a = 1$, Eq. 7.59 gives

$$-10 \log 2 = -3 \text{ dBV}$$

and the correct gain is -3 dBV. Thus the corner frequencies are the half-power frequencies. Correction factors to be applied to the gain asymptotes around the corner frequencies are given in Table 7.1, and with their use accurate curves can be obtained from the Bode asymptotic plots, as shown by the solid curve in Fig. 7.11.

TABLE 7.1 **CORNER-FREQUENCY CORRECTIONS IN DECIBELS**

Pole			Zero		
ω/ω_a	*Gain, dBV*	$\theta°$	ω/ω_a	*Gain, dBV*	$\theta°$
0.1	−0	−84	0.1	0	+6
0.5	−1	−63	0.5	+1	+27
1.0	−3	−45	1.0	+3	+45
2.0	−1	−27	2.0	+1	+63
10.0	0	−6	10.0	0	+84

The method can be extended to include complex roots, but great accuracy is required and direct plotting of the function is usually preferred.

The phase angle variation may also be plotted from the basic function forms. We note first that

$$\text{(a)} \quad F_1(j\omega) = \frac{j\omega}{\omega_a} \qquad (\theta = +90°) \qquad (7.61)$$

$$\text{(b)} \quad F_2(j\omega) = \frac{\omega_a}{j\omega} \qquad (\theta = -90°) \qquad (7.62)$$

We also have

$$\text{(c)} \quad F_3(j\omega) = 1 + \frac{j\omega}{\omega_a} \qquad \left(\theta = \tan^{-1}\frac{\omega}{\omega_a}\right) \qquad (7.63)$$

$$\text{(d)} \quad F_4(j\omega) = \frac{1}{1 + j\omega/\omega_a} \qquad \left(\theta = -\tan^{-1}\frac{\omega}{\omega_a}\right) \tag{7.64}$$

For small values of ω/ω_a, $\theta \cong 0°$ for (c). With $\omega/\omega_a = 1$, $\theta = +45°$, and for ω/ω_a large, $\theta \cong 90°$.

For (d) the angle starts at $\theta \cong 0°$ for ω/ω_a small, becomes $-45°$ at $\omega/\omega_a = 1$ at the corner frequency, and reaches $\theta \cong -90°$ for large ω/ω_a.

In either case, between $\omega/\omega_a = 0.1$ and $\omega/\omega_a = 10$, the phase angle can be approximated by a straight line, dashed in Fig. 7.12, to an accuracy

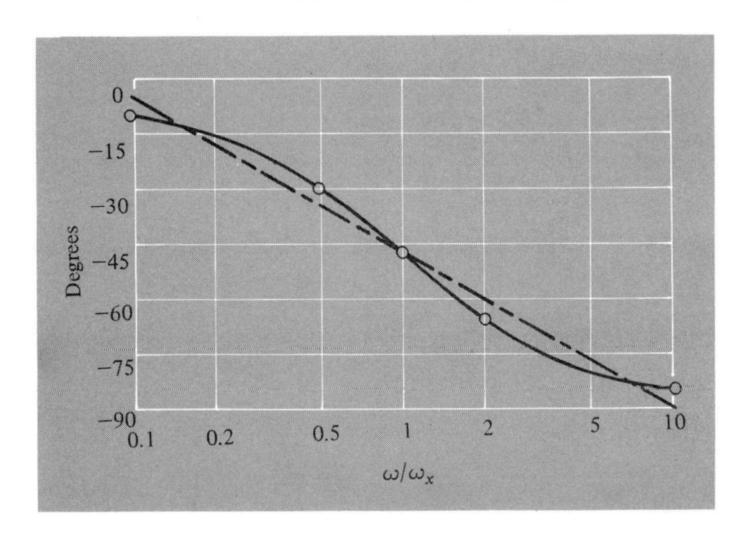

FIGURE 7.12
Phase angle corrections.

of about $\pm 5°$. More accuracy can be obtained through use of the correction factors of Table 7.1, applied in the solid curve of the figure.

We conclude that where the gain rises at $+20$ dBV/decade the phase angle approximates $+90°$ and that where the gain falls at -20 dBV/decade the phase angle nears $-90°$. Thus the phase angle is directly related to the gain in *minimum-phase shift networks* in which the poles and zeros are in the left half plane. A rate of gain change of $20n$/decade carries with it a phase shift of $90n°$, where n is the integer order of the pole or zero.

7.12 The inductively coupled, singly tuned circuit

Another solution for the problem of coupling between an amplifier and a following stage is provided by the inductively coupled, singly resonant circuit of Fig. 7.13(a). Coupling and frequency selectivity are obtained at frequencies in the range from a few hundred kilohertz to hundreds of megahertz. Greater design flexibility is possible over the direct resonant load by reason of the presence of another parameter, the mutual inductance or the coefficient of coupling between the primary and secondary coils.

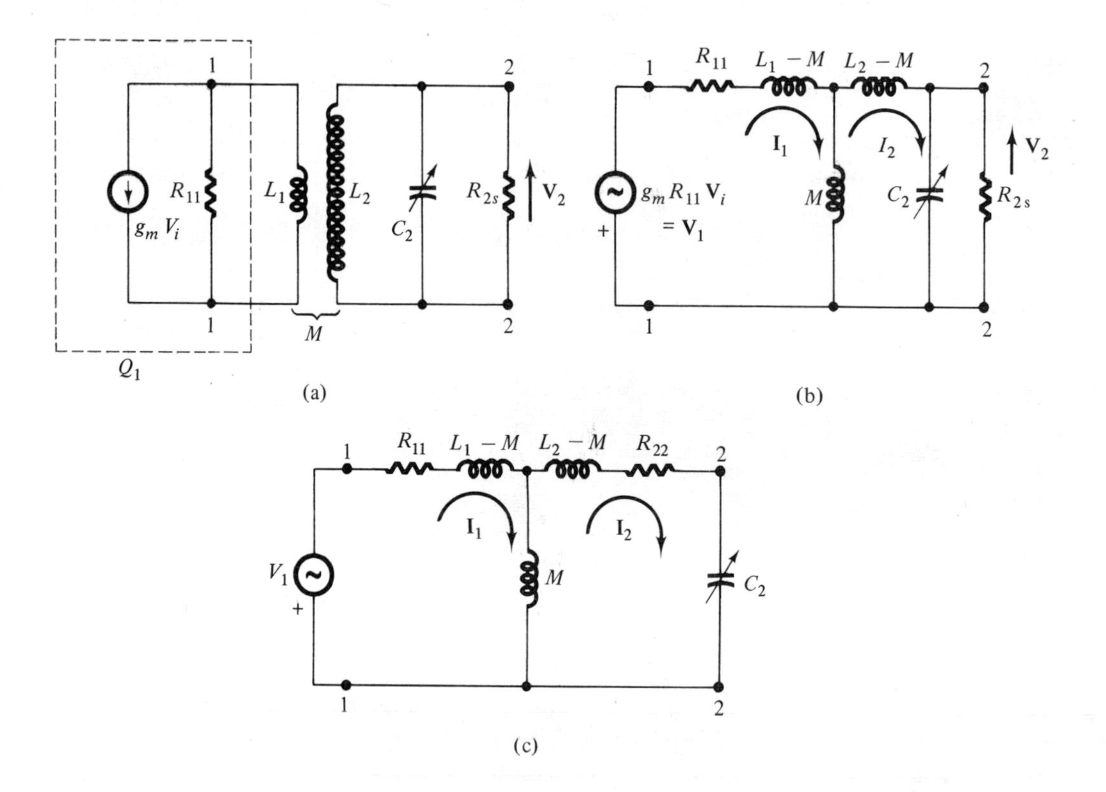

FIGURE 7.13
Singly tuned, inductively coupled circuit.

The output of a transistor amplifier, represented by a controlled current source at Q_1, is to be power-matched to a following stage, represented by the load R_{2s}. The equivalent T network of the inductive circuit is used, as derived in Section 6.7. Analysis is simplified if the parallel load R_{2s} is converted to the series value R_{22} of (c), by use of Eq. 7.19:

$$R_{22} = \frac{\omega^2 L_2^2}{R_{2s}} \tag{7.65}$$

The resistances of the transformer windings may be included in R_{11} and R_{22}, and C_2 will be considered the load $\mathbf{Z}_L$ at the 2 port.

Two adjustable parameters are needed to achieve a power match and these are available as M and C_2 of the circuit. With the Thévenin source labeled $g_m \mathbf{V}_i R_{11} = \mathbf{V}_1$ in Fig. 7.13(b) and with $R_{11} \gg \omega L_1$, the secondary current can be derived by use of the transfer admittance, Eq. 6.52, and

$$\mathbf{I}_2 = Y_T \mathbf{V}_1 = \frac{-\mathbf{z}_f \mathbf{V}_1}{\mathbf{z}_i(\mathbf{z}_o + \mathbf{Z}_L) - \mathbf{z}_f^2}$$

Also $\mathbf{V}_2 = -j\mathbf{I}_2/\omega C_2$, so that

$$\mathbf{V}_2 = \frac{1}{\omega C_2} \frac{-\omega M \mathbf{V}_1}{\omega^2 M^2 + R_{11}\{R_{22} + j[\omega L_2 - (1/\omega C_2)]\}} \tag{7.66}$$

The secondary voltage will be maximum when C_2 is adjusted for secondary resonance, and

$$\mathbf{V}_{2o} = \frac{-M\mathbf{V}_1/C_2}{\omega_o^2 M^2 + R_{11}R_{22}} \tag{7.67}$$

Power transfer will be greatest if $\mathbf{V}_{2o}$ is maximized by circuit design, and this is possible if

$$\omega M = \sqrt{R_{11}R_{22}} \tag{7.68}$$

This is the condition of *critical coupling.*

The *coefficient of coupling* was defined as

$$k = M/\sqrt{L_1 L_2} \tag{7.69}$$

and for *critical coupling*

$$k_c = \frac{\sqrt{R_{11}R_{22}}}{\omega\sqrt{L_1 L_2}} \tag{7.70}$$

The secondary voltage for critical coupling is then

$$\mathbf{V}_{2o} = \frac{-M\mathbf{V}_1/C_2}{2\sqrt{R_{11}R_{22}}} \tag{7.71}$$

Selection of C_2 for resonance provides the proper phase angle for matching R_{11} to R_{22}, and use of $k = k_c$ adjusts the magnitude.

Writing $Q_2 = \omega_o L_2/R_{22}$, at resonance

$$\begin{aligned}\mathbf{V}_{2o} &= \frac{-\mathbf{V}_1}{R_{11}} \frac{\omega_o M Q_2}{1 + (\omega_o^2 M^2/R_{11}R_{22})} \\ &= \frac{-\mathbf{V}_1}{R_{11}} \omega_o M Q_e\end{aligned} \tag{7.72}$$

where the effective Q_e for the circuit is seen to be

$$Q_e = \frac{Q_2}{1 + (\omega_o^2 M^2/R_{11}R_{22})} \tag{7.73}$$

Since the bandwidth is $BW = f_o/Q_e$, it is apparent that the choice of ωM will alter the bandwidth.

By use of the definition for k and k_c the above equation becomes

$$Q_e = \frac{Q_2}{1 + k^2/k_c^2} \tag{7.74}$$

With $k = k_c$ for the matched condition, or for maximum power transfer, then

$$BW = f_2 - f_1 = \frac{2f_o}{Q_e} \tag{7.75}$$

which is then twice the value expected from the secondary resonant circuit alone. Because of the control of bandwidth, the circuit is often designed with $k < k_c$ to obtain a specified bandwidth, and the reduction in power transfer due to mismatch is accepted. Control of $Q_2 = \omega_o L_2/R_{22}$ is possible with k which may be varied by moving L_1 relative to L_2.

Figure 7.14 demonstrates the effect of the coupling coefficient on the frequency response near resonance.

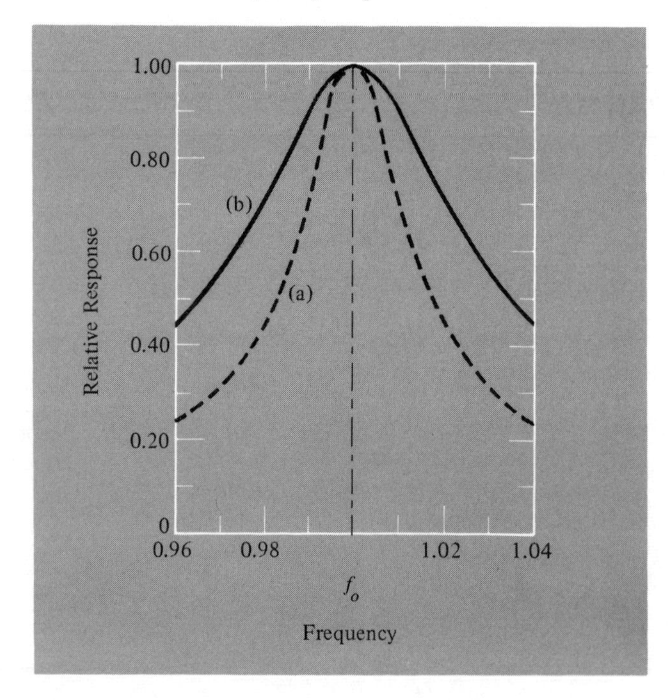

FIGURE 7.14

Effect of M on selectivity; (a) secondary alone; (b) at critical coupling.

7.13 Stagger-tuned band-pass amplifiers

For fixed-frequency use, where tuning complexities need to be resolved only once, the response band may be further broadened by staggering the resonant frequencies of several resonant circuits, with interaction of the tuned circuits eliminated by isolating amplifiers. The flatness of the overall response is then dependent on the locations of the several resonant frequencies and their respective circuit Q values; better performance is realized if slightly different circuit constants are used. An amplifier with three stages is shown in Fig. 7.15.

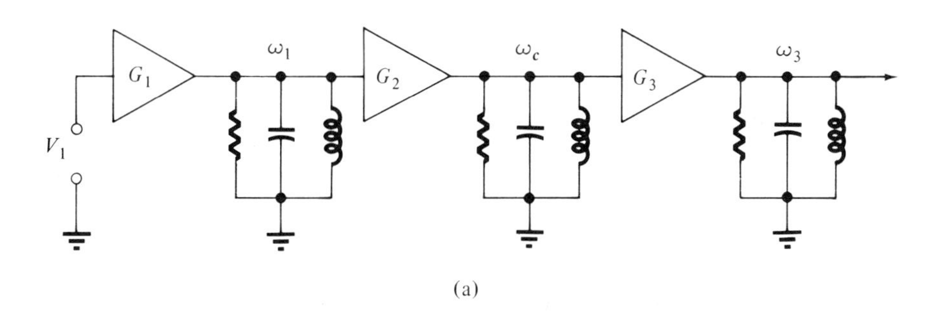

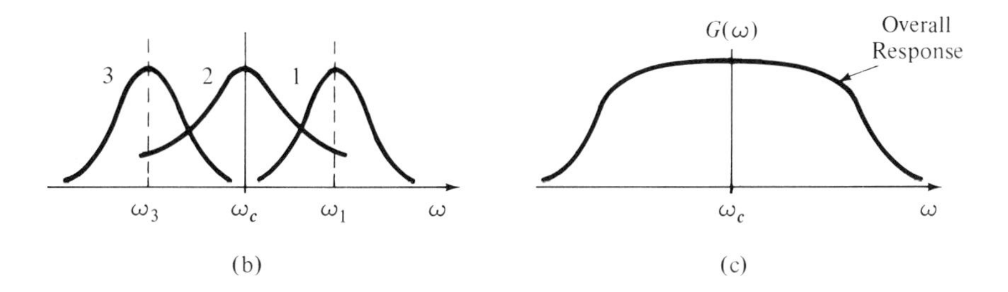

FIGURE 7.15
Stagger-tuned band-pass amplifier.

Consider the amplifiers of the figure to contain voltage-controlled current sources; the voltage transfer function of the first amplifier is

$$G_1(s) = \frac{V_2(s)}{V_1(s)} = -g_m Z(s)$$

From Eq. 7.35 we have the impedance $Z(s)$ of the resonant circuit,

$$Z(s) = \frac{1}{C} \frac{s}{(s + s_1)(s + s_1^*)} \tag{7.76}$$

where the asterisk indicates the conjugate root. The overall transfer function $G(s)$ is then found from

$$G_1(s) = \frac{V_2(s)}{V_1(s)}, \qquad G_2(s) = \frac{V_3(s)}{V_2(s)}, \qquad G_3(s) = \frac{V_4(s)}{V_3(s)}$$

giving, in general,

$$G(s) = \frac{V_{n+1}(s)}{V_1(s)}$$

$$\frac{g_{m1}g_{m2}\cdots g_{mn}}{C_1C_2\cdots C_n} \frac{S^n}{(s+s_1)(s+s_1^*)(s+s_2)(s+s_2^*)\cdots(s+s_n)(s+s_n^*)} \quad (7.77)$$

The cascaded system with n stages has n zeros and $2n$ poles. Each resonant circuit contributes a pole in the upper left half plane, a pole in the lower left half plane, and a zero. The positions of the poles can be adjusted in the left half plane by choosing the resonant frequency $\omega_1, \omega_2, \ldots, \omega_n$ and adjusting the Q of each circuit. The response is that of n isolated narrow-band circuits with high midband frequency ω_c and small frequency spacing between the circuit resonances so that the pass band is large, perhaps $0.1\omega_c$. For $n = 2$, the pole positions become similar to Fig. 7.16(a).

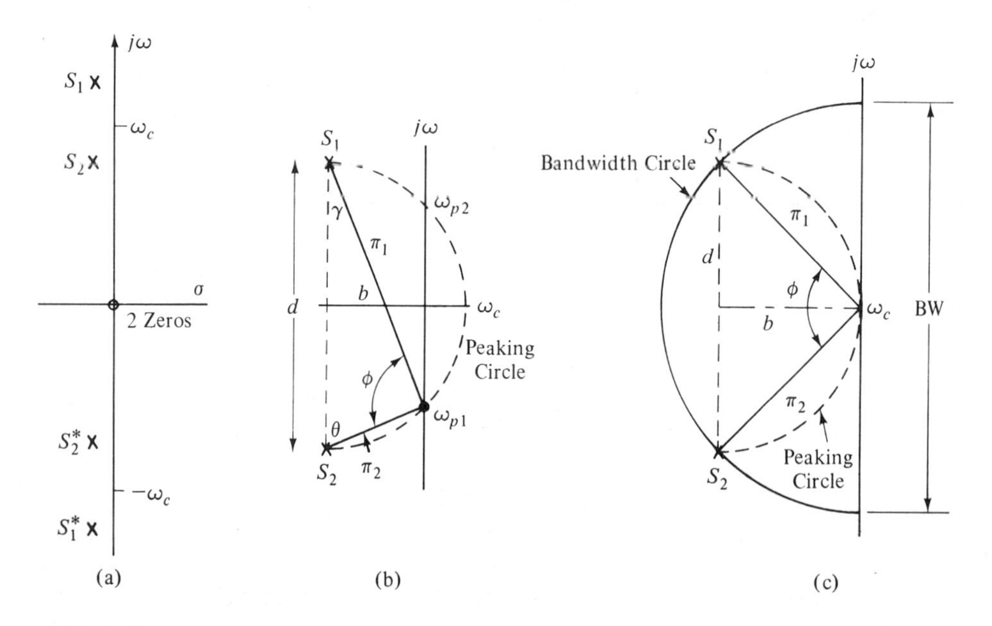

FIGURE 7.16
(a) Pole locations for $n = 3$; (b) at peak gain; (c) for maximal flatness.

Under these conditions the zeros and the lower-frequency poles resolve to an approximate constant,

$$H = \frac{g_{m1}\cdots g_{mn}}{C_1\cdots C_n} \frac{s^n}{(s+s_1^*)(s+s_2^*)\cdots(s+s_n^*)}$$

and so

$$G(s) = H\left(\frac{1}{(s + s_1)(s + s_2)\cdots(s + s_n)}\right) \tag{7.78}$$

For the two-circuit case chosen for discussion, the region around the poles s_1 and s_2 may be expanded to (b) in Fig. 7.16. As for the resonant circuit, the pole coordinates on the real axis are

$$b = \frac{-1}{2R_1C_1} = \frac{-1}{2R_2C_2} \tag{7.79}$$

The frequency ω_c is the center frequency of the two poles, with their frequencies separated by a distance d on the diagram. The gain $G(s)$ varies as

$$G(s) = H\frac{1}{\pi_1\pi_2} \tag{7.80}$$

where π_1 and π_2 are distances to ω, which moves along the frequency axis.

The triangle formed by π_1, π_2, and d can be studied by use of the law of sines, giving

$$\frac{\pi_1}{\sin\theta} = \frac{\pi_2}{\sin\gamma} = \frac{d}{\sin\varphi}$$

From Fig. 7.16(b),

$$\sin\theta = \frac{b}{\pi_2} \qquad \text{and} \qquad \sin\gamma = \frac{b}{\pi_1}$$

and so

$$\pi_1\pi_2 = \frac{bd}{\sin\varphi} \tag{7.81}$$

thus resolving the gain function into a function of one variable, $\sin\varphi$. From Eq. 7.80,

$$G(s) = H\frac{\sin\varphi}{bd} \tag{7.82}$$

In Fig. 7.16(b), at $\omega = \omega_{p1}$, the angle φ is 90° and the gain is at a maximum. As ω moves below ω_{p1}, the angle φ decreases and the gain falls; as ω moves up from ω_{p1} toward ω_c, the angle φ increases and the gain also falls. The condition of $\varphi = 90°$ requires that π_1 and π_2 intersect on the peaking circle of diameter d. However, this intersection must also occur on the $j\omega$ axis, and so we find the two frequencies, ω_{p1} and ω_{p2}, for which the gain peaks with $\varphi = 90°$. With $d/2$ as the radius of the peaking circle, the geometry of the triangle leads to

$$(\omega_p - \omega_c)^2 + b^2 = \left(\frac{d}{2}\right)^2$$

$$\omega_p - \omega_c = \sqrt{\left(\frac{d}{2}\right)^2 - b^2} \tag{7.83}$$

When the pole separation is made $d = 2b$, the peaking circle has the location in Fig. 7.16(c), just tangent to the $j\omega$ axis. The value of δ is 90° at $\omega = \omega_c$, and only one peak is obtained.

A bandwidth circle may be drawn through the poles, as shown by the solid circle in (c), with its center at ω_c. The diameter is $\sqrt{2}d$ or 1.414 times the pole separation. This diameter is the bandwidth of the doubly peaked circuit. Therefore

$$d = \frac{\text{BW}}{\sqrt{2}}$$

becomes useful in circuit design.

This discussion illustrates the method followed for a *maximally flat response* with any number of poles. A circular locus of radius BW/2 is drawn in the left half plane, with center at ω_c of the desired pass band. The poles are then symmetrically placed, as indicated in Fig. 7.17. The real part of each pole location fixes the Q of the resonant circuit as $Q = \omega_c/2b$, and the ω-axis location determines the frequency to which the particular circuit is tuned.

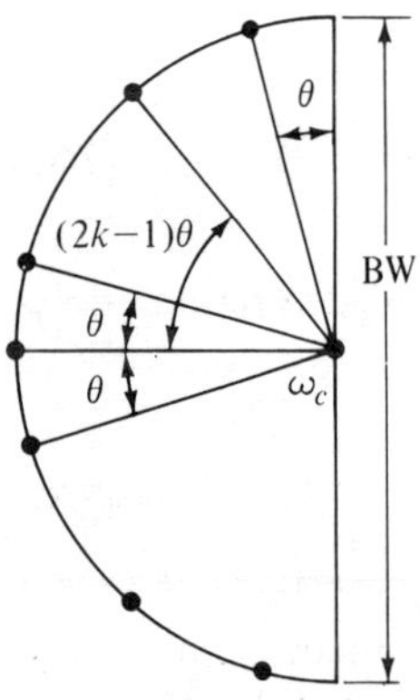

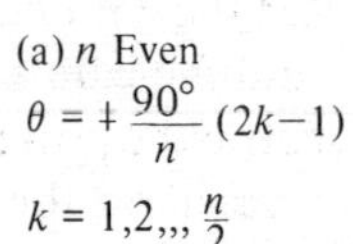

(a) n Even
$\theta = \pm \frac{90°}{n}(2k-1)$
$k = 1, 2, \ldots, \frac{n}{2}$

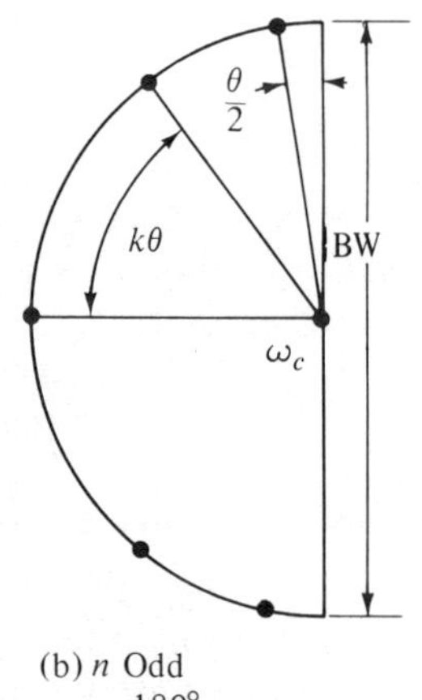

(b) n Odd
$\theta = \pm \frac{180°}{n} k$
$k = 0, 1, 2, \ldots, \frac{n-1}{2}$

FIGURE 7.17
Pole locations, maximally flat response.

Equation 7.78 can be written for the steady state, with $s = j\omega$, in the form

$$|\mathbf{G}(\omega)| = \left|\frac{\mathbf{V}_{n+1}(\omega)}{\mathbf{V}_1(\omega)}\right| = H\frac{1}{\sqrt{A_n + (\omega/\omega_b)^{2n}}}$$

We define $A_n = \omega_b^{2n}$ and can write

$$|\mathbf{G}(\omega)| = \frac{H}{\omega_b^n}\frac{1}{\sqrt{1 + (\omega/\omega_b)^{2n}}} \tag{7.84}$$

where ω_b is the frequency of a half-power point. This is the relation developed by Butterworth and known as a *maximally flat response.* Actual responses are plotted in Fig. 7.18 for $n = 2$ and $n = 5$; neither has the ideal rectangular

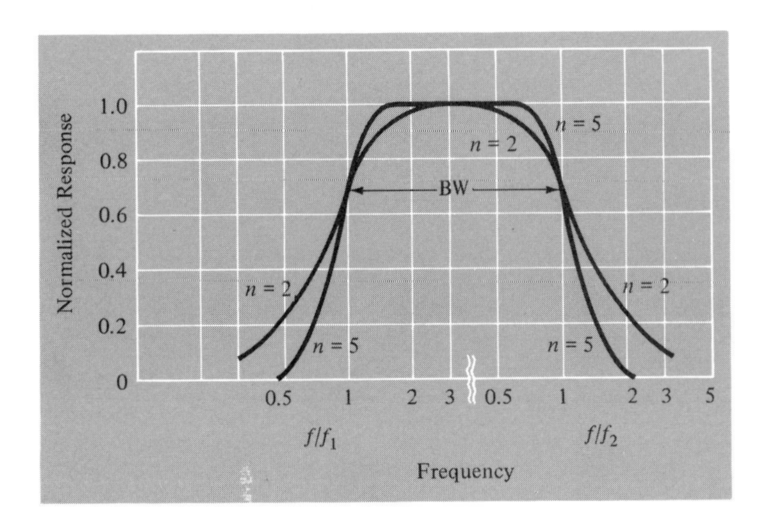

FIGURE 7.18
Effect of n on frequency response.

form, but the larger value of n gives a better approximation. Since all the curves pass through the 0.707 half-power response point, the bandwidth can be kept constant but the steepness of side frequency cutoff is improved with n; the pass-band response also flattens. However, when n exceeds 4, the oscillatory nature of the response to a step input causes considerable overshoot of such a steeply rising wave form.

The use of maximally flat networks as filters is also considered in Chapter 10.

7.14

Summary

The following are useful relations.

SERIES *RLC* CIRCUIT

$$\omega_o = \frac{1}{\sqrt{LC}}$$

$$\mathbf{Z}_o = R_s$$

$$Q = \frac{\omega_o L}{R_s} = \frac{1}{R_s}\sqrt{\frac{L}{C}}; \qquad \text{BW} = \frac{f_o}{Q}$$

The circuit behaves as a capacitance below resonant frequency; it appears as an inductance at frequencies above resonance.

PARALLEL *RLC* CIRCUIT

$$\omega_o = \frac{1}{\sqrt{LC}} \qquad \text{for } Q > 7$$

$$\mathbf{Z}_o = R_p = \frac{\omega_o^2 L^2}{R_s} = Q\omega_o L$$

$$Q = \omega_o C R_p = \frac{1}{R_s}\sqrt{\frac{L}{C}} = R_p\sqrt{\frac{C}{L}}$$

$$\text{BW} = \frac{f_o}{Q}$$

NEAR RESONANCE

$$\mathbf{Z} = \frac{R_p}{1 + j2Q\delta}$$

where $\delta = (\omega - \omega_o)/\omega_o$. The circuit behaves as an inductance below resonant frequency; it appears capacitive at frequencies above resonance.

The series *RLC* circuit develops a minimum resistance at resonance; the parallel circuit is of high resistance at resonance and is suited as a load for transistor and tube amplifiers. The quality factor of the circuit is Q, and a high resonant resistance requires a high Q. Bandwidth, as a measure of frequency selectivity obtainable with the circuit, is inversely proportional to Q; ideally the frequency response would be a rectangle, uniformly passing all desired frequencies and blocking all undesired frequencies.

Greater bandwidth, with good rejection of outside band or skirt frequencies, is often needed. The method of visualizing circuit response through study of the position of poles and zeros on the complex plane is helpful in the analysis of selective circuits. The method is here applied to the singly tuned and the stagger-tuned n-tuple. Bandwidths as large as 30 percent of the center frequency can be obtained, as used in television and in radar, with good skirt-frequency rejection.

PROBLEMS

7.1 The series circuit of Fig. 7.19(a) is in resonance at $\omega_o = 100$ r/s. Resistance $R = 100\ \Omega$, $V = 1$ V rms, and $V_o = 5$ V rms. What are the values of L and C in henrys and farads?

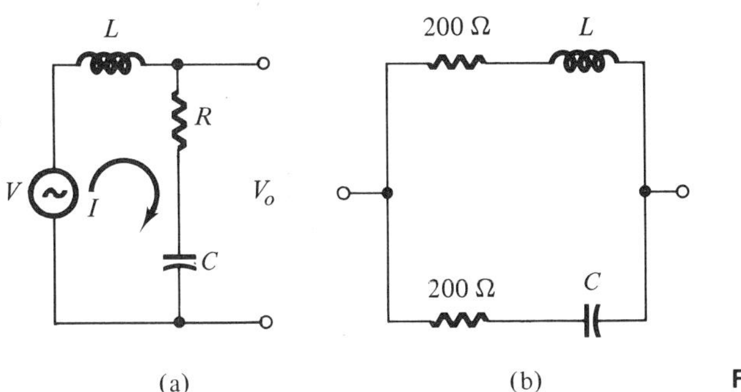

FIGURE 7.19

7.2 Find an expression for the resonant angular frequency ω_o for the circuit of Fig. 7.19(b).

7.3 An inductance of 200 μH has a resistance of 40 Ω. Connected in series is a capacitance of 0.05 μF, with negligible resistance. Find f_o, Q, and the circuit impedance at resonance.

7.4 Repeat Problem 7.3 with L and C in parallel. The coil resistance is in series with L.

7.5 A parallel-resonant circuit has a resistance R in series with L. The resonant frequency is 1200 Hz, $Q = 7.6$, and the resonant resistance is to be 2600 Ω. Find thc value of resistance R of the coil, the value of L, and the value of C.

7.6 A series-resonant circuit has a resonant resistance R_1 and a quality factor Q_1. Resistor R_2 is connected in parallel, across the resonant circuit terminals. Find an expression for the Q of the paralleled circuit.

7.7 Employing the definition of Q, find the Q of a tennis ball which rebounds to 2.8 m after being dropped from a height of 3.2 m.

7.8 A parallel-resonant circuit is assembled with $C = 500$ pF and $L =$ 10 μH and is driven at a port by a current source of 0.1 A, with a 30,000-Ω internal shunt resistance. (a) Find the resonant frequency.

(b) Find Q and the bandwidth. (c) What is the power taken by the circuit at resonance?

7.9 A radio broadcast station has an assigned center frequency of 1.550 MHz and a channel width of ± 5 kHz. (a) If a parallel-resonant circuit is tuned to this station and is to have a bandwidth equal to the channel width, what circuit Q is required? (b) The inductance is 17.8 μH; what value of current-source shunt resistance can be tolerated? (c) What is the resonant impedance of the complete circuit?

7.10 An inductance of 150 mH and a series resistance of 4 Ω is used in a series-resonant circuit, resonating at 60 Hz. With the supply from the line at 120 V, 60 Hz, and a capacitor with a voltage rating of 600 V peak, what value of series resistor must be added to protect the capacitor from voltage breakdown?

7.11 An inductance in series with a capacitance of 400 pF is resonant at 0.8 MHz. The power supplied at 0.78 MHz is half that supplied at resonance. Specify the R, L, and C values of the circuit components.

7.12 An inductor of 200 μH is paralleled with a capacitor, and resonance is reached at 1200 kHz. Plot the magnitude of the voltage, supplied by a current source of 0.01 A and 0.1 MΩ shunt resistance, across the circuit from 1100 to 1300 kHz.

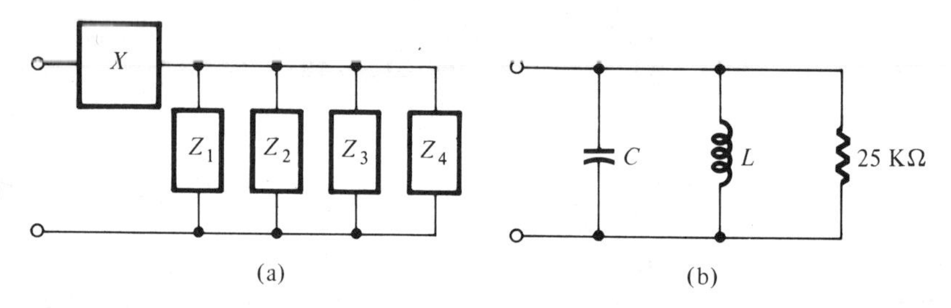

FIGURE 7.20

7.13 Four impedances are connected in parallel as in Fig. 7.20(a), with $\mathbf{Z}_1 = 100\underline{/0°}\ \Omega$, $\mathbf{Z}_2 = -j50\ \Omega$, $\mathbf{Z}_3 = 30 + j20\ \Omega$, and $\mathbf{Z}_4 = 20 - j30\ \Omega$. (a) Determine the type and magnitude of the reactance, connected as X, which will produce resonance. (b) What will be the overall Q value measured at the port?

7.14 The circuit of Fig. 7.20(b) is resonant at $\omega_o = 1000$ r/s. Find L, Q, and BW if $C = 0.015\ \mu$F.

7.15 In the circuit of Fig. 7.21(b), $R_L = R_C = \sqrt{L/C} = R$. Show that $\mathbf{Z}_{1,1} = R$ for all values of ω.

7.16 A parallel-resonant circuit has fixed C and variable L. The Q of the circuit is 14. (a) Find the values of L and C to give a circuit impedance at resonance of $\mathbf{Z}_{1,1} = 3000 + j0\ \Omega$ at 2.4 MHz. (b) What is the bandwidth?

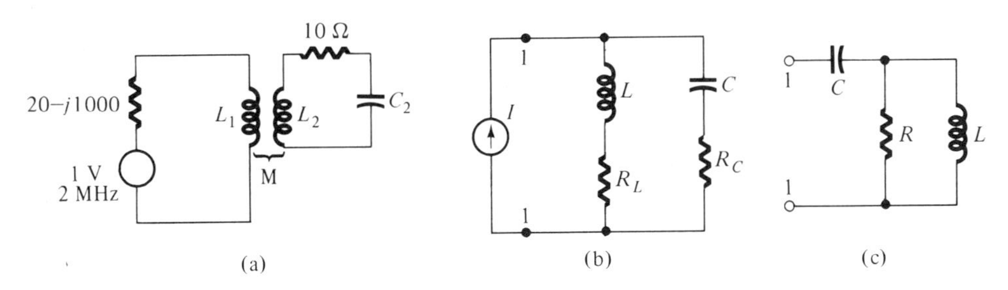

FIGURE 7.21

7.17 The Q of a series RLC circuit is 7.5 at resonance. The current from the source is 0.1 A at 1.5 V at resonance; if $L = 0.1$ H, find the value of C.

7.18 For the circuit of Fig. 7.21(c), express the series-resonant frequency as a function of the Q of the circuit. Resistance R is not negligible.

7.19 Figure 7.22(a) shows a pole-zero diagram for a series RLC circuit. Determine ω_o, Q, and BW and plot the value of $\mathbf{Z}(\omega)$ over a range from $\omega_o/3$ to $3\omega_o$ using graphical procedures.

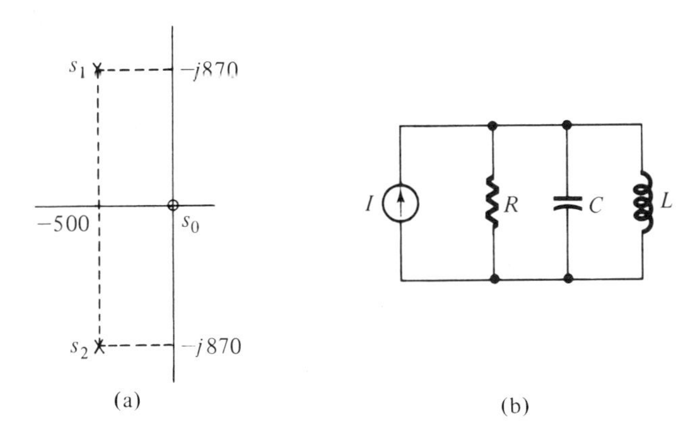

FIGURE 7.22

7.20 The impedance of a circuit is obtained as

$$\mathbf{Z}(s) = 14\frac{s + 21}{s^2 + 2s + 1000}$$

What are ω_o, Q, and BW?

7.21 In the circuit of Fig. 7.22(b), $L = 0.002$ H, $R = 2000\ \Omega$, $C = 0.05\ \mu$F, and $\omega_o = 7500$ r/s. Write the impedance function in terms of s, and find Q and BW.

7.22 For an admittance function

$$Y(s) = \frac{16s}{s^2 + 4s + 1604}$$

develop the pole-zero diagram and graphically determine the variation of $Y(\omega)$ and $\theta(\omega)$ and plot the results.

7.23 A parallel RLC circuit having $L = 120\ \mu\text{H}$ with 18-Ω series resistance is resonant at 1 MHz. The circuit is supplied by a current generator of 1 mA, $25 \times 10^4\ \Omega$ of shunt resistance. (a) What is the resonant voltage across C? (b) What is the current in C at 1.015 MHz? (c) What is the bandwidth?

7.24 Referring to (a) in Fig. 7.21, with $M = 10\ \mu\text{H}$, $L_1 = 30\ \mu\text{H}$, $L_2 = 200\ \mu\text{H}$, and $\omega L_2 = 1/\omega C_2$ at 2 MHz, find the voltage developed across C_2.

7.25 For a stagger-tuned maximally flat quadrupole (four resonant circuits) the center frequency is to be 40 MHz, and the overall bandwidth is to be 5 MHz. Specify the center frequencies of each tuned circuit and the Q values needed from a complex-plane plot.

7.26 A band-pass amplifier is to have a midband frequency of 10 MHz and a half-power bandwidth of 10 MHz ±500 kHz. Three stages seem ample for the required gain. For a maximally flat response, determine f_o, BW, and Q for each tuned circuit.

7.27 Draw the Bode asymptotes for the gain function:

$$G(s) = \frac{A_o(1 + 120s)}{(1 + 8s)(1 + 44s)(1 + 250s)}$$

7.28 Given a gain function:

$$G(s) = \frac{12}{(1 + jf/20)(1 + jf/60)}$$

What is the gain magnitude at $f = 40$ Hz?

7.29 Given a band-pass gain function:

$$A = \frac{jf/3}{(1 + jf/3)(1 + jf/12)}$$

(a) What is the frequency of maximum gain, and (b) what is the value of that gain?

REFERENCES

1. H. A. Wheeler, "The Potential Analog Applied to the Synthesis of Stagger-Tuned Filters." *IRE Trans.*, CT-2, 86 (1955).
2. J. D. Ryder, *Networks, Lines, and Fields*, 2d ed. Prentice-Hall, Inc., Englewood Cliffs, N.J., 1957.
3. E. J. Angelo, Jr., and A. Papoulis, *Pole-Zero Patterns.* McGraw-Hill Book Company, New York, 1964.
4. R. S. Sanford, *Physical Networks.* Prentice-Hall, Inc., Englewood Cliffs, N.J., 1965.

transformation of impedances

eight

The transfer of maximum power from a generator to a load requires that the load impedance match that of the generator, but in practice it is improbable that a given load will be suited to the generator from which it is to take power. For example, a radio transmitter designed for a load of 400 Ω may be required to supply power to an antenna having a resistance of 70 Ω. A circuit means of making the antenna impedance appear as a load of 400 Ω facing the output port of the generator can be designed.

In general, since a load must have an impedance conjugate to that of the generator, two variable parameters must be provided in the matching circuit to provide a power match. One adjustment will cause the magnitude of the load impedance to vary as it appears at the generator output port, and the second adjustable parameter will cause variation of the apparent load phase angle, to give the conjugate impedance and a condition of circuit resonance.

8.1 Impedance transformation with tapped resonant circuits

Coupled circuits at resonance have already been used to obtain maximum power transfer at a single frequency. In many cases the resistance value of the load will be so low as to severely reduce the Q when the load is connected in parallel with the resonant circuit. A reduced Q value causes greater bandwidth and this may be undesirable. However, by connecting the load resistance R_1 across only a portion of the resonant-circuit inductance or capacitance as in (a), Fig. 8.1, the low value of R_1 may be transformed to a higher value R_p effectively shunted across the circuit at a,c as in (c).

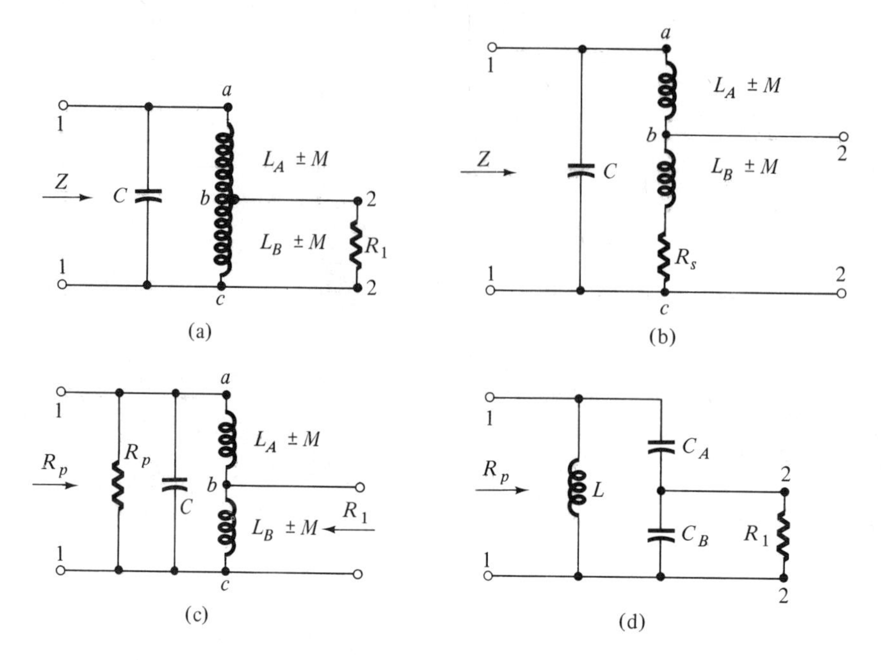

FIGURE 8.1
Impedance transformation with tapped resonant circuits.

Note that coils L_A and L_B (part of one continuous winding) have a mutual inductance M and that the total inductance is

$$L = L_A \pm M + L_B \pm M = L_A + L_B \pm 2M$$

Considering terminals a and c, resonance will occur with equal reactances; that is,

$$\omega(L_A + L_B \pm 2M) = \frac{1}{\omega C} \tag{8.1}$$

For terminals b and c, we write the reactance condition as

$$\omega(L_B \pm M) = \frac{1}{\omega C} - \omega(L_A \pm M) \tag{8.2}$$

These two equations are identical, and we generalize the result by stating that *if a circuit is resonant at one port at a given frequency, it is also resonant at any other port.*

In Fig. 8.1, the circuit of (a) is resonant across b and c, and with a load R_1 applied, the inductive-resistance combination of $L_B \pm M$ and R_1 becomes equivalent to the b,c branch circuit at (b) with series resistance R_s if

$$R_s = \frac{\omega^2(L_B \pm M)^2}{R_1} \tag{8.3}$$

by reason of Eq. 7.19. Again using this relation, we arrive at the circuit at (c), with

$$R_p = \frac{\omega^2 L^2}{R_s} = \frac{L^2}{(L_B \pm M)^2} R_1 \tag{8.4}$$

Thus a small impedance R_1 can be transformed to a parallel R_p of larger value; in reverse, an impedance R_p can be transformed to a load R_1, with the ratio of impedances determined by the inductance ratio squared.

The effective Q of the resonant circuit can be adjusted by choice of location of the tap on the inductor so that the desired R_p is established for a given R_1. This allows control of the circuit bandwidth.

If the load is placed across part of the capacitance, as in (d) in Fig. 8.1, it can be similarly shown that

$$R_p = \frac{1/\omega^2 C^2}{1/\omega^2 C_B^2} R_1 = \frac{C_B^2}{C^2} R_1$$

where C is the effective capacitance of the series capacitors. Since $C = C_A C_B/(C_A + C_B)$,

$$R_p = \frac{(C_A + C_B)^2}{C_A^2} R_1 \tag{8.5}$$

Both these methods of impedance transformation are employed in communication equipment design.

8.2 Transformation of impedances by reactive networks

A more general problem is to transform impedance values at will, so as to make a load R_{22} appear to be matched to a generator of internal resistance R_{11}, as in (a) in Fig. 8.2. Let us assume that a network can be designed to achieve that objective and represent that network by the T in (b). If the resistance at port 1 can be made to appear as R_{11}, then the generator of voltage V and internal resistance R_{11} will be power-matched at its terminals. The generator sees its *image impedance* at port 1 and delivers its maximum possible power output into the network. We now specify that the T network be purely reactive and can dissipate no power; the maximum possible power output from the generator must then be delivered to the load at the 2 port. Therefore a T network of reactances can match a load resistance R_{22} to a generator of resistance R_{11} for maximum power transfer.

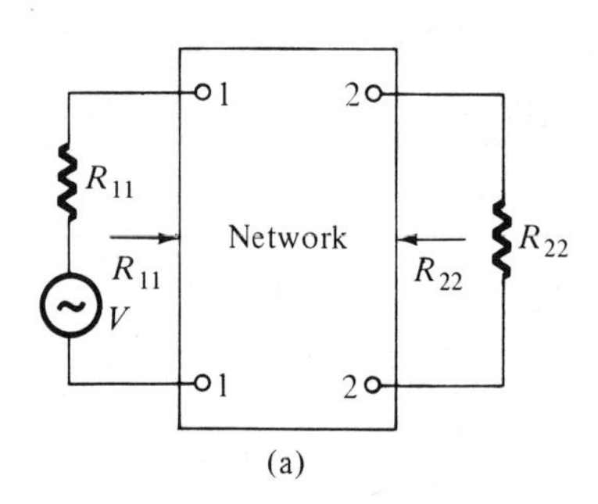

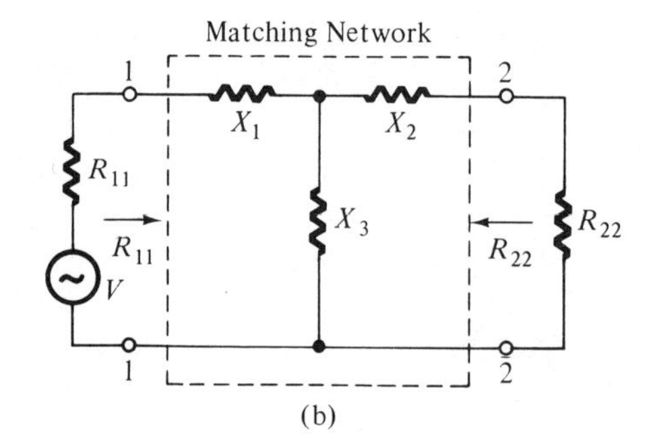

FIGURE 8.2
Matching reactance network.

We now seek to determine the needed parameters of the T network of (b). The impedance seen looking into the 1 port must be R_{11}, so we require that

$$R_{11} = jX_1 + \frac{jX_3(R_{22} + jX_2)}{R_{22} + j(X_2 + X_3)} \tag{8.6}$$

where the reactances may be either positive or negative, inductive or capacitive. Then

$$R_{11}R_{22} + jR_{11}(X_2 + X_3) = -(X_1X_2 + X_2X_3 + X_3X_1) + jR_{22}(X_1 + X_3) \tag{8.7}$$

Equating the real terms,

$$R_{11}R_{22} = -X_1X_2 - X_2X_3 - X_3X_1 \tag{8.8}$$

Both R_{11} and R_{22} are real and positive, and the right-hand side of the equation must also be positive. Therefore, one or more of the terms on the right must be positive and so one reactance must be of opposite sign to the other two, because $jX_A(-jX_B) = +X_AX_B$. To force this condition, the T network must be composed of one inductance and two capacitances or vice versa. This is the first design condition for the matching network.

The equating of the reactive terms of Eq. 8.7 leads to

$$\frac{R_{11}}{R_{22}} = \frac{X_1 + X_3}{X_2 + X_3} \tag{8.9}$$

Rewriting Eq. 8.8,

$$R_{11}R_{22} = -[(X_1 + X_3)(X_2 + X_3) - X_3^2] \tag{8.10}$$

Equations 8.9 and 8.10 relate the product and the quotient of R_{11} and R_{22}. These lead to

$$X_1 = -X_3 \pm \sqrt{\frac{R_{11}}{R_{22}}(X_3^2 - R_{11}R_{22})} \tag{8.11}$$

$$X_2 = -X_3 \pm \sqrt{\frac{R_{22}}{R_{11}}(X_3^2 - R_{11}R_{22})} \tag{8.12}$$

These equations provide values for X_1 and X_2 in terms of X_3. The plus or minus choice in Eq. 8.11 requires an equivalent choice in Eq. 8.12 and two networks can therefore be built.

We have two equations and three circuit unknowns, and so one reactance arm must be arbitrarily chosen. It is usually economic to make X_1 or X_2 equal to zero, thus requiring no equipment. Choosing $X_1 = 0$, Eq. 8.11 leads to

$$X_3 = \pm R_{11}\sqrt{\frac{R_{22}}{R_{11} - R_{22}}} \tag{8.13}$$

Our practical generators are often nonlinear and their outputs will include harmonic frequencies. It is usually best to choose X_3 as a capacitance, because the reactance of a capacitive X_3 will be low at the harmonic frequencies, shunting those components to ground and improving the output wave form.

Then

$$X_3 = -R_{11}\sqrt{\frac{R_{22}}{R_{11} - R_{22}}} \tag{8.14}$$

and we see that $R_{11} > R_{22}$ is another design requirement.

Using the expression for X_3 in Eq. 8.12, we can write X_2 as

$$X_2 = R_{22}\sqrt{\frac{R_{11}}{R_{22}} - 1} \tag{8.15}$$

Summarizing for a matching network with

$$R_{11} > R_{22} \qquad \text{and} \qquad X_1 = 0$$

we have

$$C_3 = \frac{1}{\omega R_{11}\sqrt{\frac{R_{11}}{R_{22}} - 1}} \tag{8.16}$$

$$L_2 = \frac{R_{22}}{\omega}\sqrt{\frac{R_{11}}{R_{22}} - 1} \tag{8.17}$$

A second possible network follows if X_2 is chosen to be zero. If X_3 is made capacitive as before, the design equations are

$$X_3 = -R_{22}\sqrt{\frac{R_{11}}{R_{22} - R_{11}}} \tag{8.18}$$

and so $R_{11} < R_{22}$ is a requirement of this selection. Using the result in Eq. 8.13, we obtain

$$X_1 = R_{11}\sqrt{\frac{R_{22}}{R_{11}} - 1} \tag{8.19}$$

For the matching network with

$$R_{11} < R_{22} \qquad \text{and} \qquad X_2 = 0$$

the circuit parameter for X_3 is

$$C_3 = \frac{1}{\omega R_{22}}\sqrt{\frac{R_{22}}{R_{11}} - 1} \tag{8.20}$$

and for X_1

$$L_1 = \frac{R_{11}}{\omega}\sqrt{\frac{R_{22}}{R_{11}} - 1} \tag{8.21}$$

The resultant circuits appear in Fig. 8.3. In each case the matching is the result of an L section of reactances, forming a resonant circuit with the resistance. In (a) the circuit to the right of the 1,1 terminals constitutes a parallel-resonant load on the generator. The parallel circuit produces a resistance which matches that of the generator. In (b) the L section is reversed, but the resonant action is again present.

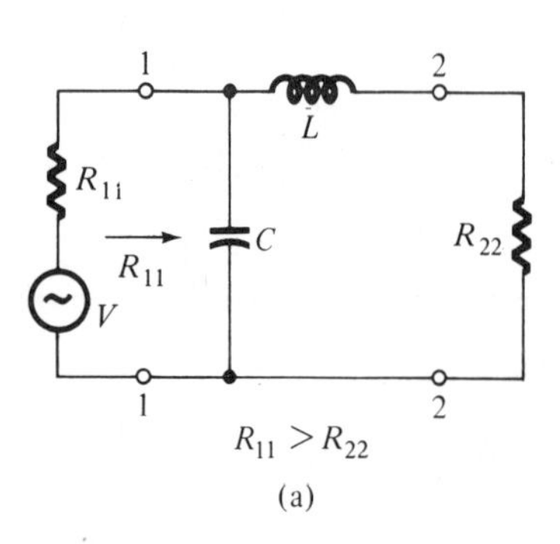

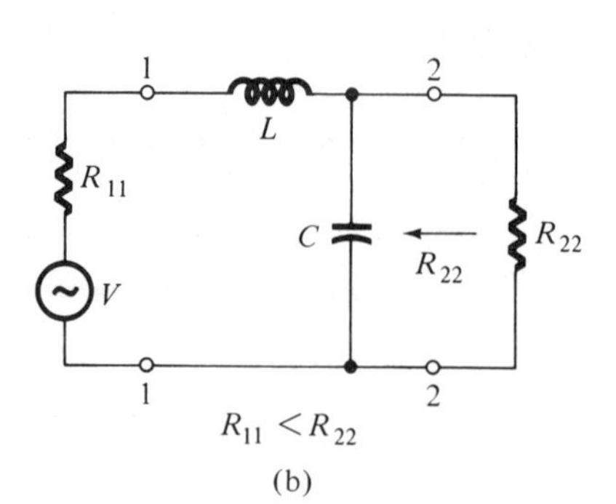

FIGURE 8.3

Designs for reactive matching networks.

When reactances appear in the load or generator, an appropriate value may be chosen for X_1 or X_2 to resonate that arm. A reactance will then be physically present, but the arm reactance would be zero in the design equations.

If $X_3^2 < R_{11}R_{22}$, the radicals in Eqs. 8.11 and 8.12 become imaginary. The circuit cannot accomplish a power match; although power is moved to the load, the amount of power is less than obtained when $X_3^2 \geqq R_{11}R_{22}$, and the coupling is said to be *insufficient*.

If X_3 is chosen so that $X_3^2 = R_{11}R_{22}$, we have the condition that

$$|X_1| = |X_2| = |X_3|$$

with X_3 usually chosen to be capacitive. This represents the condition of *critical coupling*, as previously discussed, and a power match is achieved at one frequency.

If $X_3^2 > R_{11}R_{22}$, the circuits are *overcoupled* and maximum power is transferred to the load at two frequencies. With C and L fixed, Eqs. 8.11 and 8.12 show these effects, and there will be two response peaks. Overcoupling is often employed to broaden the bandwidth, but with a response dip between the peaks.

8.3 Bandwidth and Q of the matching network

The response bandwidth of the reactive matching network for critical coupling or below can be found by determination of the circuit Q. With $R_{11} > R_{22}$, the series resistance of the parallel circuit is R_{22}, and Q can be found by the usual definition, using f_o as the design frequency and the value of L from Eq. 8.17. That is,

$$Q = \frac{\omega_o L}{R_{22}} = \sqrt{\frac{R_{11}}{R_{22}} - 1} \tag{8.22}$$

The bandwidth is then

$$\text{BW} = \frac{f_o}{Q} = f_o\sqrt{\frac{R_{22}}{R_{11} - R_{22}}} \tag{8.23}$$

For $R_{11} < R_{22}$ the bandwidth can be obtained as

$$\text{BW} = f_o\sqrt{\frac{R_{11}}{R_{22} - R_{11}}} \tag{8.24}$$

Computing the resonant impedance seen at the 1 port in (a) in Fig. 8.3 gives

$$\mathbf{Z}_{11} = Q\omega_o L = R_{11} - R_{22} \cong R_{11} \tag{8.25}$$

The approximation follows for $R_{11} \gg R_{22}$ or $Q > 7$. The result confirms the predicted impedance match at the 1 port; that is, the generator drives a matching load at the port.

8.4 The matching π network

The π network provides two shunt branches, which can be made capacitive to bypass any harmonic frequencies present, giving the π form an advantage over the T circuit. Convenience also dictates the choice of the π network, since variable capacitance input and output arms are easier to adjust than the inductances of the T circuit.

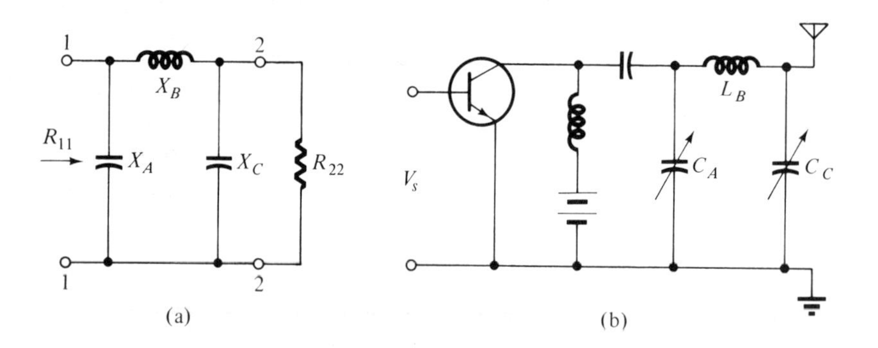

FIGURE 8.4
(a) π Network for impedance transformation; (b) as applied in a radio transmitter.

We may begin the design of a π matching network, as we did for the T circuit, by requiring that the input impedance of the circuit of Fig. 8.4 be R_{11}:

$$R_{11} = \frac{X_A\{jX_B + [jX_C R_{22}/(R_{22} + jX_C)]\}}{X_A + X_B + [X_C R_{22}/(R_{22} + jX_C)]} \tag{8.26}$$

Again equating the reals and the reactive terms, we arrive at

$$R_{11}R_{22} = \frac{-X_A X_B X_C}{X_A + X_B + X_C} \tag{8.27}$$

$$\frac{R_{11}}{R_{22}} = \frac{X_A(X_B + X_C)}{X_C(X_A + X_B)} \tag{8.28}$$

Equation 8.27 shows the necessity for one of the reactive arms to be of opposite sign to the other two, as was discussed for Eq. 8.8.

By solution of the above equations, we determine that

$$X_A = \frac{-R_{11}X_B}{R_{11} \pm \sqrt{R_{11}R_{22} - X_B^2}} \tag{8.29}$$

$$X_C = \frac{-R_{22}X_B}{R_{22} \pm \sqrt{R_{11}R_{22} - X_B^2}} \tag{8.30}$$

There are two equations and three unknowns, so one reactance must be arbitrarily selected, as was the case for the T network.

With Q chosen for a desired bandwidth, we note from Eq. 8.25 that

$$\mathbf{Z}_{11} = R_{11} = Q\omega_o L = \frac{Q}{\omega_o C_1}$$

With C_1 as the capacitive branch of the resonant circuit, making $X_B - X_C$ inductive, we can write

$$|X_A| = \frac{1}{\omega_o C_1} = \frac{R_{11}}{Q}$$

and so this is a convenient choice for the arbitrary branch. It is usual for Q to be selected in the range from 8 to 15. Then with $R_{11} > R_{22}$ and using Q from Eq. 8.22, the design equations are obtained as

$$X_A = -\frac{R_{11}}{Q} = \frac{-R_{11}}{\sqrt{(R_{11}/R_{22}) - 1}} = \frac{-R_{11}\sqrt{R_{22}}}{\sqrt{R_{11} - R_{22}}} \tag{8.31}$$

$$X_B = \frac{R_{11}}{Q}\left(1 + \sqrt{\frac{R_{22}}{R_{11}}}\right) = \frac{1}{Q}\left(R_{11} + \sqrt{R_{11}R_{22}}\right) \tag{8.32}$$

$$X_C = -\frac{R_{22}}{Q} = \frac{-R_{22}}{\sqrt{(R_{11}/R_{22}) - 1}} = \frac{-R_{22}^{3/2}}{\sqrt{R_{11} - R_{22}}} \tag{8.33}$$

The π matching network is widely used as an impedance transformation between radio transmitters and radiating antenna systems, as shown in Fig. 8.4(b).

EXAMPLE. A generator of 100 V, 500 Ω internal series resistance is to supply power to a 70 Ω antenna load at a frequency of 2.0 MHz. A π coupling network is to be used, and a power match is to occur at only one frequency.

The requirement for a matched response at only one frequency tells us that critical coupling is to be used so

$$R_{11}R_{22} - X_B^2 = 0$$

in Eqs. 8.29 and 8.30. Then

$$X_B = \sqrt{R_{11}R_{22}} = \sqrt{5 \times 10^2 \times 7 \times 10}$$

$$= \sqrt{3.5 \times 10^4} = 187\ \Omega$$

Equations 8.29 and 8.30 show that $|X_A| = |X_B| = |X_C|$ and so for the π network we have

$$X_A = -187\ \Omega$$

$$X_B = 187\ \Omega$$

$$X_C = -187\ \Omega$$

At 2.0 MHz the value of $\omega = 2\pi \times 2 \times 10^6 = 12.6 \times 10^6$ r/s, and

$$C_A = \frac{1}{12.6 \times 10^6 \times 187} = 0.425 \times 10^{-9} = 425\ \text{pF}$$

$$L_B = \frac{187}{12.6 \times 10^6} = 14.9 \times 10^{-6}\ \text{H} = 14.9\ \mu\text{H}$$

$$C_C = C_A = 425\ \text{pF}.$$

This completes the design of the π matching network.

EXAMPLE. A generator of 500 Ω internal series resistance and 10 V is power matched by a T network at 7.0 MHz to a resistive load of 100 Ω, with response at one frequency only. The network has X_2 consisting of an inductance of 10 μH and with X_3 capacitive. Find the power, current and voltage at the load.

This is a simple problem. Since the generator is matched, the generator of 500 Ω and 10 V will appear to have a load of 500 Ω resistance at its output port. The generator current and the current in the apparent load of 500 Ω is

$$\mathbf{I}_g = \frac{\mathbf{V}}{500 + 500} = \frac{10}{1000} = 0.01\ \text{A}$$

This current develops maximum power in the apparent 500 Ω load of

$$P = \mathbf{I}^2 R = 0.01^2 \times 500 = 0.05\ \text{W}$$

None of this power can be lost in the reactive matching elements, but must be transferred to the 100 Ω actual load. The current due to 0.05 W in 100 Ω is

$$\mathbf{I}_L = \sqrt{P/R} = \sqrt{\frac{5 \times 10^{-2}}{10^2}} = 0.0223\ \text{A}$$

The voltage across the load is

$$\mathbf{V}_L = \mathbf{I}_L R_L = 0.0223 \times 100 = 2.23 \text{ V}$$

We need not be concerned with the values of L or C for the matching network, because their values are such as to provide a matched load at the generator output port.

8.5 The ideal transformer

Another method of impedance transformation uses the *iron-cored transformer*, suited to broad frequency ranges from a few hertz to above 15,000 H when using a magnetic core of laminated iron. At high frequencies, powdered iron and ferrite materials are employed to reduce the losses in the magnetic material. Figure 8.5 shows a transformer and its equivalent T circuit. The latter was derived in Section 6.7, and the analysis will use the transfer functions from Section 6.6.

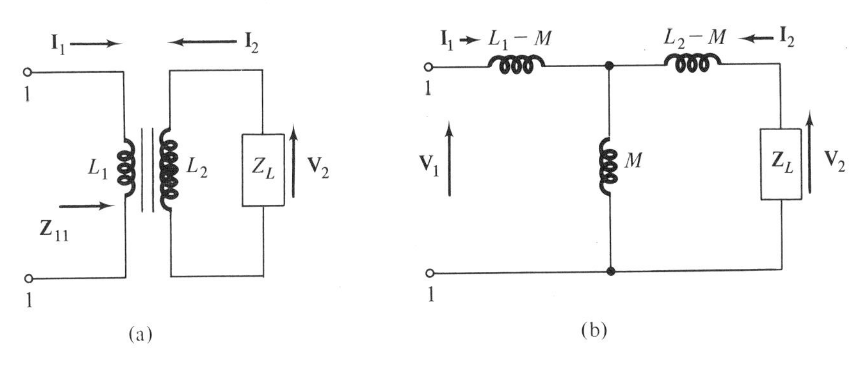

FIGURE 8.5
Ideal transformer and its equivalent circuit.

Because of the iron path for the magnetic flux between the coils, the flux is confined and nearly all flux couples the two coils. The coefficient of coupling, developed as

$$k = \frac{M}{\sqrt{L_1 L_2}}$$

is assumed to be unity, and

$$M = \sqrt{L_1 L_2} \tag{8.34}$$

We also assume that the transformer is lossless, the copper losses and the iron losses being negligible compared to the power being handled by the transformer.

The input impedance of the transformer at the 1,1 port can be written by use of Eq. 6.48 as

$$\mathbf{Z}_{11} = \mathbf{z}_i - \frac{\mathbf{z}_f^2}{\mathbf{z}_o + \mathbf{Z}_L} = \frac{-\omega^2(L_1L_2 - M^2) + j\omega L_1\mathbf{Z}_L}{j\omega L_2 + \mathbf{Z}_L}$$

$$= \frac{j\omega L_1\mathbf{Z}_L}{j\omega L_2 + \mathbf{Z}_L} \tag{8.35}$$

The first term in the numerator is zero because of Eq. 8.34. Transformers are designed with their coil reactances large with respect to the usual load impedance, and so $\omega L_2 \gg |\mathbf{Z}_L|$ in the denominator. Then we have

$$\mathbf{Z}_{11} = \frac{L_1}{L_2}\mathbf{Z}_L \tag{8.36}$$

Because of the voltage-transfer function of Eq. 6.50, we can write

$$\frac{\mathbf{V}_2}{\mathbf{V}_1} = \frac{\mathbf{z}_f\mathbf{Z}_L}{\mathbf{z}_i(\mathbf{z}_o + \mathbf{Z}_L) - \mathbf{z}_f^2} = \frac{j\omega M\mathbf{Z}_L}{-\omega^2(L_1L_2 - M^2) + j\omega L_1\mathbf{Z}_L} \tag{8.37}$$

and using Eq. 8.34 again, there results

$$\left|\frac{\mathbf{V}_2}{\mathbf{V}_1}\right| = \frac{\sqrt{L_1L_2}}{L_1}$$

and so

$$\left|\frac{\mathbf{V}_2}{\mathbf{V}_1}\right| = \sqrt{\frac{L_2}{L_1}} = \frac{n_2}{n_1} = \frac{1}{a} \tag{8.38}$$

since the inductance is proportional to the square of the coil turns, no leakage flux being present. The parameter

$$a = \frac{n_1}{n_2} \tag{8.39}$$

is the *turns ratio* of the transformer.

Writing the current-transfer function from Eq. 6.49,

$$\frac{\mathbf{I}_2}{\mathbf{I}_1} = \frac{-\mathbf{z}_f}{\mathbf{z}_o + \mathbf{Z}_L} = \frac{-j\omega M}{j\omega L_2 + \mathbf{Z}_L}$$

and again using Eq. 8.34 and the assumption that $\omega L_2 \gg |Z_L|$, the above reduces to

$$-\left|\frac{\mathbf{I}_2}{\mathbf{I}_1}\right| = \sqrt{\frac{L_1}{L_2}} = \frac{n_1}{n_2} = a \tag{8.40}$$

The minus sign is present because of the defined direction of $\mathbf{I}_2$ in the T network.

Equations 8.38 and 8.40 are the performance conditions of the *ideal transformer*. Such an operation requires that the transformer meet the conditions assumed in the analysis, namely

1. No leakage flux, or $k = 1$.
2. Zero power loss.
3. Winding reactances large with respect to the design load.

Practical transformers closely approach these conditions.

Inverting Eq. 8.38 and using Eq. 8.40, we have

$$-\left|\frac{\mathbf{V}_1}{\mathbf{V}_2}\right|\left|\frac{\mathbf{I}_2}{\mathbf{I}_1}\right| = a^2$$

from which

$$\left|\frac{\mathbf{V}_1}{\mathbf{I}_1}\right| = -a^2\left|\frac{\mathbf{V}_2}{\mathbf{I}_2}\right| \tag{8.41}$$

We note that $|\mathbf{V}_1/\mathbf{I}_1| = |\mathbf{Z}_{11}|$ and that $-|\mathbf{V}_2/\mathbf{I}_2| = |\mathbf{Z}_L|$, so that

$$|\mathbf{Z}_{11}| = a^2|\mathbf{Z}_L| \tag{8.42}$$

states the impedance-transforming property of the ideal transformer.

The analysis shows that *the transformer changes magnitude but not the phase angle of the load impedance*. The ideal transformer is a widely employed concept for impedance transformation over broad frequency bands in the audio- and low-radio-frequency ranges.

8.6 Models for the power transformer

For purposes of simplicity in analysis we adopted the ideal transformer as a model for the iron-cored transformer, with a slight loss in accuracy. Figure 8.6 is a better representation, albeit more complex for analysis since more of the internal transformer parameters are explicitly included. Thus R_1 and R_2 are the winding resistances, and L_1 and L_2 represent inductances due to leakage flux which couples only one coil; these inductances produce small series voltage drops.

The circuit in (b) in Fig. 8.6 is derived without further assumption by *reflection* of the secondary elements into the primary, using the ideal transformer impedance ratio. That is,

$$R_a = R_1 + a^2R_2 \tag{8.43}$$

$$L_a = L_1 + a^2L_2 \tag{8.44}$$

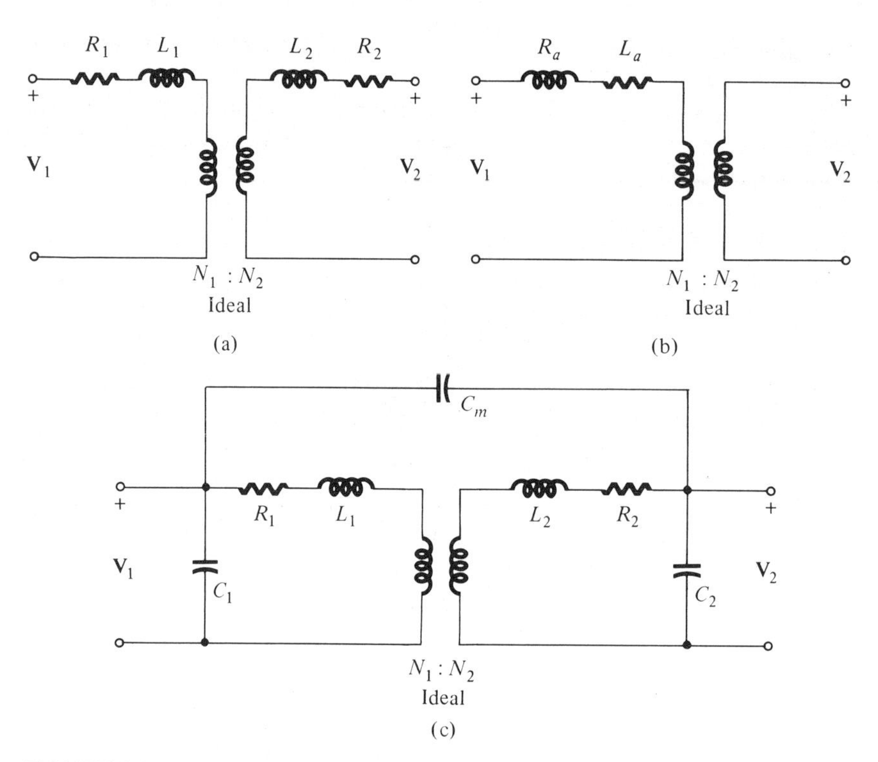

FIGURE 8.6
Nonideal models of the iron-core transformer.

This circuit is widely used in the representation of transformers operated at power frequencies.

At the upper audio frequencies the interwinding capacitances as in (c), Fig. 8.6 become appreciable reactances and introduce resonance effects. Series resonance develops between L_2 and C_2 and produces a response peak at some high frequency. By careful arrangement of the windings these reactances can be made small and the resonant frequency moved outside the band of interest; this leads to expensive construction, and for broad-band response iron-core transformers are avoided.

8.7 Summary

Tapped transformer windings are a simple means employed to raise the effective Q of resonant circuits when connected to low resistance loads. The general reactive network, however, provides means for matching loads to sources under resonance conditions, giving response over a narrow frequency band. When designed with one arm of zero reactance, the T network becomes a parallel-resonant form of circuit. The π-network form of matching network provides better filtering action for generators having harmonics present, as is common with electronic sources, and this network is very generally employed in radio transmitters.

The ideal transformer, having specifications which are rather easily met, is useful for impedance transformation over broad frequency bands. However, it can transform only impedance magnitude, and it is necessary to introduce additional reactance to correct the load phase angle.

PROBLEMS

8.1 A generator of 1 V, 0.7×10^6 Hz frequency, has 200 Ω of internal resistance. Design a T matching network of minimum cost and maximum harmonic suppression to match this generator to a 100-Ω load.

8.2 A generator of 1 V, 1.5 MHz, has an internal impedance of $50 + j50\,\Omega$. (a) Design a T network to match this generator to a resistive load of 2000 Ω, giving inductance and capacitance values. (b) Determine the current and power in the load.

8.3 If the mutual impedance is made capacitive, design a T network to match a source of 4000 Ω, resistive, 10 V, and 796 Hz to a load of $90 + j120\,\Omega$. What power is delivered to the load? What is the load voltage?

8.4 Derive the design Equations 8.29 and 8.30, starting with the network of (a) in Fig. 8.4.

8.5 A radio transmitter with an output resistance of 800 Ω is to supply power to an antenna of 70 Ω of resistance at 7 MHz. The mutual impedance is to be capacitive, and the circuit is to respond at only one frequency. Design a T network to transfer maximum power, giving inductance and capacitance values. What is the bandwidth in hertz?

8.6 Design a π network, with shunting capacitance arms, to accomplish the specifications of Problem 8.5.

8.7 An inductor is to be used with a 500-pF capacitor in a T network to match a source of 10 V, 0.7 MHz, and internal resistance of 20,000 Ω. (a) What resistive load can be matched? (b) What L is needed? (c) What is the generator current? (d) Show that the power from the generator is equal to the power delivered to the load.

8.8 A parallel circuit, resonant at 3×10^6 Hz, has $Q = 20$ and $L = 100$ μH. It is driven by a generator having 1000 Ω of internal resistance, connected to the 1 port of Fig. 8.7. Specify the capacitances to provide an impedance match for the generator. If the generator has 10 V on open circuit, what power is delivered to the resonant circuit?

8.9 An audio-frequency amplifier has an internal generator impedance (resistive) of 800 Ω. It drives a 16-Ω loudspeaker, resistive. (a) If an ideal transformer couples the amplifier to the speaker, what turns ratio is needed? (b) Find the load current and power if the amplifier-generated (open circuit) voltage is 100 V.

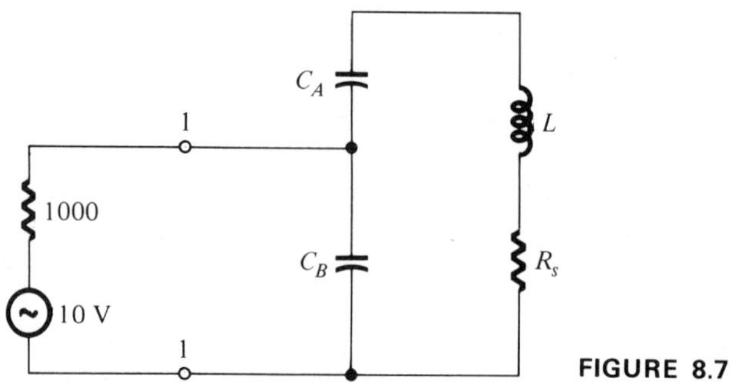

FIGURE 8.7

8.10 If the mutual impedance is an inductor, design a π network to match a generator of 4000 Ω of resistance, 100 V, 796 Hz, to a load of 900 + j1200 Ω. What power is delivered to the load?

8.11 An ideal transformer is to couple a source of 10 V, 1200 Hz, and 700 Ω of internal resistance to a load of 20 − j30 Ω. (a) For the best power transfer, what should be the turns ratio? (b) What power will be delivered to the load?

8.12 With a transformer as in Fig. 8.6(b), we short-circuit the secondary and apply a primary voltage just sufficient to cause rated currents in the two windings. The applied voltage is 27 V, 61.1 A, and the power input is 240 W. For this low voltage the core loss and magnetizing currents can be neglected. Compute the values of the primary and secondary resistances and reactances if the turns ratio is 20:1.

8.13 (a) Again neglecting the magnetizing losses, what would be the efficiency of the transformer of Problem 8.13 if it delivered 10 kVA at a leading power factor of 0.8? (b) What would be the secondary voltage at no load and full load? Secondary load = 0.045Ω.

REFERENCES

1. J. D. Ryder, *Networks, Lines, and Fields*, 2d ed. Prentice-Hall, Inc., Englewood Cliffs, N.J., 1957.
2. N. Balabanian, *Fundamentals of Circuit Theory*. Allyn and Bacon, Inc., Boston, 1961.
3. W. L. Cassell, *Linear Electric Circuits*. John Wiley & Sons, Inc., New York, 1964.

elements with gain; feedback

nine

An ideal amplifier has been characterized as a piece of wire with gain; our integrated circuit packages can be used to closely approach such an ideal "gain block," and the use of gain as a circuit element is increasing. We shall here consider how to provide gain as an additional element for the circuit designer. As the port characteristics of the gain package more closely approach the ideal, we become disinterested in the internal structure of that package. We buy packages of gain and place our emphasis on the circuit in which the gain element is embedded, rather than on the circuitry of the gain element itself.

9.1 Gain elements

The two-port active elements of Sections 6.11 and 6.12 are controlled energy sources and can represent amplifier elements. Controlled sources are characterized in general by a transfer relation between an input circuit variable and the source output. Since the controlled source can have an independent supply of energy, released under control of the input circuit variable, it is possible to have amplification of power levels between input and output ports.

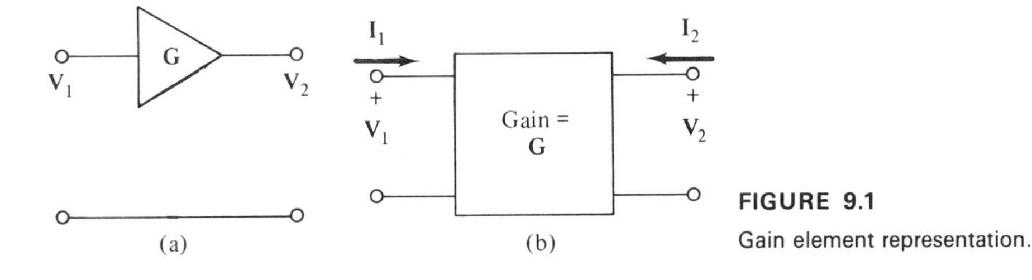

FIGURE 9.1

Gain element representation.

We are interested in the output-input relation of the *gain element* of Fig. 9.1, with the *forward voltage gain* being

$$\mathbf{G} = \frac{\mathbf{V}_2}{\mathbf{V}_1} \tag{9.1}$$

In addition to the functional definition, use of the gain element implies

1. *Linearity* between input and output in order that superposition will apply.
2. The *phase angle* of Eq. 9.1 is 0° or 180°, and the gain is a positive or negative real number.
3. *Signal transfer is in only the forward direction,* isolating the output from the input.
4. *Infinite input impedance,* to assure zero loading effect on preceding circuits.
5. *Zero output impedance,* so that gain blocks may be cascaded with other circuits.

In the symbol at (a) in Fig. 9.1, the ground line is not indicated but is implied as a necessity for a two-port element, as in the equivalent symbol at (b). The figure at (a) is used in a form of one-line diagram.

It is common to employ gain elements in cascade, as in Fig. 9.2(a). We can write the overall gain as

$$\mathbf{G}_o = \frac{\mathbf{V}_4}{\mathbf{V}_1} = \frac{\mathbf{V}_2}{\mathbf{V}_1}\frac{\mathbf{V}_3}{\mathbf{V}_2}\frac{\mathbf{V}_4}{\mathbf{V}_3} = \mathbf{G}_1\mathbf{G}_2\mathbf{G}_3 \tag{9.2}$$

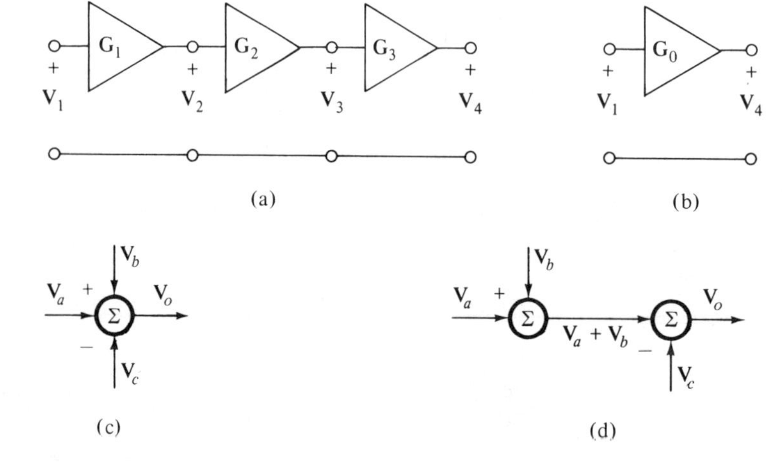

FIGURE 9.2
Gain-element algebra.

showing the overall gain as the product of the individual element gains. The single gain block at (b) is equivalent to the three cascaded gain blocks at (a).

If the individual block gains are specified in *decibels* (Section 6.13), then the overall gain is obtained as the sum of the individual decibel gains:

$$G_o\,(\text{dB}) = G_1\,(\text{dB}) + G_2\,(\text{dB}) + G_3\,(\text{dB}) \tag{9.3}$$

A summing point is illustrated in (c) in Fig. 9.2. This symbol implies that

$$\mathbf{V}_o = \mathbf{V}_a + \mathbf{V}_b - \mathbf{V}_c \tag{9.4}$$

the signs indicating addition or subtraction of the signals. Any number of inputs may be included, but only one output is given. Equation 9.4 would be unchanged if the summing operation were divided into parts, as in Fig. 9.2(d).

The summing operation may be moved past a block of gain, G_a, if gain G_a is inserted in the variable added, as shown in Fig. 9.3(a). Similarly,

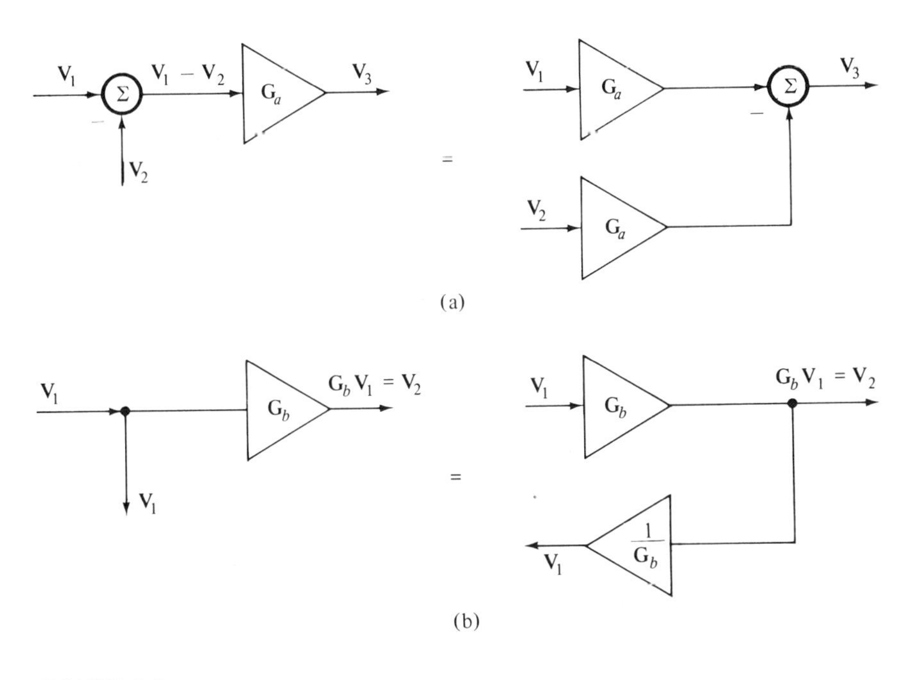

FIGURE 9.3
Shifting of summing and pickoff points.

a pickoff point for a voltage may be moved past a block of gain G_b by recognizing that the level of the signal picked off must be returned by a gain $1/G_b$, as in (b).

These are useful processes in the manipulation of gain elements.

9.2 Basic feedback concepts

Constancy of gain and other useful properties of the gain element are assured by use of a *feedback* mechanism, whereby part of the output is compared to the input, and a difference signal is used to bring the output into correspondence with the input. The principle is apparent in such simple activities as placing a pen on paper; without optical feedback of the difference between actual (output) and desired (input) locations of the pen, we could not write. The principle, while old, was only mathematically formulated by H. S. Black in 1934, and its application to amplifiers leads to our ability to transform practical but somewhat unprecise amplifiers into quasi-ideal gain elements.

In the controlled source the output is controlled by the input variable through a parameter h_f, g_m, y_f, or r_m in the circuits of Fig. 9.4. This is a *feed-forward* mechanism, and constancy of the parameter is desired. Applying

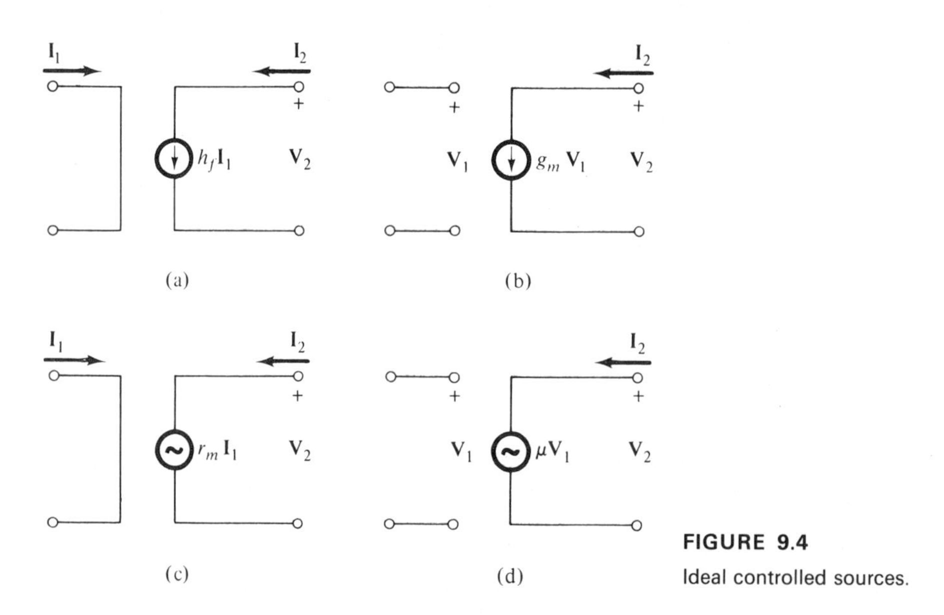

FIGURE 9.4
Ideal controlled sources.

the principle of *feedback*, we add a path with transmission function **B** from the output of the active element to its input port, as in the block diagram of Fig. 9.5. The summing point receives a portion of the output signal $\mathbf{BV}_2$ and compares it with the input signal $\mathbf{V}_i$, their difference becoming the amplifier input $\mathbf{V}_1$.

The internal gain of the amplifier is defined as

$$\mathbf{G}_{21} = \frac{\mathbf{V}_2}{\mathbf{V}_1} \tag{9.5}$$

It is assumed that the transfer function of the feedback network is unilateral,

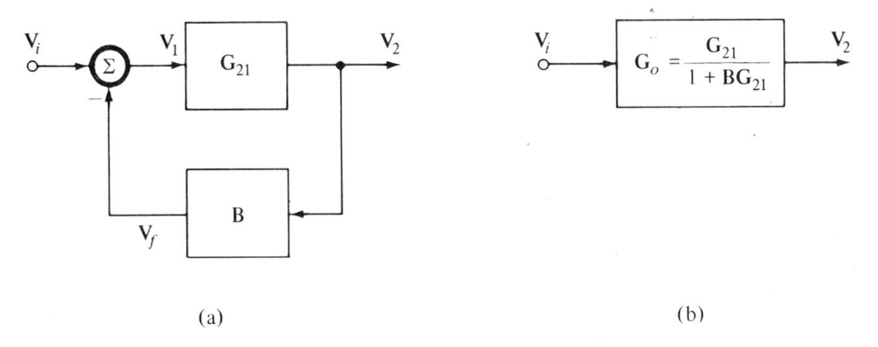

FIGURE 9.5
(a) Feedback around an active element; (b) equivalent gain element.

acting to transmit signal only toward the amplifier input, and that this network is of sufficient impedance that it does not load the amplifier at either port. The transfer function of the feedback network is defined as

$$\mathbf{B} = \frac{\mathbf{V}_f}{\mathbf{V}_2} \tag{9.6}$$

The difference signal after the summing point, and the input to the amplifier, is

$$\mathbf{V}_1 = \mathbf{V}_i - \mathbf{B}\mathbf{V}_2 \tag{9.7}$$

with the network designed to make $\mathbf{B}$ a real number.

We use Eqs. 9.5 and 9.7 to obtain the gain of the feedback system, $\mathbf{G}$, as

$$\mathbf{G}_o = \frac{\mathbf{V}_2}{\mathbf{V}_i} = \frac{\mathbf{G}_{21}}{1 + \mathbf{B}\mathbf{G}_{21}} \tag{9.8}$$

This result is the basic feedback equation.* We can replace the circuit at (a) with the equivalent active system, as drawn at (b) in Fig. 9.5.

Suppose that $\mathbf{B}$ is positive and real and that $\mathbf{G}_{21}$ is also made positive and real, such that $\mathbf{B}\mathbf{G}_{21} > 0$. Equation 9.8 shows that $|\mathbf{G}_o| < |\mathbf{G}_{21}|$ and that the *gain has been reduced* by feedback. The signals at the summing point are compared by subtraction; this condition is called *negative feedback*.

When $\mathbf{G}_{21}$ is real and negative, it is necessary to reverse the output of the feedback network so that $\mathbf{B}\mathbf{V}_2$ remains subtractive to $\mathbf{V}_i$. The gain remains less than the nonfeedback value and the feedback is still negative.

The needed gain can be restored by designing the amplifier elements for a larger value of $\mathbf{G}_{21}$, before feedback is introduced. The reduction in

*In some literature the comparison is by addition of a negative $\mathbf{V}_f$ with $\mathbf{V}_i$. For negative feedback, $\mathbf{B}\mathbf{G}_{21}$ must then carry a negative sign in Eq. 9.8, but the physical result of subtractive comparison will be unchanged.

overall system gain with negative feedback is the price paid for the favorable ends achieved, as will be shown.

A situation in which the voltage $\mathbf{BV}_2$ is so phased as to have a component additive to $\mathbf{V}_i$ is to be avoided. This is the condition of *positive feedback* and will make $|\mathbf{G}_o| > |\mathbf{G}_{21}|$. Such circuits tend to instability and oscillation. This is a matter affecting the internal design of the amplifier and will not concern us here.

Gain $\mathbf{G}_{21}$ is the *internal amplifier gain*, and $\mathbf{G}_o$ is the *closed-loop gain* of the entire system. The feedback network will ordinarily employ a stable resistive voltage or current divider to make $\mathbf{B}$ a real number, and gain $|\mathbf{G}_{21}|$ will be large, perhaps greater than 10^4. Then if $|\mathbf{BG}_{21}| \gg 1$, Eq. 9.8 becomes

$$\mathbf{G}_o = \frac{1}{B} \tag{9.9}$$

The gain is dependent on B and independent of G_{21}. The internal gain of the amplifier may change due to variation of active device parameters, but, with a large internal gain, the total system gain will remain substantially constant.

The *sensitivity* of the overall gain $\mathbf{G}_o$ to changes in internal gain $\mathbf{G}_{21}$ can be determined with BG_{21} made real. Starting with Eq. 9.8,

$$\begin{aligned} dG_o &= d\left(\frac{G_{21}}{1 + BG_{21}}\right) \\ &= \frac{1}{1 + BG_{21}} \frac{G_{21}}{1 + BG_{21}} \frac{dG_{21}}{G_{21}} \\ &= \frac{1}{1 + BG_{21}} \frac{G_o}{G_{21}} dG_{21} \end{aligned}$$

This result allows us to write the sensitivity S as

$$S = \frac{dG_o/G_o}{dG_{21}/G_{21}} = \frac{1}{1 + BG_{21}} \tag{9.10}$$

That is, the percentage change in G_o to the percentage change in G_{21} is the same as the ratio $1/(1 + BG_{21})$. Again the value of $BG_{21} \gg 1$ becomes a desirable condition.

Since gain and phase angle are related, the phase angle will also be regulated. This performance cannot be expected at all frequencies, but we are able to design gain elements with constant gain and phase angle over bands of frequencies used in most problems. With constant gain the transfer parameter will be independent of signal level; this is confirmed by the linearity assumption.

EXAMPLE. Suppose that we have an amplifier element with gain $G_{21} = 1000$ and add a negative feedback network with $B = 0.1$. The gain with feedback is

$$G_o = \frac{1000}{1 + 1000 \times 0.1} = \frac{1000}{101} = 9.90$$

The internal gain is now observed to change by 10 percent due to transistor parameter change. The gain of the feedback gain element becomes

$$G_o = \frac{900}{1 + 900 \times 0.1} = \frac{900}{91} = 9.89$$

With feedback the change in gain is from 9.90 to 9.89, or only 0.1 percent, as compared to 10 percent for the gain element without feedback.

9.3 Impedances of the feedback element

Practical gain circuits are developed by the addition of negative feedback to the active circuits of Section 6.12, and two useful forms are reproduced in Fig. 9.6. To allow control of the feedback, we require that the signal fed back be transmitted entirely by way of the feedback network. That is, the internal feedback $\mathbf{y}_r\mathbf{V}_2$ or $\mathbf{h}_r\mathbf{V}_2$ of the active network is made zero. Removal of this element ensures that changes in the loading will not be reflected back to cause changes in the input circuit.

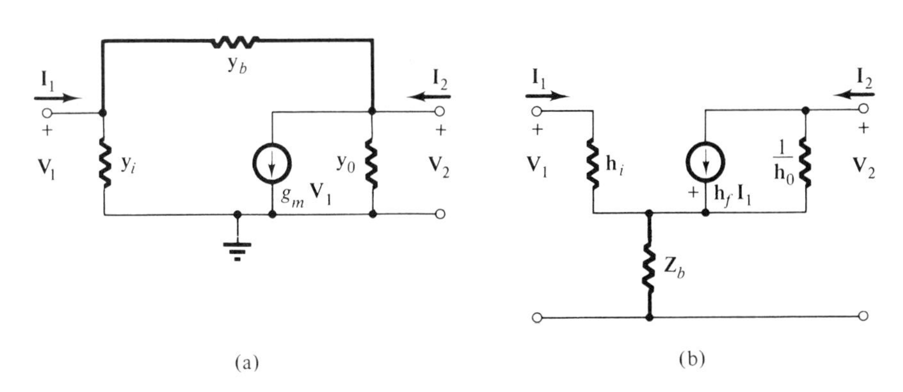

FIGURE 9.6
(a) Shunt feedback in the **y**-parameter active circuit; (b) series feedback in the **h**-parameter active circuit.

This requirement does not unduly idealize our practical circuits; active devices with suitably small values of $\mathbf{h}_r$ or $\mathbf{y}_r$ are available.

With three-terminal circuits, it is possible to introduce feedback networks in only two ways—either in shunt from output terminal to input terminal, as $\mathbf{y}_b$ in Fig. 9.6(a), or in series to the common terminal, as $\mathbf{z}_b$ in Fig. 9.6(b). The shunt element samples the output voltage and gives *shunt-derived* or voltage feedback; the series element samples the output current and introduces *series-derived* or current feedback.

Shunt-derived feedback causes the circuit to maintain constant $\mathbf{V}_2$, making the output voltage independent of load. But this is the characteristic of a current source with large $\mathbf{y}_o$, so regulation for constant output voltage appears to raise the output admittance $\mathbf{y}_o$ or lower the output impedance $\mathbf{z}_o$. It can be shown that

$$\mathbf{Z}_o = \frac{\mathbf{z}_o}{1 + \mathbf{BG}_{21}} \tag{9.11}$$

and that the output impedance is reduced by the same factor as the gain is reduced. With $|BG_{21}| \gg 1$ and shunt-derived feedback, the output impedance of a feedback gain element can be very low.

In Fig. 9.6(b) the feedback voltage is

$$\mathbf{V}_f = \left(\frac{1}{\mathbf{h}_f} + 1\right)\mathbf{I}_2\mathbf{z}_b$$

and being proportional to load current, this control voltage tends to maintain the load current constant. This is the performance of a current source if $1/\mathbf{h}_o$ is very large. Therefore series-derived feedback provides a current source with raised output impedance. The output impedance can be found to be

$$\mathbf{Z}_o = \mathbf{z}_o(1 + \mathbf{BG}_{21}) \tag{9.12}$$

and the output impedance, with series-derived feedback, is raised by the same factor as the gain is reduced. We normally employ shunt-derived feedback around practical amplifier elements in order to obtain a gain element with very small output impedance.

9.4 The gain element in the operational amplifier

An important application of the gain element is in the *operational amplifier*, so called because of its versatility in performing mathematical and other operations. The general circuit of Fig. 9.7 employs a shunt feedback impedance for the reasons just discussed, to which is added a series input impedance $Z_i(s)$. These are written in the s domain for generality.

The internal gain is made very large, perhaps $|G_{21}| > 10^4$, and is negative such that

$$G(s) = \frac{-V_2(s)}{V_1(s)} \tag{9.13}$$

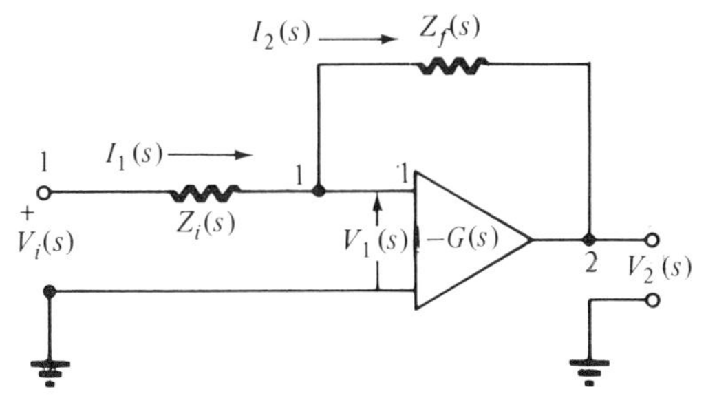

FIGURE 9.7
Operational gain element.

The output voltage $V_2(s)$ usually approximates 10 V. With very large internal gain the input $V_i(s)$ is so small that terminal 1 is *virtually at ground potential.* With the input impedance high by design and with negligible applied potential, essentially zero current enters the gain element at 1.

Then we can say that

$$I_1(s) = I_2(s) \tag{9.14}$$

and can write

$$\frac{V_i(s) - V_1(s)}{Z_i(s)} = \frac{V_1(s) - V_2(s)}{Z_f(s)}$$

Dividing by $V_2(s)$ and using Eq. 9.13, we have

$$G_o(s) = \frac{V_2(s)}{V_i(s)} = -\frac{Z_f(s)}{Z_i(s)} \frac{1}{1 + [1/G(s)]\{1 + [Z_f(s)/Z_i(s)]\}} \tag{9.15}$$

With the magnitude of G very large, the denominator terms become

$$1 \gg \frac{1}{G(s)}\left[1 + \frac{Z_f(s)}{Z_i(s)}\right]$$

and the overall system gain is

$$G_o(s) = \frac{V_2(s)}{V_i(s)} = -\frac{Z_f(s)}{Z_i(s)} \tag{9.16}$$

Again we have found a condition in which a practical amplifier may be used, yet the overall gain $G_o(s)$ of the system is independent of that of the gain element, $-G(s)$. The overall gain is dependent only on circuit impedances.

Equation 9.16 is basic for the operational amplifier. The accuracy with which the various operations are carried out by the circuit will depend on large internal gain of the gain element. Note that we have also implied a value of zero output impedance, so that loading effects on the circuit may be neglected.

9.5

Uses of the operational gain element

If the impedances in the gain expression, Eq. 9.16, are made equal and resistive, then

$$V_2(s) = -V_i(s) \tag{9.17}$$

and the circuit is an *inverter*, with unity gain magnitude. If $Z_i(s) = R_1$ and $Z_f(s) = R_2$, then

$$V_2(s) = -\frac{R_2}{R_1} V_i(s) \tag{9.18}$$

and the output represents the input voltage multiplied by a negative constant; the circuit appears in Fig. 9.8(a).

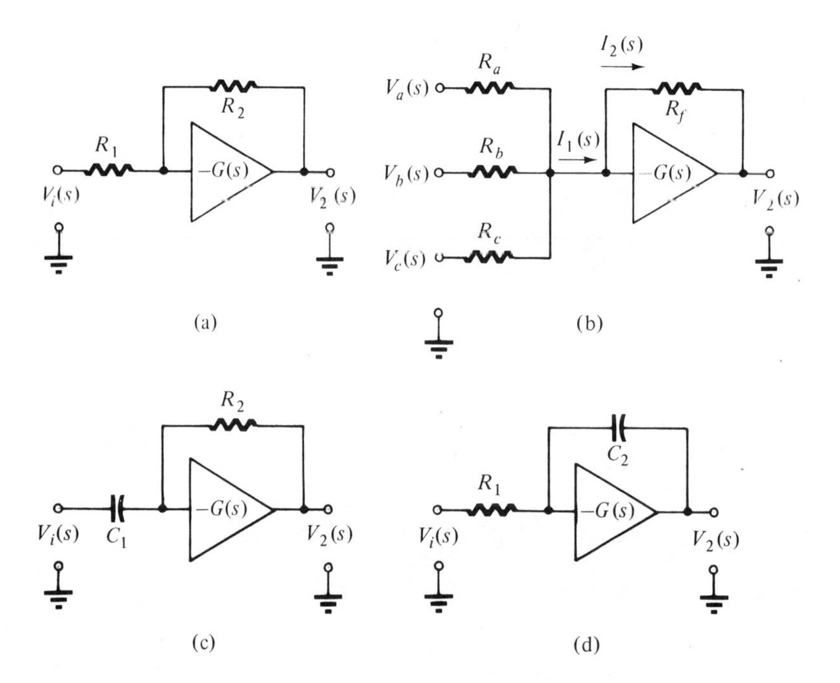

FIGURE 9.8
(a) Multiplication by a negative constant; (b) summation of signals; (c) differentiation; (d) integration.

Several inputs may be combined in the circuit of Fig. 9.8(b). Current $I_1(s)$ is

$$I_1(s) = \frac{V_a(s)}{R_a} + \frac{V_b(s)}{R_b} + \frac{V_c(s)}{R_c}$$

and

$$V_2(s) = -\left[\frac{R_f}{R_a}V_a(s) + \frac{R_f}{R_b}V_b(s) + \frac{R_f}{R_c}V_c(s)\right] \tag{9.19}$$

The output voltage is the sum of the input voltages, with each input multiplied by a negative constant.

If we make $Z_i(s)$ capacitive, as $1/sC_1$, and $Z_f(s) = R_2$, then

$$V_2(s) = -R_2C_1sV_i(s) \tag{9.20}$$

and the circuit output is the time *derivative* of the input voltage, multiplied by a negative constant $-R_2C_1$. If $R_2 = 1\,\text{M}\Omega$ and $C_1 = 1\,\mu\text{F}$, the multiplying factor is unity in value.

Making $Z_f(s) = 1/sC_2$ and $Z_i(s) = R_1$, then

$$V_2(s) = -\frac{1}{R_1C_2}\frac{V_i(s)}{s} \tag{9.21}$$

The input voltage is *integrated* and multiplied by a negative factor.

Combinations of R and C may be employed for $Z_i(s)$ and $Z_f(s)$, giving variety to the transfer functions which can be simulated by the operational amplifier. Nonlinear operations may also be undertaken, for example, when a diode with an exponential voltage-current relation is used as Z_f. The amplifier output-input relation then becomes logarithmic.

The literature on applications of the operational amplifier is prolific.

9.6 Operational amplifier limitations

A summing amplifier should have a uniform response at all frequencies present in a given application. The use of resistive feedback with large $\mathbf{BG}_{21}$ values widens the frequency band by a factor $1 + \mathbf{BG}_{21}$, but the gain is reduced by the same factor. Thus a summing amplifier of wide bandwidth may be an amplifier of reduced overall gain.

If we substitute $j\omega = s$ in Eq. 9.21 for the integration operation, we find that unity overall gain is reached when the angular frequency is given by

$$\omega = \frac{1}{R_1C_2}$$

Going to lower frequencies, the gain rises at 20 dBV/frequency decade, reaching infinite gain at zero frequency. Practical gain limits will prevent this theoretical growth to infinity and cause the gain to level off at a finite value, but as long as $G(s) \gg [1 + Z_f(s)/Z_i(s)]$ is satisfied at the lowest frequency of interest in the problem, then satisfactorily accurate integration can be obtained.

Due to capacitance in the circuits of the gain element, the rate of rise of output voltage in response to a step input change is limited to a maximum, known as the *slew rate*. The full input step appears at V_1, since the feedback from V_2 cannot instantly respond. Output βV_2 increases as fast as the several capacitances permit, according to

$$\rho = \frac{dV_2}{dt}$$

where ρ is the slew rate, a specification supplied by the operational amplifier manufacturer. For sinusoidal input the maximum output voltage at a given frequency is related to the slew rate as

$$V_2 = \frac{\rho}{\omega} \tag{9.22}$$

9.7 Analog computation

The analog computer performs summing, sign changing, integration, and differentiation and undertakes some nonlinear simulations by use of operational amplifiers. In practice these are small packaged amplifiers with voltage gains ranging up to 10^5; the appropriate input and feedback elements are added in accordance with the requirements of the problem being solved. The operational amplifier then becomes a gain element.

Analog computers ordinarily are used to solve differential equations. Consider

$$\frac{d^2x}{dt^2} + 8\frac{dx}{dt} + 30x = 4 \sin \omega t \tag{9.23}$$

This might represent the familiar mass-spring-damper mechanical problem with a sinusoidal forcing function.

We shall simulate the equation in the computer by representing displacement x in the physical system by a voltage av in the electrical computer. The coefficient a relates the maximum expected physical displacement to the maximum voltage capability of the amplifiers, usually 10 V. Time may be represented directly in *real time* or scaled as previously discussed in Section 4.10.

The first step in establishing the analog circuit is solution for the highest-order derivative, giving

$$\frac{d^2x}{dt^2} = 4 \sin \omega t - 8\frac{dx}{dt} - 30x \tag{9.24}$$

The sum of the three terms on the right represents d^2x/dt^2. An integrating amplifier is arranged at (1) in Fig. 9.9 to receive the three right-hand terms as inputs. Its output will be the negative of the integral of d^2x/dt^2, or $-dx/dt$. A following sign changer at (2) yields dx/dt. Another integration at (3) gives $-x$, and the sign is changed by (4) to furnish an output voltage representing x, which may be plotted against time or displayed as a time function on a cathode-ray screen.

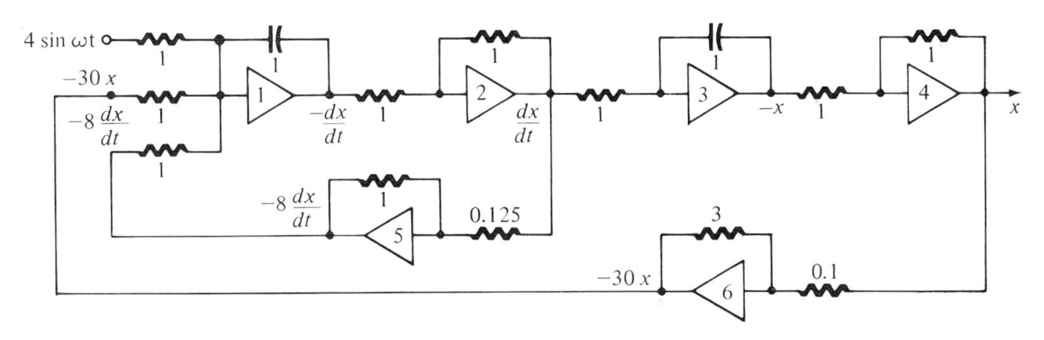

FIGURE 9.9

Analog computation of Eq. 9.24.

At (5) we have a coefficient multiplier with input dx/dt. Multiplied by $1/0.125 = 8$, the output of (5) is $-8\,dx/dt$ and this output provides one of the input quantities for (1). The displacement x from (4) is also multiplied by $3/0.1 = 30$ in amplifier (6), giving $-30\,dx/dt$, which is a second input quantity for (1). The driving function is $4 \sin \omega t$, obtained as an external signal. Its connection to the input starts the solution of the problem. More sophisticated programming could be used to consolidate some of the functions and reduce the number of amplifiers required.

Initial conditions may be introduced into the integrators by adding a potential and switch for each integrating capacitor, as in Fig. 9.10. Prior

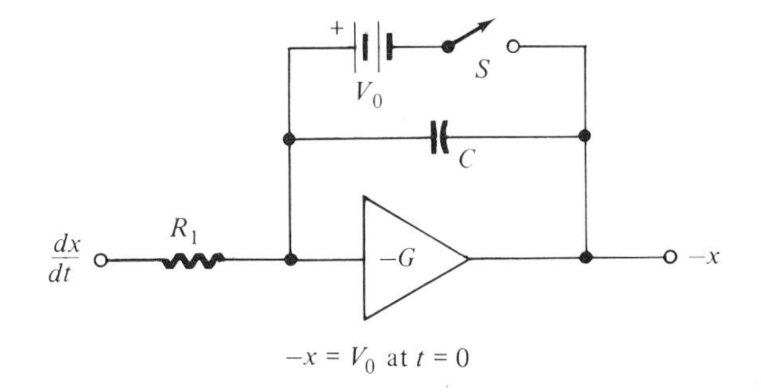

FIGURE 9.10

Introduction of an initial condition.

to the start of the solution, S is closed and voltage V_o applied to the capacitor to represent the value of the initial displacement of the variable. A similar circuit is added for each integrator, and the relays simultaneously open all contacts as the solution starts at time zero.

Solution by differentiation is not usually undertaken because a derivative of a sinusoidal function is multiplied by ω. Thus the outputs of differentiators increase with frequency, and unwanted high-frequency noise voltages become a source of error. Integration of random noise voltages leads to zero output, and the use of the integration process makes the solution error-free from that source.

Real-time solution may be too fast for the recording equipment employed, and time may be scaled by setting $t = t'/a$, where t' represents the computer time. If 10 actual seconds are to equal 1 s in the problem, then $t' = 10t$ and

$$\frac{dx}{dt} = \frac{dx}{d(t'/a)} = a\frac{dx}{dt'} = 10\frac{dx}{dt'} \tag{9.25}$$

This will slow the solution by a factor of 10.

9.8 Impedance converters

Consider the ideal controlled current-source in Fig. 9.11(a), with an impedance $\mathbf{Z}$ at its output terminals. The current in $\mathbf{Z}$ is $(1 - \alpha)\mathbf{I}$ and the input voltage is $\mathbf{V} = (1 - \alpha)\mathbf{IZ}$. The input impedance is then

$$\mathbf{Z}_{in} = \frac{\mathbf{V}}{\mathbf{I}} = (1 - \alpha)\mathbf{Z}$$

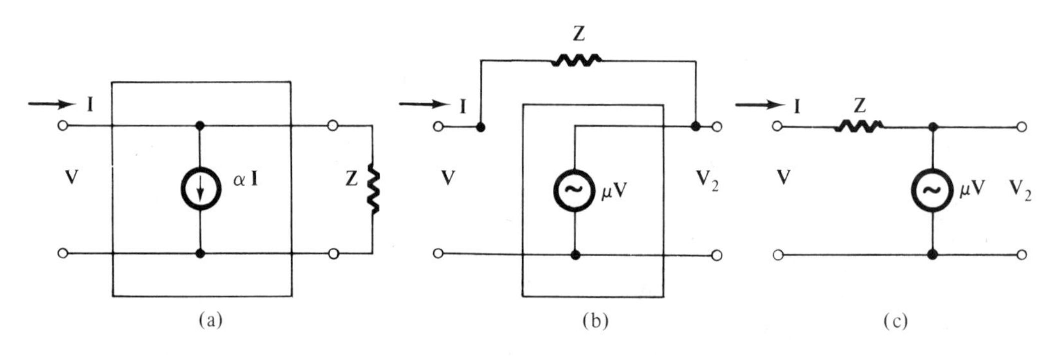

FIGURE 9.11
(a) Negative-impedance converter; (b), (c) Miller effect.

If α is much greater than unity, as is usual with controlled sources,

$$\mathbf{Z}_{in} \cong -\alpha\mathbf{Z} \tag{9.26}$$

and the input impedance is negative. The circuit is known as a *negative impedance converter* (NIC).

If $\mathbf{Z} = R$, we have a means of generating a negative resistance as a circuit element. A negative resistance implies that power may be supplied from the element rather than being absorbed in a positive resistance. This is consistent with our concept of a gain element as an amplifier, or more properly a converter of energy.

Another result of use of a gain element is shown in Fig. 9.11(b). It is actually a simplified version of the previously discussed shunt feedback circuit. With a voltage $\mathbf{V}$ applied in the resultant circuit at (c), the current is

$$\mathbf{I} = \frac{(\mathbf{y}_f + 1)\mathbf{V}}{\mathbf{Z}}$$

and the input impedance is

$$\mathbf{Z}_{in} = \frac{\mathbf{Z}}{\mathbf{y}_f + 1} \tag{9.27}$$

The input impedance is reduced by the gain property of the gain element. In amplifiers this result is known as the *Miller effect*.

A *gyrator* is a two-port network, realized with a gain element, which presents an input impedance which is proportional to the terminating admittance, with a positive proportionality constant. That is,

$$\mathbf{Z}_i = r^2\mathbf{Y}_L \tag{9.28}$$

where $r = 1/g$ is called the *gyration resistance*. The gyrator symbol appears in Fig. 9.12(a).

At the output port

$$\mathbf{I}_2 = -\mathbf{Y}_L\mathbf{V}_2$$

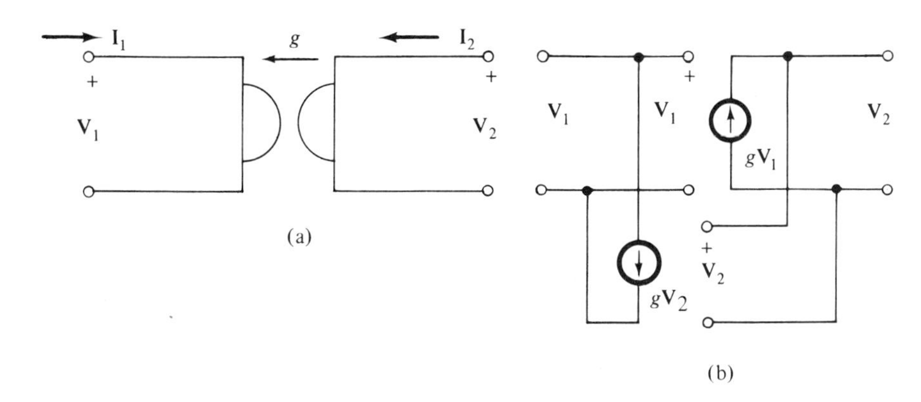

FIGURE 9.12
(a) Gyrator symbol; (b) gyrator realization.

and the input admittance is

$$\mathbf{Y}_i = \frac{\mathbf{I}_1}{\mathbf{V}_1} = \mathbf{y}_i - \frac{\mathbf{y}_r\mathbf{y}_f}{\mathbf{y}_o + \mathbf{Y}_L}$$

If we require that

$$\mathbf{y}_i = 0 \qquad \mathbf{y}_r = g = 1/e$$

$$\mathbf{y}_o = 0 \qquad \mathbf{y}_f = -g = -1/r$$

then

$$\mathbf{I}_1 = -g\mathbf{V}_2 \tag{9.29}$$

$$\mathbf{I}_2 = g\mathbf{V}_1 \tag{9.30}$$

An applied voltage $\mathbf{V}_2$ is transduced to a current *into* the device; this is analogous to a property of the gyroscope from which the gyrator derives its name.

We can take the ratio of the above equations,

$$\mathbf{Y}_i = \frac{\mathbf{I}_1}{\mathbf{V}_1} = -g^2\frac{\mathbf{V}_2}{\mathbf{I}_2} \tag{9.31}$$

but $-\mathbf{V}_2/\mathbf{I}_2 = \mathbf{Z}_L$. Let us use $Z_L = -j/\omega C$, and we have

$$\mathbf{Y}_i = \frac{-jg^2}{\omega C} \qquad \text{and} \qquad \mathbf{Z}_i = \frac{j\omega C}{g^2} \tag{9.32}$$

and so a capacitive load appears as an inductive input impedance. Likewise, an inductive load would be transformed to appear as a capacitance at the input.

The gyrator employs active gain elements in its physical realization; one example is indicated in Fig. 9.12(b), using voltage-controlled current generators in an active feedback link.

9.9 Transformation of time constants

Feedback and a gain element may be used to alter the time constant of a passive circuit. The transfer function of the RC circuit of (a) in Fig. 9.13 is

$$\frac{V_b(s)}{V_a(s)} = \frac{I(s)/sC}{I_s(R + 1/sC)} = \frac{1}{1 + sRC} \tag{9.33}$$

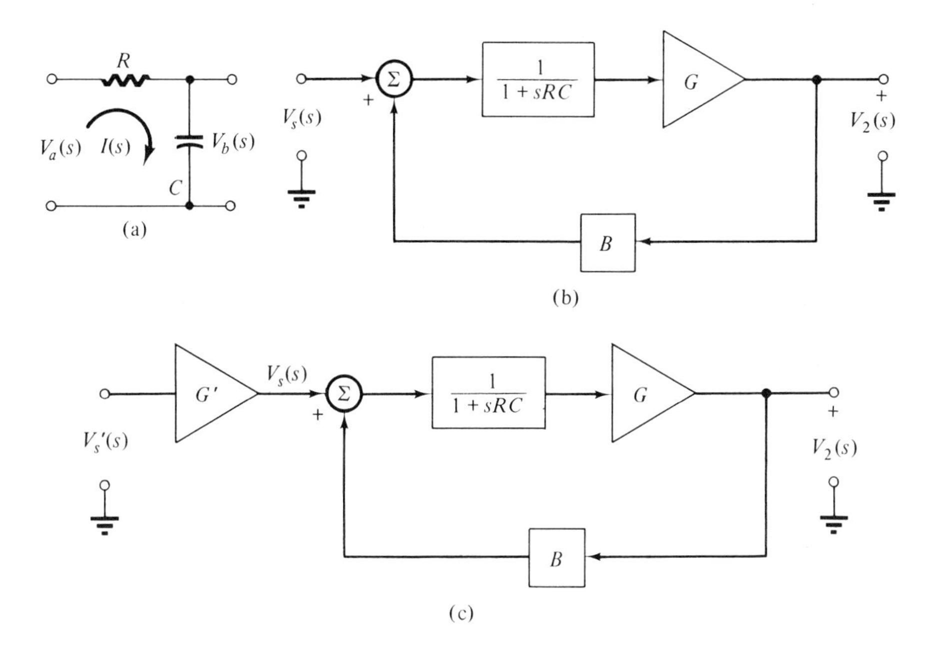

FIGURE 9.13

Transformation of the time constant.

If we place this RC circuit inside a feedback loop, as in Fig. 9.13(b), the internal gain becomes

$$\frac{V_2(s)}{V_1(s)} = \frac{G}{1 + sRC} \tag{9.34}$$

and the transfer ratio for the overall feedback system can be written as

$$G_{2s} = \frac{V_2(s)}{V_s(s)} = \frac{G/(1 + sRC)}{1 + [BG/(1 + sRC)]} = \frac{G}{1 + BG + sRC} \tag{9.35}$$

If the total gain is restored by adding an amplifier element of gain G' ahead of the summing point with

$$G' = \frac{1 + BG}{G} \tag{9.36}$$

we have the result of (c) in Fig. 9.13. Then the overall gain is

$$G'_{2s} = \frac{V_2(s)}{V'_s(s)} = \frac{G}{1 + BG + sRC}\,\frac{1 + BG}{G} = \frac{1}{1 + s[RC/(1 + BG)]} \tag{9.37}$$

and the time constant of the transfer function is reduced by the factor $1 + BG$, after comparing this result with Eq. 9.33. Since such time constants represent pole or zero locations in RC networks, we have a means of translating such locations by use of gain elements.

9.10 Summary

An ideal gain element has constant gain, has a linear transfer relation, transmits signals in only one direction, and has infinite input and zero output impedances. By use of negative feedback our practical amplifiers can be made to closely approximate these ideal conditions, and we are able to employ active elements for gain in our circuits. As long as the internal gain is large, we need not concern ourselves with the internal circuit arrangements which produce that gain.

Gain elements are broadly applied as operational amplifiers. In the analog computer these "op amps" perform differentiation, integration, linear combination, and mathematical operations which are combinations of these basic operations. Another extensive application of gain elements will be encountered in Chapter 11.

PROBLEMS

9.1 An amplifier gives an output of 15 V with an input of 0.15 V. (a) Negative feedback is added and it is found that 2.75 V are now needed as input to give the same output as before. What is the value of B? (b) For a change in internal gain to 90, what is the percentage change in overall gain with the feedback added?

9.2 A precision gain element with feedback is to have an overall voltage gain of 60. (a) If the gain must be constant within 0.5 percent but internal gain changes as large as 12 percent are expected, determine the value of B which should be used. (b) What was the value of G_{21} before feedback was added?

9.3 A negative feedback three-stage cascaded amplifier has an overall voltage gain of 50 with feedback. The overall gain changes 0.5 percent when the internal gain changes by 15 percent. Determine the internal gain G.

9.4 An amplifier has three cascaded stages with gains of -100, -100, and -75. The feedback network B is a resistance voltage divider giving $v_f = -[R_1/(R_1 + R_2)]V_2 = -0.01V_2$. Predict the overall gain with and without feedback.

9.5 A gain element, before feedback is added, has a gain $K_o = 1150$. Its upper frequency limit is $\omega_2 = 10^5$ rad/s, and its gain vs. frequency equation without feedback is

$$G = \frac{K_o}{1 + j\omega/\omega_2}$$

(a) Derive an expression for the high-frequency gain response with feedback. (b) With $B = 0.01$, what is the upper frequency limit half-power frequency of the gain element with feedback?

9.6 By block-diagram algebra, reduce the diagram of Fig. 9.14(a) to the form of the basic feedback system of Fig. 9.5(a) and label the elements with their transfer relations. Determine the gain of the system, with the diagram shown as in Fig. 9.5(b).

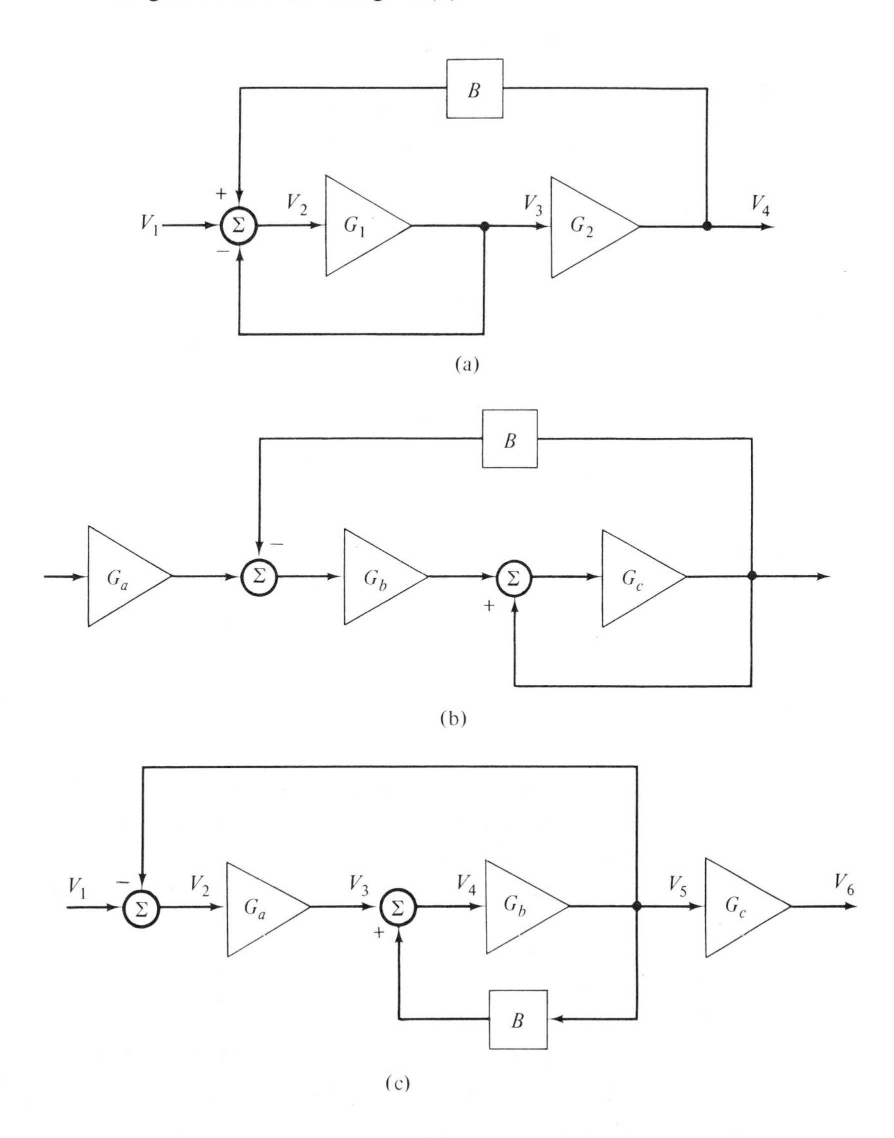

FIGURE 9.14

9.7 Reduce the block diagram of the system of Fig. 9.14(b) to the basic feedback system of Fig. 9.5(a). Then further reduce the system to that at (b) in Fig. 9.5, determining the overall gain expression.

9.8 (a) Determine the internal gain G_{21} of the system of Fig. 9.14(c). (b) What is the value of B used for the feedback transfer function?

9.9 The gain element of Fig. 9.15(a) has a gain $K = 3$. (a) Find the transfer function V_2/V_1 and plot its poles in the complex s plane. (b) Does the pole location show that the circuit is stable?

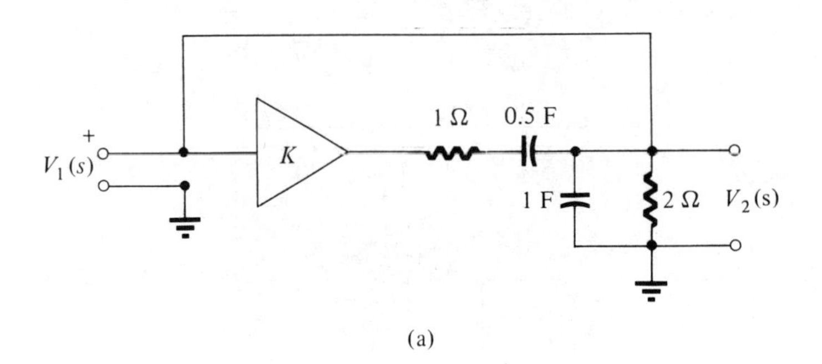

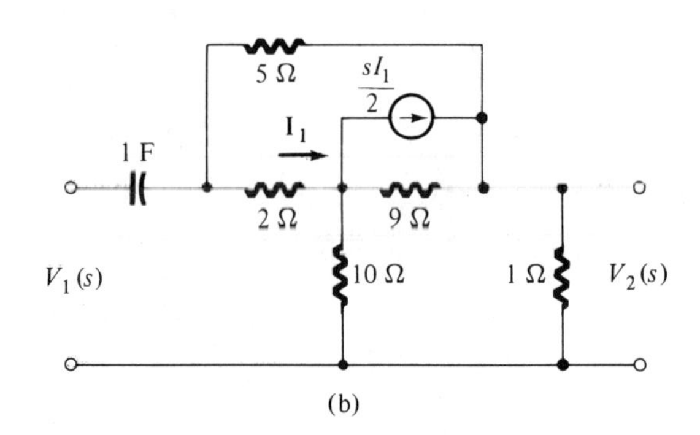

FIGURE 9.15

9.10 Find an expression for the transfer function V_2/V_1 as a function of s for the circuit of Fig. 9.15(b). The current source is an ideal gain element giving a current of $sI_1/2$.

9.11 Draw an analog computer program diagram to solve

$$\frac{d^3x}{dt^3} + 2\frac{d^2x}{dt^2} - 4\frac{dx}{dt} + x = y(t)$$

where $x(0) = 0$, $dx/dt = -2$, and $d^2x/dt^2 = 0$, all at $t = 0$.

9.12 With an operational amplifier having $Z_i = 1\ \text{M}\Omega$ and $Z_f = 0.2\ \mu\text{F}$, determine the output as a function of time for an input signal $v_i = 0.3t$.

9.13 A vibration problem with

$$\frac{d^2x}{dt^2} + \frac{dx}{dt} + 4x = 2$$

has $x(0) = 0$ and $dx/dt = 0$ at $t = 0$. One volt of output corresponds to $5x$ in the problem; draw the analog program.

9.14 Program an analog computer to solve

$$3\frac{d^3v}{dt^3} + 2\frac{d^2v}{dt^2} + \frac{dv}{dt} + 4v = 0$$

with $v = 0$, $dv/dt = 2$, and $d^2v/dt^2 = 5$ at $t = 0$.

9.15 Determine the relation of v_2 to v_1 in the s plane for each of the operational circuits in Fig. 9.16. Gain K is very large.

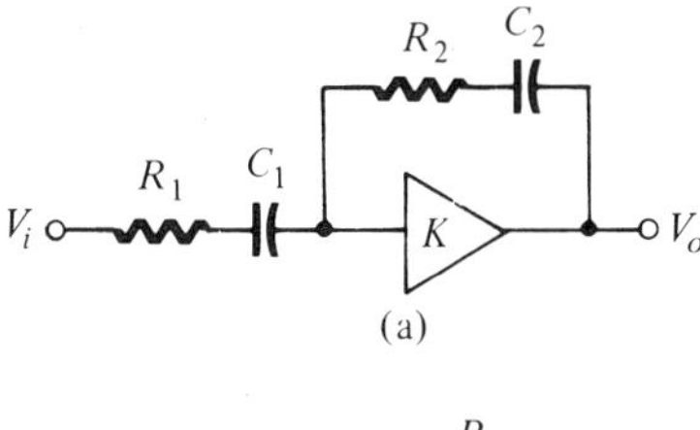

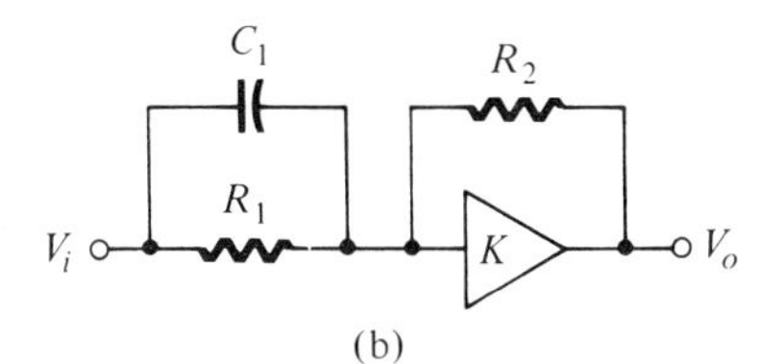

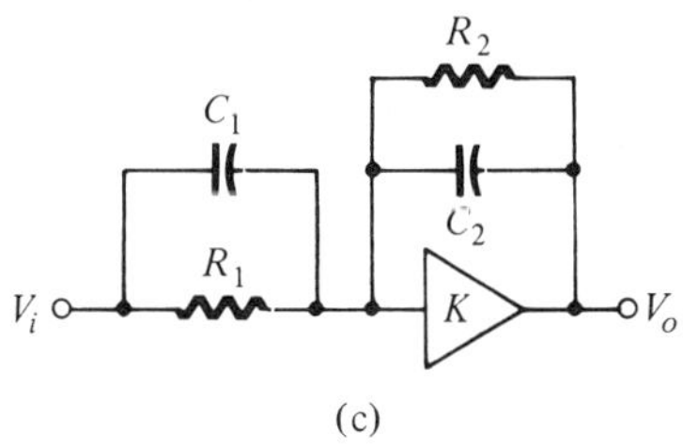

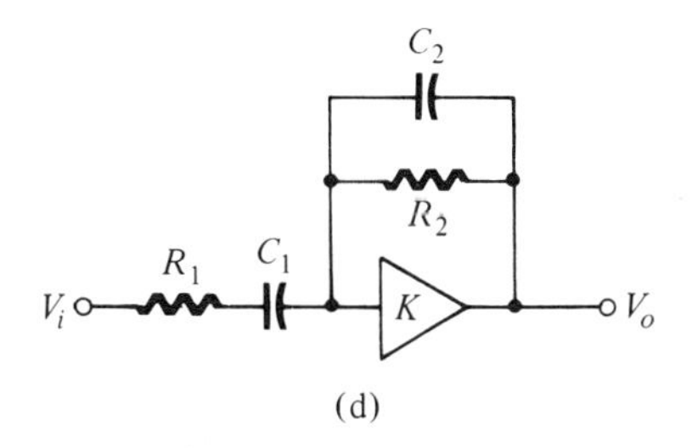

FIGURE 9.16

REFERENCES

1. W. W. Soroka, *Analog Methods in Computation and Simulation.* McGraw-Hill Book Company, New York, 1954.
2. W. A. Lynch and J. G. Truxal, *Signals and Systems in Electrical Engineering.* McGraw-Hill Book Company, New York, 1962.
3. J. R. Ashley, *Introduction to Analog Computation.* John Wiley & Sons, Inc., New York, 1963.
4. M. Kahn, *The Versatile Op-Amp.* Holt, Rinehart and Winston, Inc., New York, 1970.
5. J. G. Graeme, G. E. Tobey, and L. P. Huelsman, *Operational Amplifiers.* McGraw-Hill Book Company, New York, 1971.

passive filters

ten

Whenever ac signals in different frequency bands are to be separated, we employ *filter circuits.* An ideal filter would pass all frequencies in a given band without reduction in magnitude and would prevent the passage of energy at all other frequencies. Actual filters do not perform as well.

In this chapter the filters will employ linear, passive, bilateral components. These will be reactances, without loss. Since the filter networks are designed with reactances, any energy accepted will be transmitted to the load in the pass band of frequencies; in the stop band no net energy can be accepted by the filter and none appears at the load.

The *constant-k filter* was historically first, being developed by G. A. Campbell in 1922. While relatively imperfect, it will serve to introduce the study; it will be followed by a more modern treatment.

10.1 Characteristic impedance of symmetrical networks

Filter networks are ordinarily designed as symmetrical two-port T or π networks, as in Fig. 10.1. The branch notation is customary, coming from the view that a T or π is the result of a series connection of identical L half-sections, as shown.

We have used reactive T or π networks to transform a load $\mathbf{Z}_{oT}$ into a suitable input value $\mathbf{Z}_{1\,in}$ (Fig. 10.2). The network obtained was dependent on $\mathbf{Z}_{oT} > \mathbf{Z}_{1\,in}$ or $\mathbf{Z}_{oT} < \mathbf{Z}_{1\,in}$. Suppose we now require that $\mathbf{Z}_{oT}$ be a load such that $\mathbf{Z}_{1\,in} = \mathbf{Z}_{oT}$, or that the transformation ratio of the network be unity. Such a load $\mathbf{Z}_{oT}$ is called the *iterative or characteristic impedance,* and when used to terminate a passive two-port network will cause the input impedance also to be $\mathbf{Z}_{oT}$.

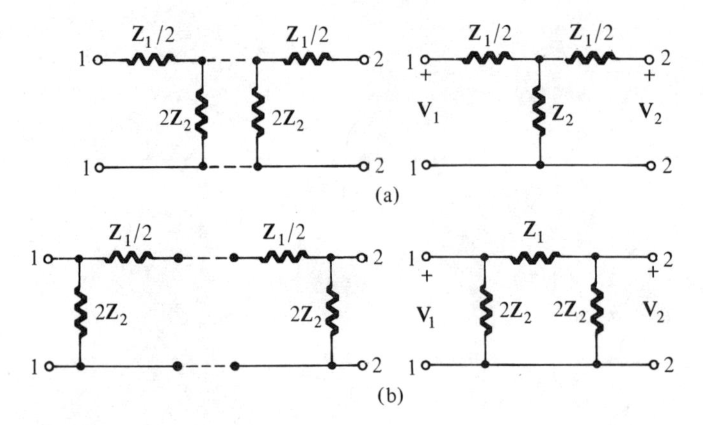

FIGURE 10.1

T and π networks derived from *L* sections.

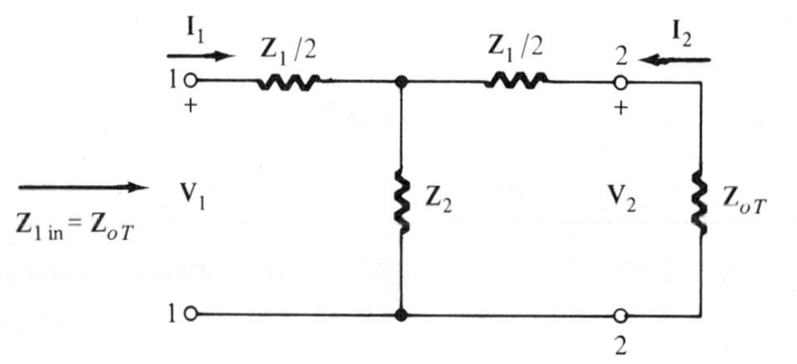

FIGURE 10.2

Terminated *T* network.

The value of $\mathbf{Z}_{oT}$ for the *T* network of Fig. 10.2 can be written by requiring $\mathbf{Z}_{1\,in}$ to equal $\mathbf{Z}_{oT}$, as

$$\mathbf{Z}_{1\,in} = \mathbf{Z}_{oT} = \frac{\mathbf{Z}_1}{2} + \frac{\mathbf{Z}_2(\mathbf{Z}_1/2 + \mathbf{Z}_{oT})}{\mathbf{Z}_1/2 + \mathbf{Z}_2 + \mathbf{Z}_{oT}} \tag{10.1}$$

$$\mathbf{Z}_{oT}^2 = \mathbf{Z}_1\mathbf{Z}_2 + \frac{\mathbf{Z}_1^2}{4}$$

and

$$\mathbf{Z}_{oT} = \sqrt{\mathbf{Z}_1\mathbf{Z}_2 + \frac{\mathbf{Z}_1^2}{4}} = \sqrt{\mathbf{Z}_1\mathbf{Z}_2\left(1 + \frac{\mathbf{Z}_1}{4\mathbf{Z}_2}\right)} \tag{10.2}$$

is the characteristic impedance for the symmetrical *T* network. In a similar manner we can find

$$\mathbf{Z}_{o\pi} = \sqrt{\frac{\mathbf{Z}_1\mathbf{Z}_2}{1 + \dfrac{\mathbf{Z}_1}{4\mathbf{Z}_2}}} \tag{10.3}$$

for the symmetrical π network.

From the general two-port network theory the value of $\mathbf{z}_i$ for Fig. 10.2 is

$$\mathbf{z}_i = \frac{\mathbf{Z}_1}{2} + \mathbf{Z}_2$$

using an open termination at the 2 port. Likewise, using a short-circuit termination,

$$\mathbf{y}_i = \frac{2}{\mathbf{Z}_1 + [\mathbf{Z}_1\mathbf{Z}_2/(\mathbf{Z}_1/2 + \mathbf{Z}_2)]}$$

Combining these expressions,

$$\frac{\mathbf{z}_i}{\mathbf{y}_i} = \mathbf{Z}_1\mathbf{Z}_2 + \frac{\mathbf{Z}_1^2}{4} = \mathbf{Z}_{oT}^2 \tag{10.4}$$

and so we have

$$\mathbf{Z}_{oT} = \sqrt{\frac{\mathbf{z}_i}{\mathbf{y}_i}} \tag{10.5}$$

This result is a general relation. It provides an experimental means of determining the $\mathbf{Z}_o$ of any symmetrical network.

10.2 Properties of the characteristic impedance; pass and stop bands

We have previously required that our transformation networks be composed of pure reactances and will impose the same condition on these filter networks. With reactive elements the characteristic impedance must be either pure resistance or pure reactance, depending on whether the radicand of Eq. 10.2 or Eq. 10.3 is a positive or a negative real number. The product $\mathbf{Z}_1\mathbf{Z}_2 = -X_1X_2$ and $\mathbf{Z}_1^2/4 = -X_1^2/4$, the sign of the radicand depending on the reactance types of X_1 and X_2 and their magnitudes.

We study filter action by assuming the network to be terminated in a real $\mathbf{Z}_{oT} = R_k$. If the value of $\mathbf{Z}_{1\,in} = \mathbf{Z}_{oT}$ is real and positive, then the network will accept power from the source at the 1 port; being internally reactive, the network must deliver this power to the R_k load without loss. The condition of $\mathbf{Z}_{oT}$ real and positive constitutes a *pass band*. If the value of $\mathbf{Z}_{1\,in} = \mathbf{Z}_{oT}$ is reactive, the network represents a pure reactance at the 1 port and can accept no power; therefore a reactive $\mathbf{Z}_{oT}$ at the 1 port is a condition for a *stop band* of frequencies.

With reactive elements, Eq. 10.2 becomes

$$\mathbf{Z}_{oT} = \sqrt{-(X_1X_2)\left(1 + \frac{X_1}{4X_2}\right)} \tag{10.6}$$

with the reactances carrying their own signs. A *pass band* will exist if

1. X_1 and X_2 are of opposite reactance type, making X_1X_2 negative above,
2. $-1 < X_1/4X_2 < 0$, since $X_1/4X_2$ will have a negative sign.

Placing these conditions in Eq. 10.6 makes the radicand positive, $Z_{1\,in} = Z_{oT} = R_k$, and the filter network will accept power at the 1 port.

A *stop band* of frequencies exists with

1. X_1 and X_2 of either reactance type,
2. $X_1/4X_2 < -1$.

The radicand is negative and the impedance $Z_{1\,in}$ is reactive; no power can be accepted by the filter at the 1 port.

Similar reasoning may be applied to the π network, where

$$\mathbf{Z}_{o\pi} = \frac{\mathbf{Z}_1\mathbf{Z}_2}{\mathbf{Z}_{oT}} \tag{10.7}$$

With $\mathbf{Z}_1\mathbf{Z}_2$ real for $\mathbf{Z}_1$ and $\mathbf{Z}_2$ as reactances, the conditions for real or reactive $\mathbf{Z}_{o\pi}$ values follow as for the T networks.

Reactances X_1 and X_2 are not necessarily single elements but may be the net values of series or parallel combinations of L and C. Because of the dependence of these filters on $\mathbf{Z}_1\mathbf{Z}_2 = k^2$, where k is independent of frequency, these filters have become known as *constant-k filters.*

10.3 The attenuation and phase constants

Using the two-port variables for the T network terminated in $\mathbf{Z}_o$, we have

$$\mathbf{V}_1 = \mathbf{Z}_{oT}\mathbf{I}_1$$
$$\mathbf{V}_2 = -\mathbf{Z}_{oT}\mathbf{I}_2$$

and the gain functions are equal:

$$\frac{\mathbf{V}_1}{\mathbf{V}_2} = -\frac{\mathbf{I}_1}{\mathbf{I}_2} \tag{10.8}$$

When terminated in $\mathbf{Z}_o$, the gain functions are phasor ratios which may be written in exponential form,

$$\frac{\mathbf{I}_1}{\mathbf{I}_2} = \varepsilon^{\gamma} = \varepsilon^{\alpha + j\beta} \tag{10.9}$$

where γ is known as the *propagation constant*, α is the *attenuation constant* in *nepers*,* and β is the *phase constant* in radians. Filter performances will be compared in terms of attenuation of current or voltage and the concurrent phase shift.

With the current division factor used on the circuit of Fig. 10.2, we can write

$$-\mathbf{I}_2 = \frac{\mathbf{Z}_2}{(\mathbf{Z}_1/2) + \mathbf{Z}_2 + \mathbf{Z}_{oT}}\mathbf{I}_1$$

$$\varepsilon^{\gamma} = \frac{\mathbf{I}_1}{\mathbf{I}_2} = \frac{(\mathbf{Z}_1/2) + \mathbf{Z}_2 + \mathbf{Z}_{oT}}{\mathbf{Z}_2} = 1 + \frac{\mathbf{Z}_1}{2\mathbf{Z}_2} + \sqrt{\frac{\mathbf{Z}_1}{\mathbf{Z}_2} + \left(\frac{\mathbf{Z}_1}{2\mathbf{Z}_2}\right)^2} \tag{10.10}$$

after using Eq. 10.2 and neglecting the minus sign on $\mathbf{I}_2$.

We can take the logarithm of both sides and evaluate the propagation constant as

$$\gamma = \ln\left[1 + \frac{\mathbf{Z}_1}{2\mathbf{Z}_2} + \sqrt{\frac{\mathbf{Z}_1}{\mathbf{Z}_2} + \left(\frac{\mathbf{Z}_1}{2\mathbf{Z}_2}\right)^2}\right] \tag{10.11}$$

where the radicand may be negative. We note that the logarithm of a complex quantity is

$$\gamma = \ln|\mathbf{Z}|\varepsilon^{j\beta} = \ln|\mathbf{Z}| + j\beta = \alpha + j\beta \tag{10.12}$$

10.4 The constant-*k* low-pass filter

Let us consider the reactance network of Fig. 10.3; then

$$\mathbf{Z}_1\mathbf{Z}_2 = k^2 = \frac{L_1}{C_2}$$

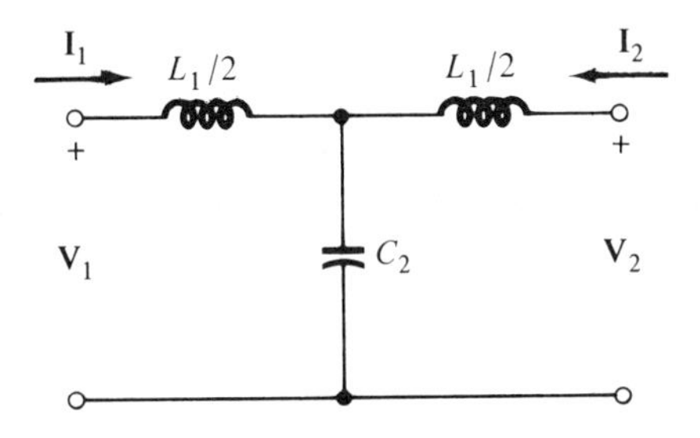

FIGURE 10.3

Low-pass filter of *T* form.

*The relation of nepers to decibels is apparent, since both are logarithmic power ratios, differing only in the logarithmic base. Two voltages differ by 1 neper when one of them is ε times the other. Thus, 1 neper = 8.69 decibels.

Since this result is independent of frequency, we have a constant-k filter.

For a pass band with a filter of opposite reactance types we must have

$$-1 < \frac{\mathbf{Z}_1}{4\mathbf{Z}_2} < 0 \tag{10.13}$$

Inserting the reactances at the zero limit,

$$-\frac{\omega_c^2 L_1 C_2}{4} = 0 \tag{10.14}$$

and so $f_c = 0$, or the pass band starts at zero frequency. Using the other limit of Eq. 10.13,

$$-\frac{\omega_c^2 L_1 C_2}{4} = -1$$

$$\omega_c = \frac{2}{\sqrt{L_1 C_2}} \qquad f_c = \frac{1}{\pi\sqrt{L_1 C_2}} \tag{10.15}$$

we have the upper cutoff frequency of the low-pass filter.

Inserting reactance values for the low-pass filter in Eq. 10.10 gives

$$\varepsilon^\gamma = 1 - \frac{\omega^2 L_1 C_2}{2} + \sqrt{-\omega^2 L_1 C_2 + \left(\frac{\omega^2 L_1 C_2}{2}\right)^2}$$

Using Eq. 10.15 for the cutoff frequency, this becomes

$$\varepsilon^\gamma = 1 - \frac{2f^2}{f_c^2} + \frac{2f}{f_c}\sqrt{\frac{f^2}{f_c^2} - 1} \tag{10.16}$$

In the pass band $f < f_c$, and the radicand will be negative; the propagation constant then has real and quadrature parts. That is,

$$\varepsilon^\gamma = \varepsilon^{\alpha + j\beta} = 1 - \frac{2f^2}{f_c^2} + j\frac{2f}{f_c}\sqrt{1 - \frac{f^2}{f_c^2}} \tag{10.17}$$

The logarithm of the magnitude of this expression is α; taking the magnitude,

$$\varepsilon^\alpha = \left[1 - \frac{4f^2}{f_c^2} + \frac{4f^4}{f_c^4} + \frac{4f^2}{f_c^2} - \frac{4f^4}{f_c^4}\right]^{1/2} = 1 \tag{10.18}$$

Then

$$\alpha = \ln 1 = 0$$

for the attenuation in the pass band.

While we have zero attenuation in the pass band, there is a progressive phase shift across the band. The phase shift β can be found by noting that the cosine of the phase angle of Eq. 10.17 is given by the real part, divided by the magnitude, which is unity. Then

$$\beta = \cos^{-1}\left(1 - \frac{2f^2}{f_c^2}\right) \tag{10.19}$$

The phase shift changes from $\beta = 0$ at zero frequency to $\beta = \pi$ at f_c or cutoff. In the stop band the phase angle remains at π radians.

In the stop band $f/f_c > 1$, the radical in Eq. 10.16 becomes real, and the equation is

$$\varepsilon^{\gamma} = 1 - \frac{2f^2}{f_c^2} + \frac{2f}{f_c}\sqrt{\frac{f^2}{f_c^2} - 1}$$

and being real

$$\alpha = \ln\left[1 - \frac{2f^2}{f_c^2} + \frac{2f}{f_c}\sqrt{\frac{f^2}{f_c^2} - 1}\right] \tag{10.20}$$

which increases in magnitude as f increases. Theoretical curves of attenuation and phase shift for the low-pass filter are shown in Fig. 10.4.

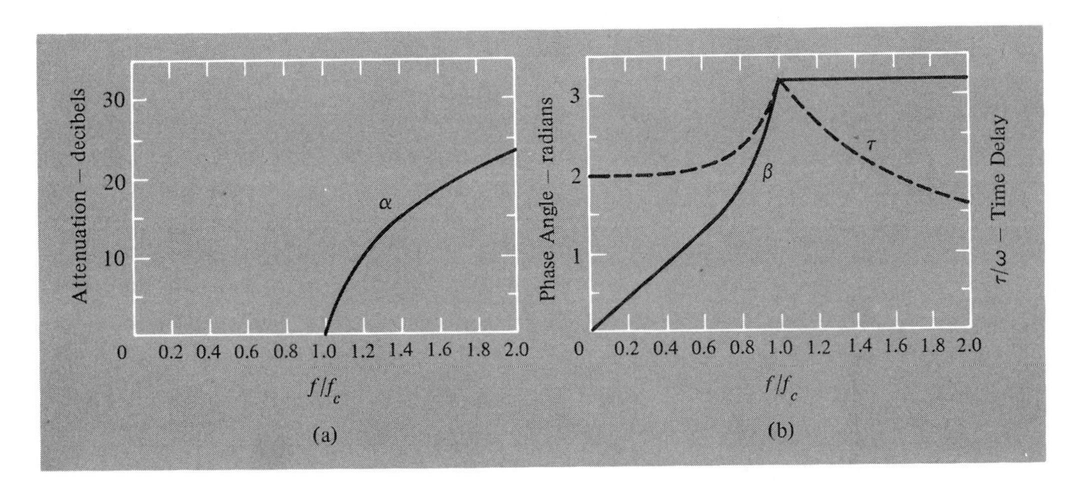

FIGURE 10.4
(a) Attenuation of low-pass constant-k filter; (b) phase shift and time delay.

A phase shift β, increasing linearly with f, will not cause wave-form distortion since all frequency components will have a time delay $\tau = \beta/\omega$, which is independent of frequency since β and ω increase linearly with f.

With reactances written for the low-pass case, as functions of frequency, the values of Z_o become

$$Z_{oT} = \sqrt{\frac{L_1}{C_2}}\sqrt{1 - \frac{f^2}{f_c^2}} \tag{10.21}$$

$$Z_{o\pi} = \sqrt{\frac{L_1}{C_2}}\sqrt{\frac{1}{1 - \dfrac{f^2}{f_c^2}}} \tag{10.22}$$

At zero frequency we designate a terminating resistance R_k as

$$R_k = \sqrt{\frac{L_1}{C_2}} \tag{10.23}$$

From this relation and the cutoff frequency we develop the low-pass filter design relations:

$$L_1 = \frac{R_k}{\pi f_c} \qquad C_2 = \frac{1}{\pi f_c R_k} \tag{10.24}$$

It is apparent from Eqs. 10.21 and 10.22 that the characteristic impedance is not a constant over the pass band and that the filter design conditions are not met when the filter is terminated in a constant resistance $- R_k$, except at zero frequency. This situation will be further discussed in Section 10.6.

10.5 Normalized models

The normalizing of filter networks to desired parameter levels and cutoff frequencies leads to general models which can be adjusted to any design specifications by use of the scaling factors of Section 4.10.

The constant-k low-pass filter might be normalized to a cutoff frequency of 1 r/s $= \omega_c$ and to an R_k design value of 1 Ω. Then, writing Eqs. 10.15 and 10.23,

$$\omega_c = \frac{2}{\sqrt{L_1 C_2}} = 1 \tag{10.25}$$

$$R_k = \sqrt{\frac{L_1}{C_2}} = 1 \tag{10.26}$$

Solution of these equations for the two filter parameters gives

$$L_1 = 2 \text{ H}$$

$$C_2 = 2 \text{ F}$$

and the normalized model of a low-pass constant-k filter appears in Fig. 10.5(b).

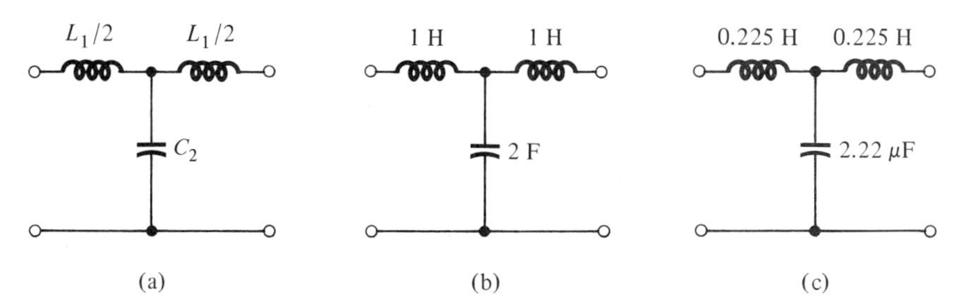

FIGURE 10.5

(a) Low-pass filter; (b) low-pass filter normalized at $\omega_c = 1$ r/s, $R_k = 1\ \Omega$; (c) denormalized to $\omega_c = 2000$ r/s, $R_k = 450\ \Omega$.

Conversion to a specific design follows from Section 4.10, where we designated scaled network parameters by primes, as

$$\omega' = q\omega$$

$$R' = pR$$

$$L' = \frac{p}{q}L$$

$$C' = \frac{1}{pq}C$$

Use of these factors is shown by an example.

EXAMPLE. We need a constant-k filter for low-pass use, with a cutoff angular frequency of 2000 r/s, to operate into a resistance load of 450 Ω. The model is that of Fig. 10.5(b).

With the primed quantities referring to the normalized model,

$$q = \frac{\omega'}{\omega} = \frac{1}{2000}$$

$$p = \frac{R'}{R} = \frac{1}{450}$$

Then

$$\frac{L_1}{2} = \frac{q}{p}\frac{L_1'}{2} = \frac{1}{2000}\frac{450}{1} \times 1\ \text{H}$$

$$= 0.225\ \text{H}$$

$$C_2 = pqC'_2 = \frac{1}{450}\frac{1}{2000} \times 2\text{ F}$$

$$= 2.22 \times 10^{-6}\text{ F} = 2.22\ \mu\text{F}$$

and we have the design parameters needed for the required filter characteristics in the circuit of Fig. 10.5(c).

10.6 Frequency transformations for other filter characteristics

By interchange of the position of L and C, the low-pass filter can be transformed to a high-pass filter. A development similar to that of Section 10.4 could be carried through, but a transformation of frequency may be employed instead.

Equation 10.20 shows that attenuation is an even function and Fig. 10.6(a) illustrates this, the pass band being extended to a negative frequency

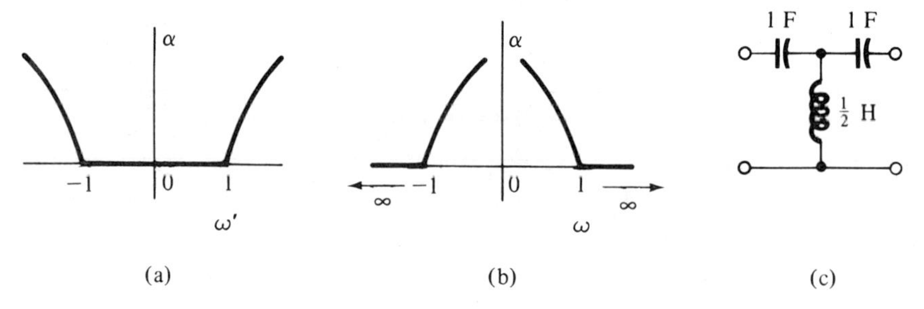

FIGURE 10.6

(a) Attenuation of normalized low-pass filter as an even function; (b) transformation to the high-pass filter; (c) normalized high-pass circuit.

cutoff, normalized at −1. Physically, the negative frequencies have no meaning, but the band is mathematically important in the transformations to be employed.

Quantities with primes will refer to the normalized low-pass model from which we work. The low-pass circuit can be transformed to a high-pass circuit by inverting frequencies, using

$$j\omega = \frac{1}{j\omega'} \tag{10.27}$$

The cutoff of the pass band of the low-pass filter inverts to

$$j\omega_c = \frac{1}{j\omega'_c} = \frac{-j}{1} = -j1$$

and the cutoff of the negative band transforms to

$$j\omega_c = \frac{1}{j\omega_c'} = \frac{-j}{-1} = j1$$

We can transform the zero frequency point in the pass band of Fig. 10.6(a) as

$$j\omega = \frac{-j}{0} = -j\infty$$

and

$$j\omega = \frac{-j}{-0} = j\infty$$

The pass bands of the normalized low-pass filter of (a) transform into the pass bands for the normalized high-pass filter at (b); we then reject operation in the negative frequencies.

To obtain the transformed filter elements, we employ the transformation of Eq. 10.27 for inductive elements as

$$j\omega' L' = \frac{L'}{j\omega} = \frac{-j}{\omega(1/L')} = \frac{-j}{\omega C}$$

which represents the reactance of a capacitance:

$$C = \frac{1}{L'} \qquad F \tag{10.28}$$

Capacitive reactance of the $1/j\omega' C'$ form becomes

$$\frac{1}{j\omega' C'} = \frac{j\omega}{C'} = j\omega L$$

which represents the reactance of an inductance of value

$$L = \frac{1}{C'} \qquad H \tag{10.29}$$

The low-pass to high-pass transformation yields the normalized high-pass filter of Fig. 10.6(c).

Band-pass filter characteristics can be derived by use of the transformation

$$j\omega' = j\omega + \frac{\omega_o^2}{j\omega} \tag{10.30}$$

or

$$j\omega = \frac{j\omega'}{2} \pm \sqrt{\left(\frac{j\omega'}{2}\right)^2 - \omega_o^2} \tag{10.31}$$

where ω_o will be identified. The normalized low-pass model has frequency cutoffs at $\omega_c = \pm 1$. The corresponding cutoff frequency of the transformed filter for $\omega_c' = -1$ will be

$$\begin{aligned} j\omega_c &= \frac{j(-1)}{2} \pm \sqrt{-\left(\frac{1}{2}\right)^2 - \omega_o^2} \\ &= j(-0.5 \pm \sqrt{0.25 + \omega_o^2}) \end{aligned} \tag{10.32}$$

The other cutoff occurs for $\omega_c = +1$, and this yields

$$j\omega_c = j(0.5 \pm \sqrt{0.25 + \omega_o^2}) \tag{10.33}$$

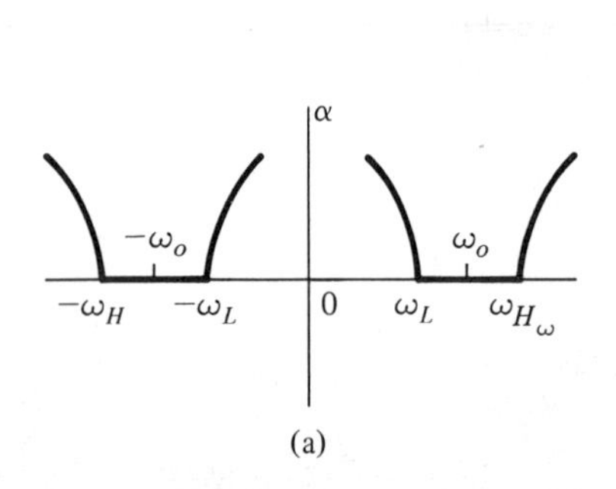

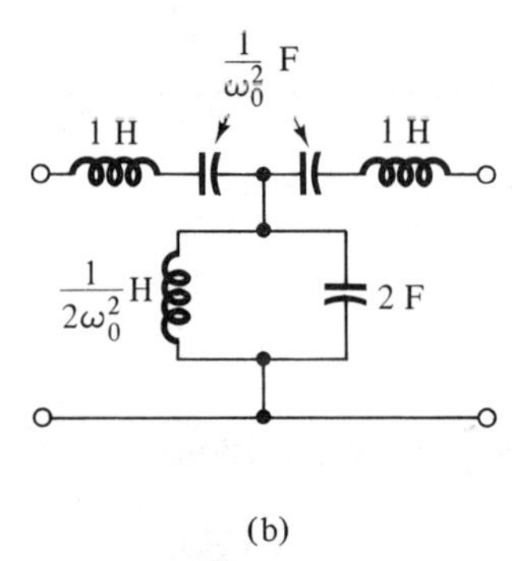

FIGURE 10.7
(a) Attenuation of normalized band-pass filter; (b) normalized band-pass circuit; BW = 1 r/s, $R_k = 1\ \Omega$.

These cutoff frequencies bound the two pass bands of Fig. 10.7(a). Again rejecting the negative band, we can write the lower cutoff frequency as ω_L,

$$\omega_L = (0.5 - \sqrt{0.25 + \omega_o^2}) \tag{10.34}$$

and the higher cutoff frequency as

$$\omega_H = (0.5 + \sqrt{0.25 + \omega_o^2}) \tag{10.35}$$

The frequency ω_o can now be identified by multiplying Eqs. 10.34 and 10.35, yielding

$$\omega_o = \pm\sqrt{\omega_L \omega_H} \tag{10.36}$$

The frequency ω_o is the geometric mean of the cutoff frequencies. The bandwidth of the filter is found as

$$\omega_H - \omega_L = 1 \tag{10.37}$$

and the attenuation characteristics of the band-pass filter appear in Fig. 10.7(a), but disregarding the negative-frequency region.

An impedance of $j\omega' L'\ \Omega$ in the low-pass model must be replaced in the band-pass model by a series-connected branch

$$j\omega' L' = j\omega L + \frac{\omega_o^2 L}{j\omega} = j\omega L - \frac{j}{(\omega/\omega_o^2)(1/L)} = j\omega L - \frac{j}{\omega C}$$

where

$$L = L' \qquad \text{H} \tag{10.38}$$

$$C = \frac{1}{\omega_o^2 L'} \qquad \text{F} \tag{10.39}$$

as a series impedance.

Also, an admittance of $j\omega' C'$ mhos in the shunt branch of the low-pass model must be replaced in the band-pass model by

$$j\omega' C' = j\omega C + \frac{\omega_o^2 C}{j\omega} = j\omega C - \frac{j}{\omega(1/\omega_o^2 C)} = j\omega C - \frac{j}{\omega L}$$

Then we have a parallel circuit for the shunt branch of the filter, comprising

$$C = C' \qquad \text{F} \tag{10.40}$$

$$L = \frac{1}{\omega_o^2 C'} \qquad \text{H} \tag{10.41}$$

and the normalized band-pass circuit of Fig. 10.7(b) follows.

10.7 Variation of Z_o over the pass-band

In Eqs. 10.21 and 10.22 we gave the values of $\mathbf{Z}_{oT}$ and $\mathbf{Z}_{o\pi}$ for the low-pass constant-k filter:

$$\mathbf{Z}_{oT} = \sqrt{\frac{L_1}{C_2}}\sqrt{1 - \frac{f^2}{f_c^2}} = R_k\sqrt{1 - \frac{f^2}{f_c^2}} \tag{10.42}$$

$$\mathbf{Z}_{o\pi} = \sqrt{\frac{L_1}{C_2}}\sqrt{\frac{1}{1-\dfrac{f^2}{f_c^2}}} = R_k\sqrt{\frac{1}{1-\dfrac{f^2}{f_c^2}}} \tag{10.43}$$

The variation of these parameters over the pass band is shown in Fig. 10.8. Neither type of network is able to produce the theoretical impedance match

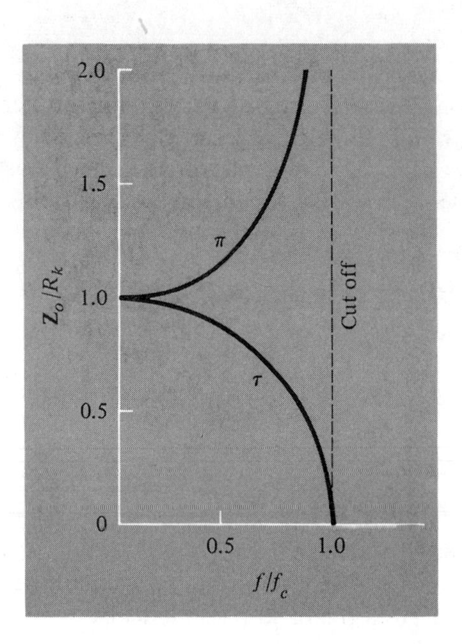

FIGURE 10.8

Variation of $\mathbf{Z}_o$ over the pass band, constant-*k* section.

to a constant resistive load which underlies this filter theory, except at zero frequency. The resultant cutoff is not sharp, and the attenuation does not reach the values expected.

The problem seems to be that of finding a matching network which will change its characteristics with frequency so as to give an approximate match to a filter over the whole of the pass band. In 1923, Otto Zobel proposed a matching network with desirable terminal characteristics by using a network model in which the series arm was scaled to the original constant-*k* *T* filter as

$$\frac{\mathbf{Z}_1}{2} = \frac{m\mathbf{Z}_1'}{2} \tag{10.44}$$

in which the primed elements refer to the original, or *prototype*, network. The multiplier m had a value $0 < m < 1$; it was also required that the so-called *m-derived network* have the same input impedance as the modeled constant-k network. This made possible the use of constant-k and m-derived sections in cascade.

Using Eq. 10.2 for $\mathbf{Z}_{1\,in} = \mathbf{Z}_{oT}$, we meet the latter requirement by making

$$\mathbf{Z}_1'\mathbf{Z}_2' + \frac{\mathbf{Z}_1'^2}{4} = \mathbf{Z}_1\mathbf{Z}_2 + \frac{\mathbf{Z}_1^2}{4} \tag{10.45}$$

Then using $m\mathbf{Z}_1'/2 = \mathbf{Z}_1/2$, we obtain the shunt arm impedance:

$$\mathbf{Z}_2 = \frac{(1 - m^2)}{4m}\mathbf{Z}_1' + \frac{\mathbf{Z}_2'}{m} \tag{10.46}$$

The shunt arm of the m-derived section must be composed of two series elements, representing the series element of the prototype scaled by $(1 - m^2)/4m$, in series with the shunt element of the prototype scaled by $1/m$.

Zobel found that half-sections of m-derived form, as in Fig. 10.9, had $\mathbf{Z}_{1\,in}$ as a function of m and with $m = 0.6$ provided an approximation to the

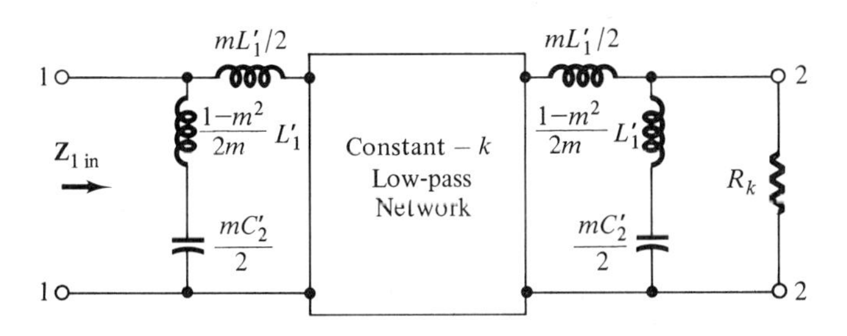

FIGURE 10.9

Use of m-derived half-sections for matching.

desired constant input resistance R_k over most of the pass band. For a low-pass filter, the input impedance of the circuit in Fig. 10.9 can be derived as

$$\mathbf{Z}_{1\,in} = \frac{R_k[1 - (1 - m^2)f^2/f_c^2]}{\sqrt{1 - f^2/f_c^2}} \tag{10.47}$$

This relation is plotted in Fig. 10.10 for several values of m and a π form of network, as the circuit of Fig. 10.9 has become when viewed from port to port. If we select $m = 0.6$, a value of $Z_{1\,in} \cong R_k$ can be obtained over about 85 percent of the pass band, and a source impedance of $R_k\,\Omega$ is matched over that range. A similar variation with m could be found for a high-pass π network.

However, the frequencies at which the m-derived matching section fails to match the load are those most critical to filter performance, near the frequency of cutoff. Even when the two-element arm of the m-derived section is chosen to provide a frequency of high attenuation at series resonance and near cutoff, the attenuation in the stop band does not reach the levels desired,

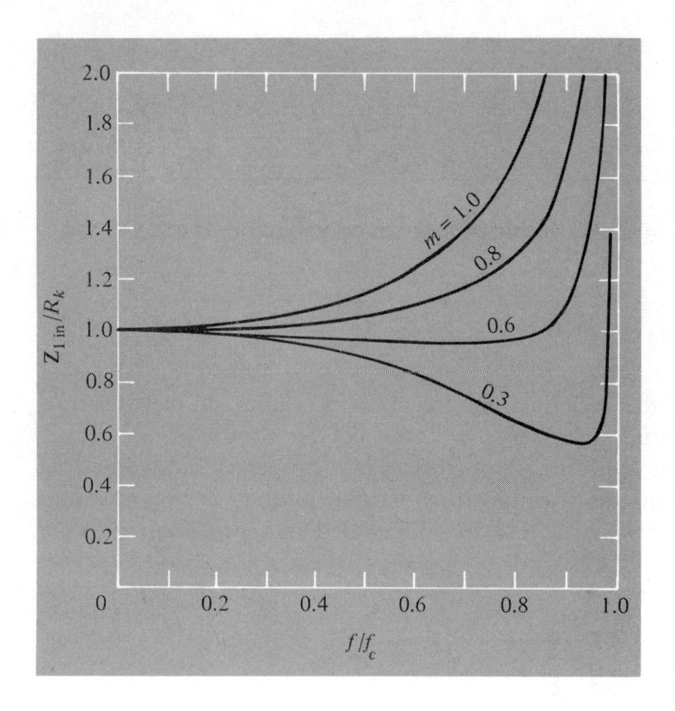

FIGURE 10.10

Dependence of $\mathbf{Z}_{o\pi}$ on m of an m-derived section.

considering the large number of circuit elements employed. A theoretical curve of attenuation is shown in Fig. 10.11 for the m-derived section, when cascaded at each end of a constant-k prototype network.

To achieve results closer to those desired, new design methods have been developed.

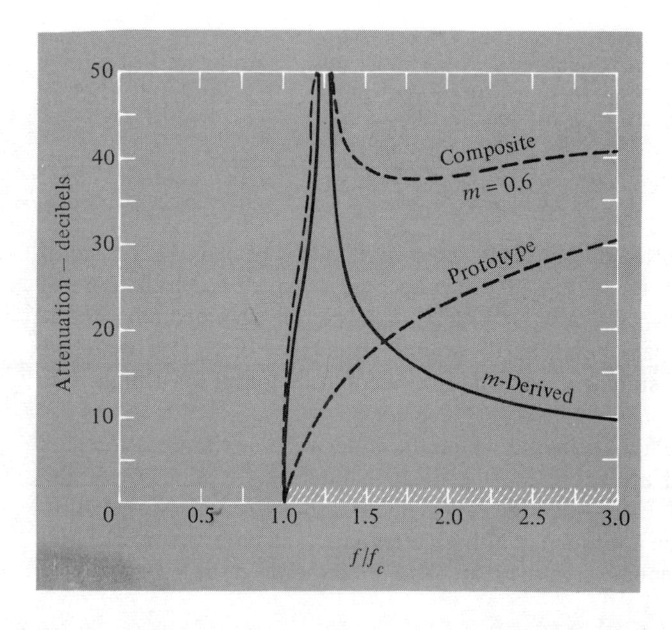

FIGURE 10.11

Attenuation for composite filter.

10.8

Modern filter theory

Constant-k filter theory was based on the concept of iterated networks, as found in telephone practice, with the load on any section provided by the iterated impedance of all the following sections. In practical situations, sections terminated with a fixed resistive load did not meet the design criteria and performance fell below expectations. Modern filter design is based on the writing of a transfer function which approximates that of an ideal filter as nearly as desired. We find the left half-plane poles of the chosen filter function and synthesize a circuit having the same pole locations. We are able to specify the terminations and to include them in the design process.

In Chapter 7 we employed transfer functions in the form of ratios of polynomials in s:

$$G(s) = A\frac{s^p + a_1 s^{p-1} + \cdots + a_p}{s^n + b_1 s^{n-1} + \cdots + b_n} \tag{10.48}$$

This function may be factored to give

$$G(s) = A\frac{(s + p_1)(s + p_2)\cdots(s + p_p)}{(s + n_1)(s + n_2)\cdots(s + n_n)} \tag{10.49}$$

Each numerator term represents a zero as $s_1 = p_1, s_2 = p_2 \cdots$ with p zeros present. The roots of the denominator lead to poles of the function, with n poles present.

All roots of a polynomial with real coefficients are real or occur in complex conjugate pairs. As discussed, all poles lie in the left half of the s plane or on the imaginary axis. Thus the choice of pole locations for our circuits is bounded.

For the transfer function of Eq. 10.48, there are $p + n + 1$ choices of coefficients which serve to establish the values of the circuit elements. These choices are p for the numerator, n for the denominator, and 1 to establish A. We would like to achieve the ideal rectangular response of (a) in Fig. 10.12 for the low-pass case, but since the number of circuit elements must be limited for economic reasons, such an ideal response can only be approximated, and we have a number of alternatives.

The *maximally flat* or Butterworth response of (b) in Fig. 10.12 is obtained by making as many derivatives of $|G(j\omega)|$ go to zero at $\omega = 0$ as possible; this choice favors uniformity of response at the lower frequencies. Design for a *maximally flat phase response* leads to a Bessel filter, but it has poorer pass-band characteristics. The *Chebyshev approximation* of (c) in Fig. 10.12 allows ripples in the pass-band attenuation, but the amplitude of the ripples can be controlled. *Elliptic function filters* are similar, but the design produces ripples in attenuation in both pass and stop bands. It yields even faster transition from pass band to stop band at cutoff.

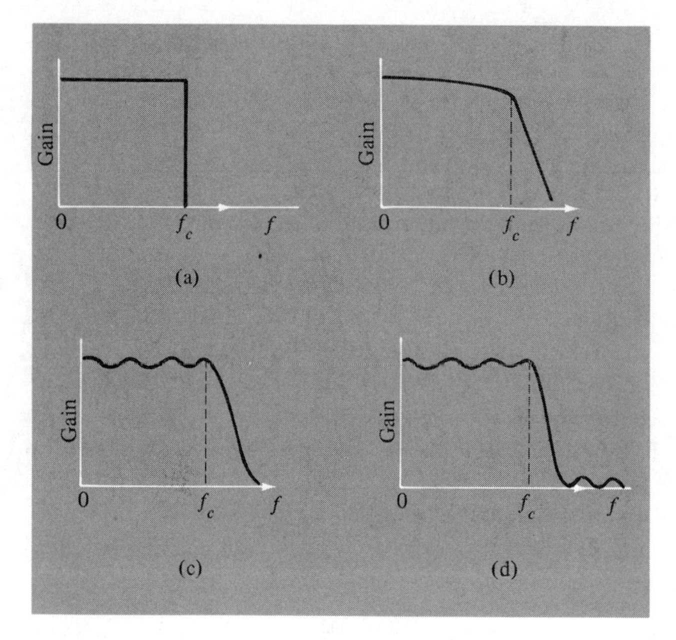

FIGURE 10.12

(a) Ideal filter response; (b) maximally flat response (Butterworth); (c) Chebyshev response; (d) elliptical function response.

We shall concentrate here on the maximally flat and the Chebyshev forms, as the most generally applied. The previously discussed filter transformations can be applied to generate high-pass and band-pass forms from basic low-pass circuit designs, and only the latter will be discussed.

10.9 First-order filters

We begin our analysis with the simplest networks yielding transfer functions of the form of Eq. 10.48; these are the one-pole networks of Fig. 10.13, where R_1 and R_2 represent the terminating resistors. The voltage transfer functions are

$$\text{(a)} \quad \frac{V_2(s)}{V_1(s)} = \frac{1}{R_1C}\left(\frac{1}{s + 1/R_pC}\right) \qquad \text{where } R_p = \frac{R_1R_2}{R_1 + R_2} \tag{10.50}$$

$$\text{(b)} \quad \frac{V_2(s)}{V_1(s)} = \frac{R_2}{L}\left(\frac{1}{s + (R_1 + R_2)/L}\right) \tag{10.51}$$

$$\text{(c)} \quad \frac{V_2(s)}{V_1(s)} = \frac{R_2}{R_1 + R_2}\left(\frac{s}{s + 1/C(R_1 + R_2)}\right) \tag{10.52}$$

$$\text{(d)} \quad \frac{V_2(s)}{V_1(s)} = \frac{R_2}{R_1 + R_2}\left(\frac{s}{s + R_p/L}\right) \qquad \text{where } R_p = \frac{R_1R_2}{R_1 + R_2} \tag{10.53}$$

The poles lie on the negative real axis. With zeros at infinity, the poles in (a) and (b) are dominant at low frequencies and these networks are low-pass in

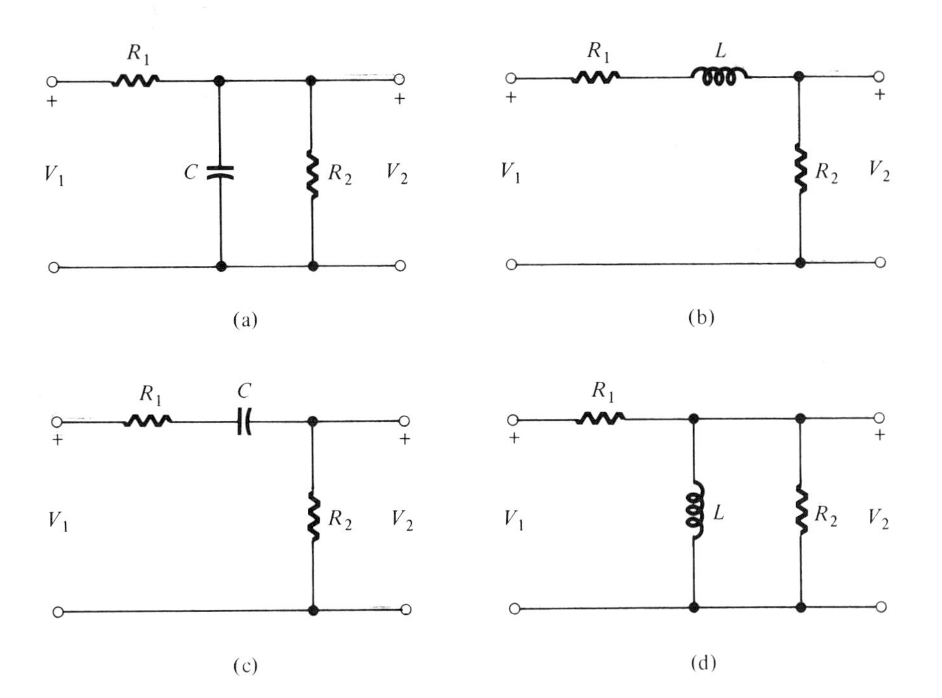

FIGURE 10.13
Single-pole filters: (a), (b) low-pass; (c), (d) high-pass.

nature. For (c) and (d) with a zero at the origin, the zero controls at low frequencies, while at high frequencies the zero and pole reduce to a constant, independent of frequency. The action is that of a high-pass circuit.

By placing $j\omega = s$ and designating the reciprocal of the time constant as ω_o, we can write the above expressions in more general form:

$$\text{Low-pass:}\quad G(\omega) = K_1 \frac{1}{1 + j\omega/\omega_o} = |G(\omega)|\varepsilon^{j\theta(\omega)} \qquad (10.54)$$

$$\theta = \tan^{-1}\frac{-\omega}{\omega_o}$$

$$\text{High-pass:}\quad G(\omega) = K_2 \frac{1}{1 - j\omega_o/\omega} = |G(\omega)|\varepsilon^{j\theta(\omega)} \qquad (10.55)$$

$$\theta = \tan^{-1}\frac{\omega_o}{\omega}$$

The symmetry of the low-pass and high-pass expressions is illustrated by the gain and phase-shift curves of Fig. 10.14.

For practical use, the attenuation well beyond cutoff should be greater than the indicated rate of -20 dB/frequency decade of the first-order low-pass circuit.

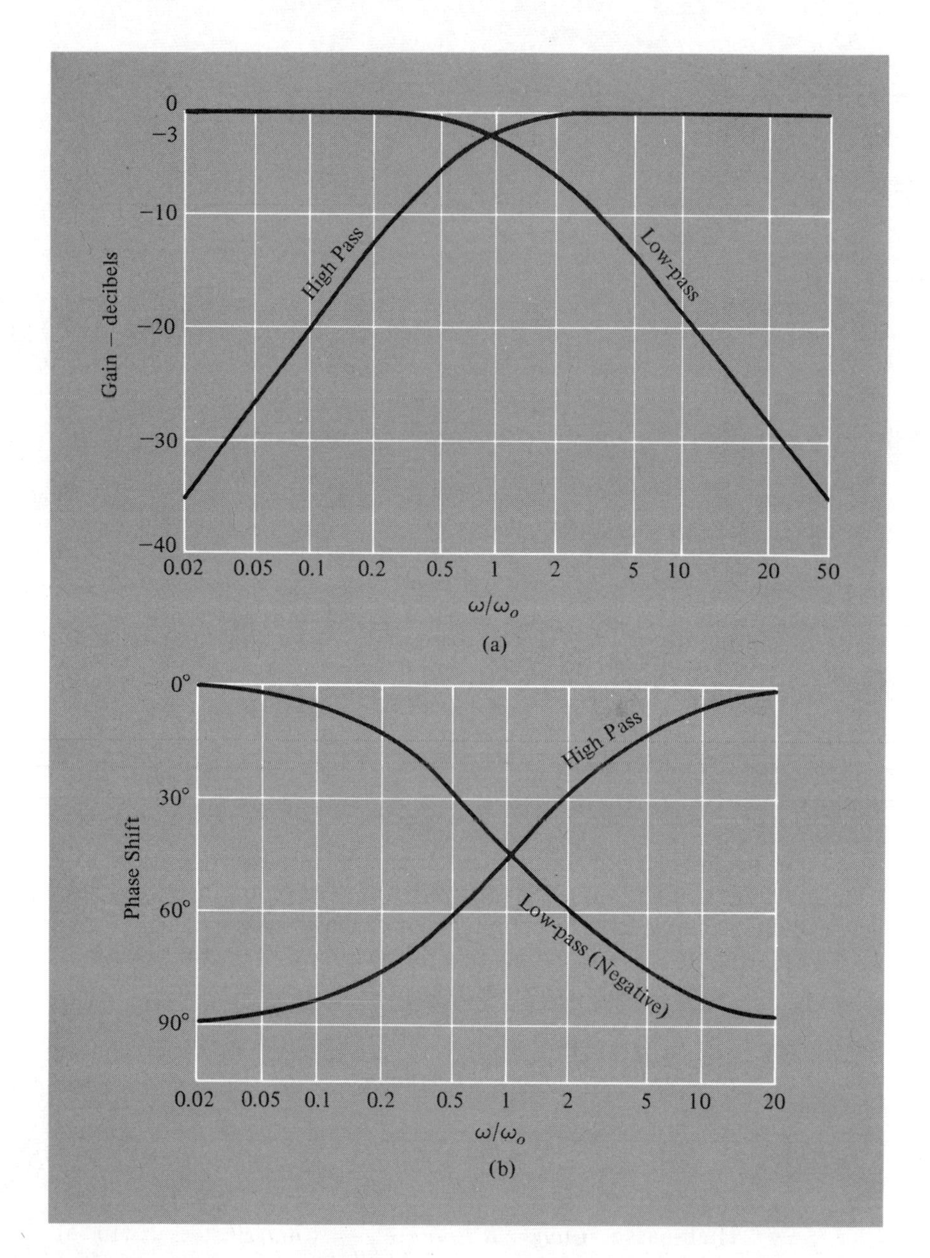

FIGURE 10.14

Gain and phase shift for one-pole networks (θ negative for low-pass case).

10.10

Second-order response

We may add an inductor to the network of (a) in the previous section and obtain the second-order low-pass filter of Fig. 10.15, having two poles. With an underdamped circuit the poles become complex conjugates

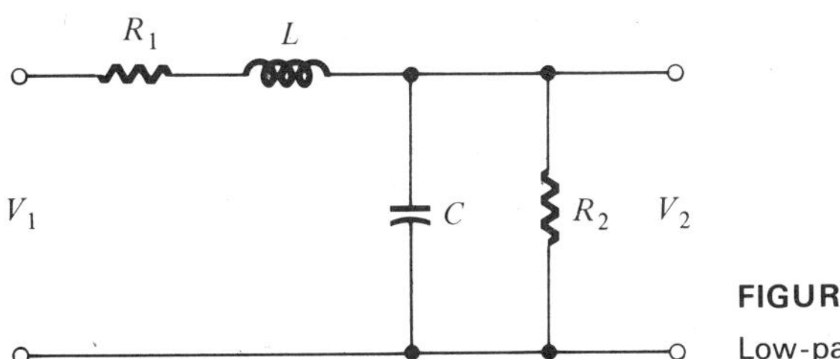

FIGURE 10.15
Low-pass second-order network.

and can be located as desired in the left half plane. From the circuit,

$$V_1(s) = \left(R_1 + sL + \frac{R_2}{1 + sCR_2}\right) I_1$$

$$V_2(s) = \left(\frac{R_2}{1 + sCR_2}\right) I_1$$

We can write

$$\frac{V_2(s)}{V_1(s)} = \frac{1}{s^2 LC + s[(L/R_2) + CR_1] + [(R_1 + R_2)/R_2]} \tag{10.56}$$

Setting $K = R_2/(R_1 + R_2)$, we have the low-pass transfer function:

$$G(s) = K\left[\frac{1/LCK}{s^2 + s[(1/CR_2) + (R_1/L)] + (1/LCK)}\right] \tag{10.57}$$

The function has a double zero at infinity and two poles in complex conjugate form.

In Section 7.9 we demonstrated that for a second-order network the coefficient of the s^0 term is ω_o^2, where ω_o is the undamped natural frequency of the network. From Eq. 10.57

$$\omega_o = \sqrt{1/LCK}$$

and

$$\omega_o^2 LCK = 1 \tag{10.58}$$

for the second-order filter. It was also shown that the coefficient of the s^1 term is ω_o/Q, and from Eq. 10.57

$$\frac{1}{CR_2} + \frac{R_1}{L} = \frac{\omega_o}{Q} \tag{10.59}$$

That is, Eq. 10.57 can be written in typical form for a *low-pass second-order network* as

$$G(s) = \frac{V_2(s)}{V_1(s)} = \frac{K\omega_o^2}{s^2 + (\omega_o/Q)s + \omega_o^2} \tag{10.60}$$

with K as a magnitude factor.

We may write $j\omega = s$ and find the steady-state amplitude and phase characteristics:

$$G(\omega) = K\left[\frac{1}{(\omega/\omega_o)^4 + (\omega/\omega_o)^2[(1 - 2Q^2)/Q^2] + 1}\right]^{1/2} \tag{10.61}$$

$$\theta = \tan^{-1}\frac{(\omega/\omega_o)(1/Q)}{1 - (\omega/\omega_o)^2} \tag{10.62}$$

The magnitude relation readily translates into decibels as

$$\begin{aligned} G_{\text{dBV}} &= -20\log\left[\left(\frac{\omega}{\omega_o}\right)^4 + \left(\frac{\omega}{\omega_o}\right)^2\frac{1 - 2Q^2}{Q^2} + 1\right]^{1/2} \\ &= -10\log\left[\left(\frac{\omega}{\omega_o}\right)^4 + \left(\frac{\omega}{\omega_o}\right)^2\frac{1 - 2Q^2}{Q_2} + 1\right] \end{aligned} \tag{10.63}$$

The parameter Q determines the manner in which the response behaves near the normalized frequency $\omega/\omega_o = 1$ and provides the usual second-order response curves in Fig. 10.16. At high frequencies all the curves approach a common asymptote, falling at -40 dBV/frequency decade. The frequency ω_o is the corner frequency and will be identified as the *cutoff frequency* of the filter; it is the -3-dBV frequency and the frequency at which the phase curves reach $-90°$. The curve for $Q = 0.5$ is the response of a second-order network of two cascaded RC sections and comparison of this curve with those for the second-order LC network illustrates the limitation in the rate of cutoff inherent in RC circuits, without other energy elements.

The frequency at which the maximum amplitude occurs is ω_m, found as

$$\omega_m = \omega_o\sqrt{1 - \frac{1}{2Q^2}} \tag{10.64}$$

The peaking curve is shown in Fig. 10.17(a), in expanded form. The peak occurs at a frequency below ω_o and exists only if $Q > 0.707$. Higher peaks are associated with higher Q values, and the peak magnitude can be calculated by

$$M_m = \frac{Q}{\sqrt{1 - (1/4Q^2)}} \tag{10.65}$$

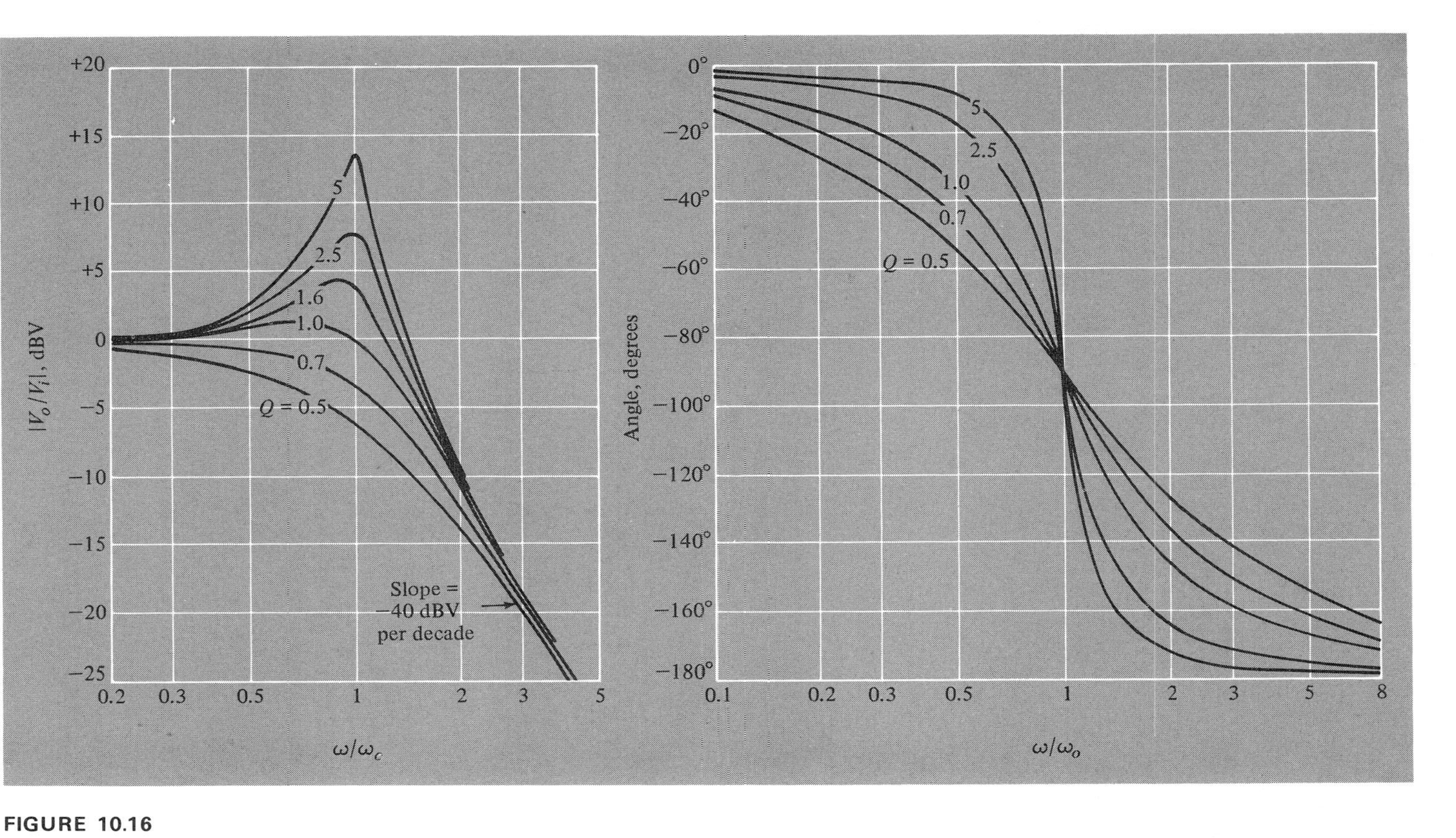

FIGURE 10.16

dBV and phase response for the second-order low-pass circuit.

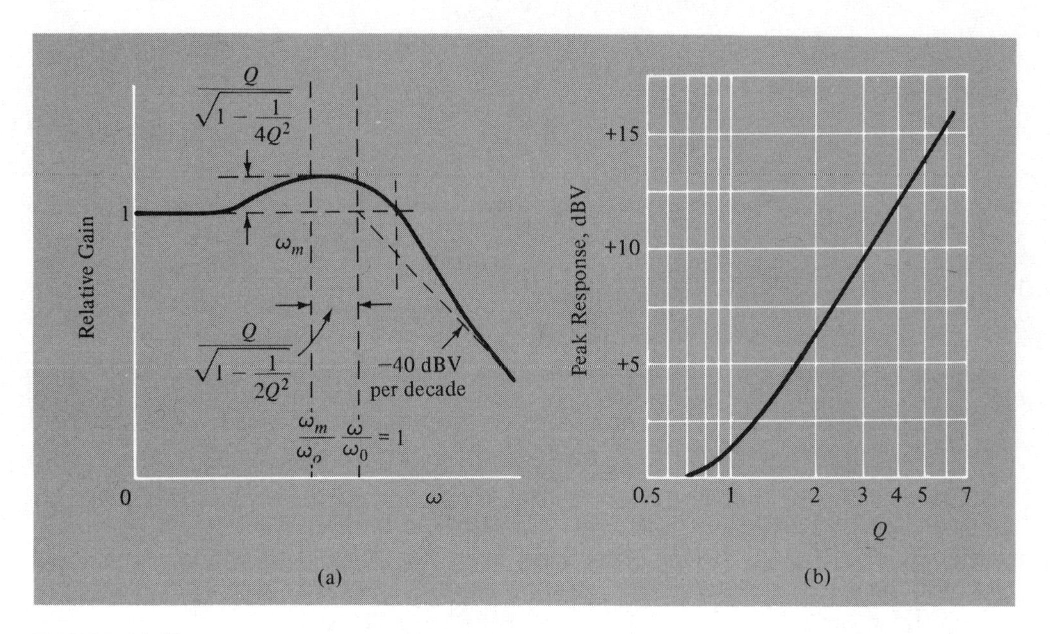

FIGURE 10.17

Study of the second-order response curves.

Higher Q values give more rapid cutoff, as well as a more rapid change of phase near cutoff but distort the pass-band response. The rate of cutoff can be increased by use of higher-order networks without such sacrifice of flatness of response in the pass band as will be shown.

10.11 The maximally flat response

The frequency response predicted by the second-order equation, Eq. 10.60, can be made maximally flat and monotonic in form. Mathematically this result can be achieved by taking successive derivatives of the magnitude expression with respect to ω^2, and setting all except the highest-order term equal to zero at $\omega = 0$ for the low-pass case. The resulting equations give specifications for the circuit parameters to meet the maximally flat requirement.

A simpler and equivalent process is to equate the numerator and denominator coefficients of equal powers of the variable ω, except for the highest-order term. This term is ω^4 in the denominator of Eq. 10.60. The coefficient of ω^2 in the denominator is $(1 - 2Q^2)/Q^2$ and the coefficient of ω^2 in the numerator is zero. We equate these coefficient values

$$\frac{1 - 2Q^2}{Q^2} = 0$$

from which

$$Q = \sqrt{\tfrac{1}{2}} = 0.707 \tag{10.66}$$

for the maximally flat response; this is confirmed in Fig. 10.16(a) as the maximally flat curve.

It is now possible to determine the necessary circuit parameters for the second-order maximally flat filter.

The terminations can be introduced through K, and we here assume $R_1 = R_2$, so that $K = \frac{1}{2}$. Using this value in Eq. 10.58 gives

$$\omega_o^2 LC = 2$$

from which

$$C = \frac{2}{\omega_o^2 L}$$

Using this relation in Eq. 10.59 with $R_1 = R_2$ and $Q = \sqrt{\frac{1}{2}}$ for maximal flatness leads to

$$\omega_o^2 L^2 - 2\sqrt{2}R_1\omega_o L + 2R_1^2 = 0$$

Solving the quadratic we have

$$L = \frac{\sqrt{2}R_1}{\omega_o} \tag{10.67}$$

and using the relation for C gives

$$C = \frac{\sqrt{2}}{\omega_o R_1} \tag{10.68}$$

Also

$$R_1 = R_2 = \sqrt{L/C} \tag{10.69}$$

For the maximally flat case we have previously shown that the poles are on the bandwidth circle at $\pm 45°$ from the negative real axis. The argument for this was developed in Section 7.13, but here the origin of the pole plot is at $\omega = 0$ and the radius ω_o is the low-pass bandwidth. High-pass circuits can be designed by interchange of L and C and by writing ω_o/ω for the variable in the response equation.

With the coefficient of $(\omega/\omega_o)^2$ made zero as required, the maximally flat response function of Eq. 10.61 is

$$|G(\omega)| = \frac{K}{[1 + (\omega/\omega_o)^{2n}]^{1/2}} \tag{10.70}$$

with n as the order of the network. This result is also known as the *Butterworth response*. It is plotted for $n = 2$ as the solid curve of Fig. 10.18. The frequency

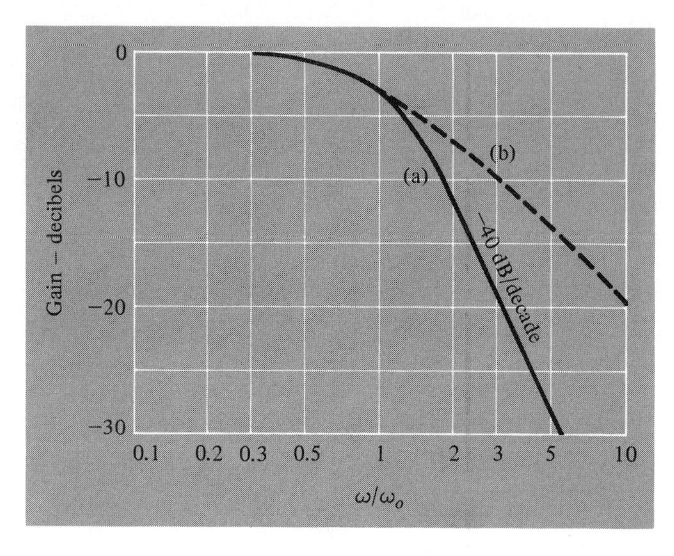

FIGURE 10.18

(a) Second-order filter, low-pass; (b) first-order low-pass filter.

$\omega = \omega_o$ is recognizable as the -3-dBV frequency or the defined *cutoff frequency* of the filter. The rate of attenuation at frequencies well above ω_o is -40 dBV/frequency decade and twice that obtainable with the first-order filter, shown for comparison; circuits as shown in Fig. 10.15 are used in wide-band amplifiers.

10.12 The third-order maximally flat filter

While it is possible to assemble a filter of order n by cascading the proper number of first- and second-order filters with proper isolation, we shall still consider here the design of a third-order filter. It is obtained from a second-order circuit by adding another circuit element, as in Fig. 10.19(a). For this circuit we can write current

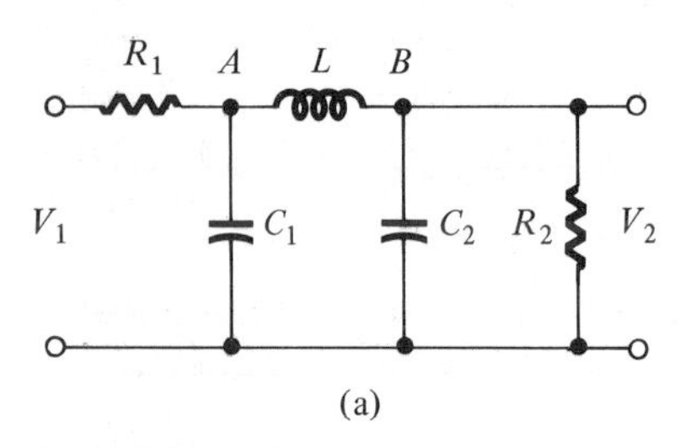

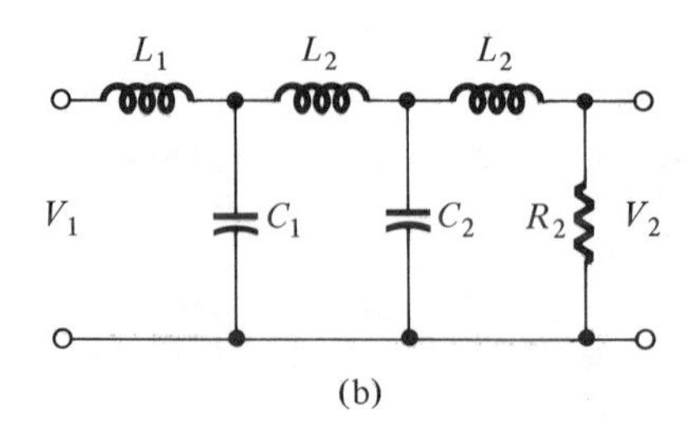

FIGURE 10.19

(a) Third-order low-pass filter; (b) fifth-order network.

summations at A and B:

$$\frac{V_1(s) - V_A(s)}{R_1} = \frac{V_A(s) - V_2(s)}{sL} + sC_1V_A(s) \tag{10.71}$$

$$\frac{V_A(s) - V_2(s)}{sL} = V_2(s)\left(sC_2 + \frac{1}{R_2}\right) \tag{10.72}$$

After some labor we can obtain

$$G(s) = \frac{V_2(s)}{V_1(s)} = \frac{R_2}{R_1 + R_2}\left\{\frac{1}{s^3LC_1C_2R_p + s^2L[(C_1R_1 + C_2R_2)/(R_1 + R_2)] + s\{(C_1 + C_2)R_p + [L/(R_1 + R_2)]\} + 1}\right\} \tag{10.73}$$

where

$$R_p = \frac{R_1R_2}{R_1 + R_2}$$

With $s = j\omega$ the steady-state amplitude and phase relations are

$$|G(\omega)| = \frac{K}{A^{1/2}} \tag{10.74}$$

where the denominator A is

$$\omega^6(LC_1C_2R_p)^2 + \omega^4\left\{\frac{L^2(C_1R_1 + C_2R_2)^2}{(R_1 + R_2)^2} - 2LC_1C_2R_p\left[(C_1 + C_2)R_p + \frac{L}{R_1 + R_2}\right]\right\} + \omega^2\left[\left(C_1 + C_2 + \frac{L}{R_1R_2}\right)^2 R_p^2 - 2L\left(\frac{C_1R_1 + C_2R_2}{R_1 + R_2}\right)\right] + 1 \tag{10.75}$$

$$\theta = \tan^{-1}\frac{-\omega\{\omega^2LC_1C_2R_p - (C_1 + C_2)R_p - [L/(R_1 + R_2)]\}}{1 - \omega^2L[(C_1R_1 + C_2R_2)/(R_1 + R_2)]} \tag{10.76}$$

The above is a formidable expression but it provides the needed design information for a maximally flat response characteristic by equating the coefficients of ω^4 and ω^2 to zero since the numerator lacks terms in ω^4 and ω^2. To demonstrate the ease with which this method handles variations in terminating resistance, let us take $R_2 = \infty$. Numerous terms in the coefficients vanish and we have $K = 1$, $R_p = R_1$. From the ω^2 coefficient we have

$$(C_1 + C_2)^2 R_1^2 = 2LC_2 \tag{10.77}$$

From the coefficient of ω^4 we have

$$L^2C_2^2 - 2LC_1C_2R_1^2(C_1 + C_2) = 0$$

which reduces to

$$2C_1R_1^2(C_1 + C_2) = LC_2 \tag{10.78}$$

Division of the first by the second gives

$$C_1 + C_2 = 4C_1$$

and so

$$C_2 = 3C_1$$

We can then use Eq. 10.78 and obtain

$$L = \frac{2C_1(C_1 + 3C_1)R_1^2}{3C_1} = \frac{8}{3}C_1R_1^2 \tag{10.79}$$

We also have

$$\frac{1}{\omega_o^6} = (LC_1C_2R_p)^2 \tag{10.80}$$

from the sixth-order term. Substituting Eq. 10.79, we have

$$\omega_o^6\left(\frac{64}{9}C_1^2R_1^4\right)(C_1^2)(9C_1^2)(R_1^2) = 1$$

or

$$\omega_o = \frac{1}{2C_1R_1} = \frac{3}{2C_2R_1} \tag{10.81}$$

This completes the design relations.

With the coefficients of ω^4 and ω^2 set to zero, Eq. 10.74 becomes of the form

$$|G(\omega)| = \frac{1}{[1 + (\omega/\omega_o)^6]^{1/2}} \tag{10.82}$$

which is again the general form of the Butterworth response. This third-order response ($n = 3$) is plotted in Fig. 10.20(a) and shows the response to fall at a rate of -60 dBV/frequency decade when well above ω_o. The better

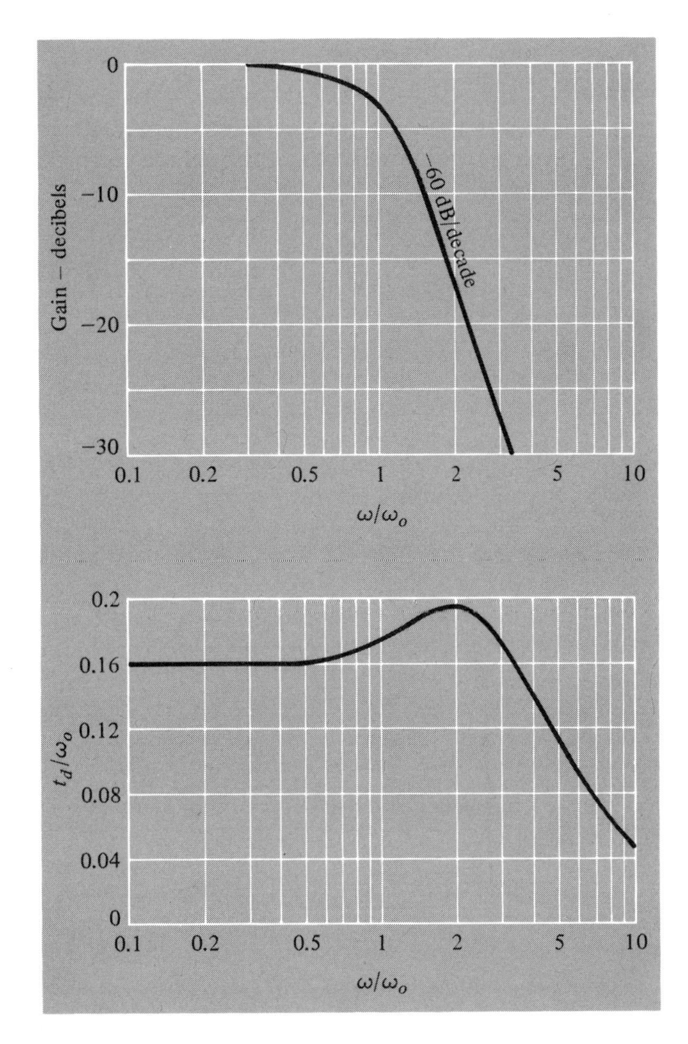

FIGURE 10.20

Third-order low-pass filter: (a) amplitude response; (b) time delay.

approximation to the ideal rectangular response is apparent. When used in a video amplifier, the third-order flat-response filter is said to provide "series compensation," since the inductor is in series with capacitances present in the amplifier.

The phase shift produced by the network, when designed for maximal flatness of amplitude, reduces to

$$\theta = \tan^{-1} \frac{\omega R_1(C_1 + C_2)\{[1 - \omega^2 R_1^2(C_1 + C_2)^2]/8\}}{1 - \omega^2\{[R_1^2(C_1 + C_2)^2]/2\}} \tag{10.83}$$

and a time-delay characteristic is plotted in Fig. 10.20(b). The values are normalized on $(C_1 + C_2)R_1$; the result is not constant with frequency, but maximal flatness of amplitude response was the objective of the circuit design.

10.13 The poles for the maximally flat response

The normalized maximally flat requirement has been shown to reduce to an amplitude response function:

$$|G(\omega)| = \frac{1}{[1 + (\omega/\omega_o)^{2n}]^{1/2}} \tag{10.84}$$

This response can be obtained from a network having a transfer characteristic of the form

$$G(s) = \frac{\omega_o^n}{(s + s_1)(s + s_2)\cdots(s + s_n)} \tag{10.85}$$

We have previously shown that the poles are located on a circle locus of radius ω_o in the left half of the complex frequency plane.

If n is odd, there must be one root on the negative real axis at $s = -\omega_o$ and one or more pairs of complex conjugate roots on the circle. These pairs can be located by setting

$$s^{2n} = \pm 1$$

depending on whether n is odd or even, so that $G(s)$ is infinite at its poles. The poles are then located at

$$s_k = \varepsilon^{j(\pi/2)(2k/n)} \qquad \text{for odd } n \tag{10.86}$$

$$s_k = \varepsilon^{j(\pi/2)[(2k-1)/n]} \qquad \text{for even } n \tag{10.87}$$

TABLE 10.1 THE POLES OF THE BUTTERWORTH RESPONSE

n	s
1	$\pm 1\underline{/0^\circ}$
2	$\pm 1\underline{/45^\circ}$; $\pm 1\underline{/-45^\circ}$
3	$\pm 1\underline{/0^\circ}$; $\pm 1\underline{/60^\circ}$; $\pm 1\underline{/-60^\circ}$
4	$\pm 1\underline{/22.5^\circ}$; $\pm 1\underline{/-22.5^\circ}$; $\pm 1\underline{/67.5^\circ}$; $\pm 1\underline{/-67.5^\circ}$
5	$\pm 1\underline{/0^\circ}$; $\pm\underline{/36^\circ}$; $\pm 1\underline{/-36^\circ}$; $\pm 1\underline{/72^\circ}$; $\pm 1\underline{/-72^\circ}$

However, these expressions locate twice as many poles as there are in our function. Since a passive real network cannot have poles in the right half plane, we need retain only those in the left half plane; an example appears in Fig. 10.21 for $n = 3$ and $n = 4$. The results are consolidated in Table 10.1.

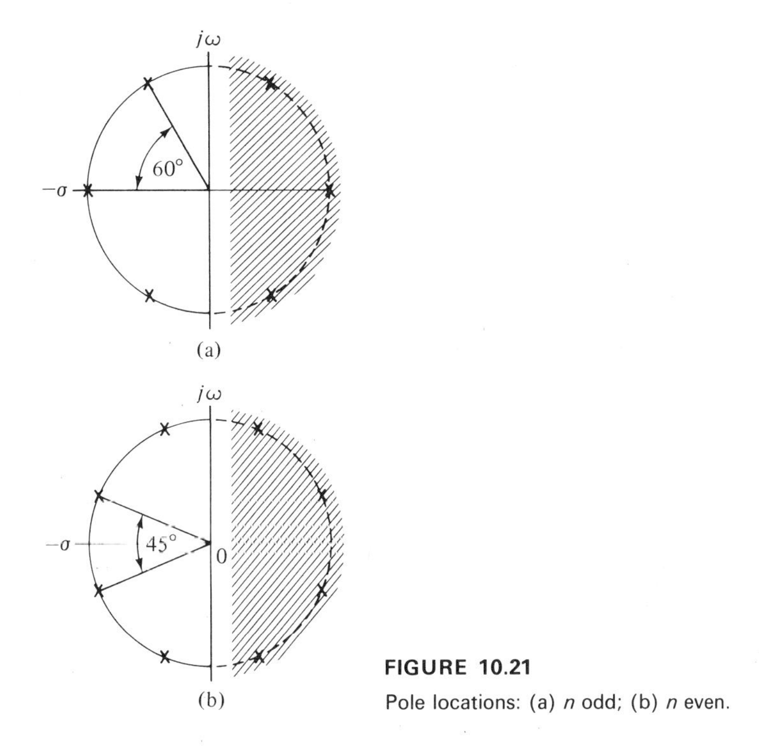

FIGURE 10.21
Pole locations: (a) n odd; (b) n even.

Having located the poles for any n, we can use Eq. 10.85. For example, with $n = 3$,

$$G_3(s) = \frac{1}{(s + 1\angle 0°)(s + 1\angle 60°)(s + 1\angle -60°)} \tag{10.88}$$

$$= \frac{1}{(s + 1)(s + 0.5) + j0.866(s + 0.5) - j0.866}$$

$$= \frac{1}{s^3 + 2s^2 + 2s + 1} \tag{10.89}$$

Similarly, we could obtain a denominator polynomial in s for any n. These are the Butterworth polynomials given in Table 10.2; having the desired polynomial in s, one need merely equate the coefficients to the polynomial in s written from the circuit, that is, Eq. 10.60 for the second order and Eq. 10.73 for the third order, thereby saving the labor of squaring the function in ω.

TABLE 10.2 **BUTTERWORTH POLYNOMIALS (NORMALIZED TO $\omega_o = 1$)**

n	$G_n(s)$
1	$\frac{1}{s+1}$
2	$\frac{1}{s^2 + 1.41s + 1}$
3	$\frac{1}{s^3 + 2s^2 + 2s + 1}$
4	$\frac{1}{s^4 + 2.61s^3 + 3.41s^2 + 2.61s + 1}$
5	$\frac{1}{s^5 + 3.24s^4 + 5.24s^3 + 5.24s^2 + 3.24s + 1}$

At high frequencies, well above ω_o, the gain amplitude expression of Eq. 10.84 becomes

$$G(\omega) = \frac{1}{[(\omega/\omega_o)^{2n}]^{1/2}} \tag{10.90}$$

The decibel loss in this region beyond cutoff is

$$\text{dBV} = -20n \log \frac{\omega}{\omega_o}$$

We have already had attenuation rates of -40 dBV/frequency decade and -60 dBV/decade for two- and three-pole filters. An eight-pole filter ($n = 8$) would reach an attenuation rate of -160 dBV/frequency decade.

We cascaded constant-k filters to increase attenuation in the stop band, but for the Butterworth filter it is better to design for the attenuation needed. An eight-pole filter will have less loss in the pass band than will two four-pole filters in cascade. The filter complexity increases with the number of poles, as shown in Fig. 10.19(b), but the Butterworth filter approximates the ideal response rectangle more and more closely, as shown by Fig. 7.18.

The delay variation is small, and designs for maximal flatness or for minimal delay error do not differ greatly.

10.14 The Chebyshev response

The Butterworth maximally flat response is often satisfactory, but we have no assurance that it is the most efficient circuit design. For comparison purposes we divide the response into the pass region, the transition region, and the stop region. We consider the pass region as that in which the attenuation is ± 1 dB and the stop region as that in which the attenuation exceeds 20 dB. In the transition region between, the signal is too small to be useful, yet too large to be overlooked.

The Butterworth approximation fixed the amplitude at the origin, as well as the values of the first $n - 1$ derivatives there. This response preferentially weights the frequencies near the origin and is more stringent than is often necessary. While having unit response at the origin, the response still falls uniformly to -3 dB at $\omega = \omega_o$, so that the response is not uniform in the pass band, and the transition band is wide.

Returning to the pole-zero plot for $n = 2$, we might consider a further separation of the poles parallel to the j axis. This will reduce the response at zero frequency and will emphasize a frequency ω nearer cutoff. Various frequencies in the pass band would rise or fall, but proper pole location can control these gain *ripples* to a specified peak-to-peak amplitude. Such a response curve, while having a varying response in the pass band, has a steeper fall at cutoff and a narrower transition region. By experiment with pole locations, we are led to the *Chebyshev approximation*, which allows amplitudes in the pass band to rise or fall by a specified amount, but with control of the amplitude of these *error ripples*. This result is illustrated in Fig. 10.22(a).

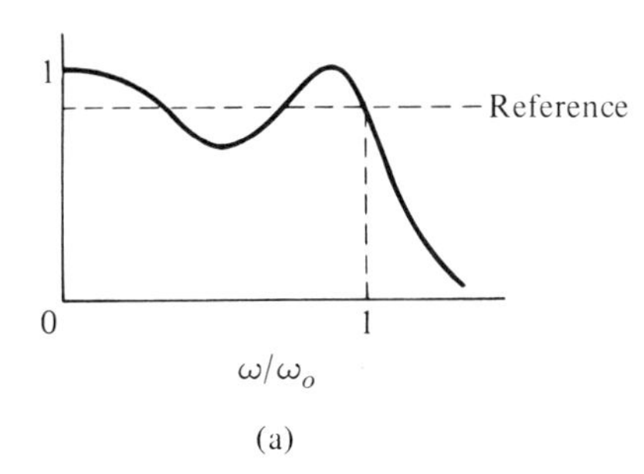

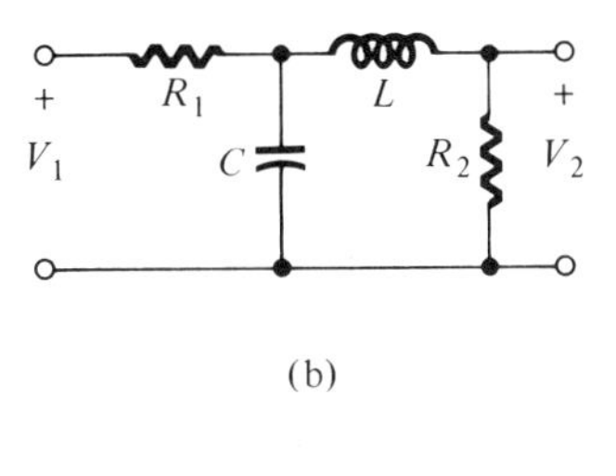

FIGURE 10.22
(a) Chebyshev equal-ripple low-pass response; (b) second-order network.

Chebyshev functions are usually developed in terms of a normalized ω, and the amplitude is an even function of that variable. If the reciprocal of $G(\omega)$ differs from the ideal flat response by a peak-to-peak error ε^2, then

$$\frac{1}{G(\omega)} - 1 = \frac{\varepsilon^2}{2} C_{2n}(\omega) \tag{10.91}$$

Solving for the response function,

$$|G(\omega)| = \frac{1}{1 + (\varepsilon^2/2) C_{2n}(\omega)} \tag{10.92}$$

The term $C_{2n}(\omega)$ is an even-order Chebyshev polynomial.* This is defined as

$$C_{2n}(\omega) = \cos(2n \cos^{-1} \omega)$$

Using the identity

$$\cos 2n\omega = 2 \cos^2 n\omega - 1$$

we then have

$$|G(\omega)| = \frac{1}{1 - (\varepsilon^2/2) + \varepsilon^2 C_n^2(\omega)} \tag{10.93}$$

It being possible to neglect the $\varepsilon^2/2$ term as a small constant, the Chebyshev polynomials vary between zero and 1 in the pass band $0 \leqq \omega \leqq 1$ so that the amplitude of the response will vary between $(1 - \varepsilon^2/2)^{-1}$ and $(1 + \varepsilon^2/2)^{-1}$. The first few Chebyshev polynomials are given in Table 10.3.

TABLE 10.3 **CHEBYSHEV POLYNOMIALS**

Order	*Polynomial*
0	1
1	ω
2	$2\omega^2 - 1$
3	$4\omega^3 - 3\omega$
4	$8\omega^4 - 8\omega^2 + 1$
5	$16\omega^5 - 20\omega^3 + 5\omega$
6	$32\omega^6 - 48\omega^4 + 18\omega^2 - 1$
8	$128\omega^8 - 256\omega^6 - 160\omega^4 - 32\omega^2 + 1$

It is found that the steepness of cutoff of the function of Eq. 10.93, in the transition band, exceeds that possible with the Butterworth maximally flat approximation; this is the improvement purchased by acceptance of the equal-ripple response in the pass band.

The magnitude of the response oscillates with ε as the ripple-scaling parameter, where

$$\text{Ripple (dB)} = 10 \log(1 + \varepsilon^2) \tag{10.94}$$

The value of the ripple may be specified, and for a ripple of ± 0.5 dB we find $\varepsilon^2 = 0.125$.

*This dates from 1875. The development of these forms had nothing to do with filter networks but resulted from a study of valve linkage for steam engines.

The second-order circuit of Fig. 10.22(b) requires use of the fourth-order polynomial,

$$8\omega^4 - 8\omega^2 + 1$$

and with $\varepsilon^2 = 0.125$ we have

$$|G(\omega)| = \frac{1}{1 - (0.125/2) + 0.125(8\omega^4 - 8\omega^2 + 1)}$$

$$\cong \frac{1}{\omega^4 - \omega^2 + 1} \tag{10.95}$$

We note that $0.125/2 \neq 0.125$, but this introduces only a small constant gain error.

Equation 10.95 represents the form of the second-order magnitude response, which will meet the specification for a ripple of ± 0.5 dB. We make the second-order gain magnitude of Eq. 10.61 conform by equating coefficients. That is,

$$\frac{K}{(\omega/\omega_o)^4 + (\omega/\omega_o)^2[(1 - 2Q^2)/Q^2] + 1} = \frac{1}{\omega^4 - \omega^2 + 1}. \tag{10.96}$$

where $K = R_2/(R_1 + R_2)$. Assuming $R_2 = \infty$ again, we find $K = 1$. The first coefficient identity reveals

$$1 = \frac{1}{\omega_o^4}$$

from which

$$\omega_o^2 LCK = \omega_o^2 LC = 1 \tag{10.97}$$

From the second identity

$$-1 = \frac{1 - 2Q^2}{Q^2}$$

and we find that

$$Q = 1 \tag{10.98}$$

is the required parameter. Using Eq. 10.59, we have

$$\frac{1}{Q} = 1 = \frac{1}{\omega_o C R_2} + \frac{R_1}{\omega_o L} = \frac{R_1}{\omega_o L}$$

the last equality resulting from $R_2 = \infty$. We then find

$$L = \frac{R_1}{\omega_o} \tag{10.99}$$

and with Eq. 10.97 we have

$$C = \frac{1}{\omega_o R_1} \tag{10.100}$$

and also

$$R_1 = \sqrt{\frac{L}{C}} \tag{10.101}$$

These design results may be compared with the values for the maximally flat second-order circuit from Eqs. 10.67 and 10.68:

$$L = \frac{\sqrt{2}R_1}{\omega_o} \qquad \text{and} \qquad C = \frac{\sqrt{2}}{\omega_o R_1}$$

These indicate the design differences. Comparison of the Chebyshev response of Fig. 10.23 with the maximally flat response of Fig. 10.18(a) may be made, and we see that the pass band within the ± 0.6-dBV limits has increased in

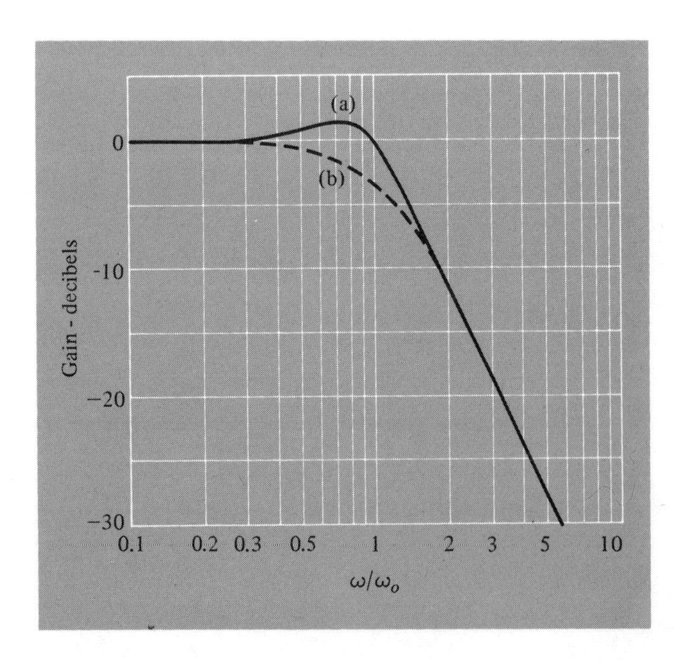

FIGURE 10.23

(a) Chebyshev response, $\varepsilon^2 = 0.625$; (b) maximally flat response.

the ratio of 0.7 to 1 in favor of the Chebyshev design. At higher frequencies the gain falls at the rate of $-20n$ dBV/decade, so that the major difference between the Butterworth and the Chebyshev design is the faster attenuation in the transition region.

For the Butterworth design we use $Q = 0.707$ and obtain the response given for that value of the parameter in Fig. 10.16(a). Using the same circuit but changing Q to unity gives a Chebyshev response for $Q = 1$ in the same figure, for a gain ripple of ± 0.625 dBV.

The design of such complex filters, especially those of higher orders, is normally carried out by use of tables of design constants or by computer. While pole location is not then of importance, the poles for the Chebyshev filter could be shown to be located on an elliptical locus, with the major axis equal to $2\omega_o$. This confirms the concept of spreading the poles, introduced at the beginning of this section.

10.15 Summary

The constant-k filter was designed with the telephone system in mind, where the attenuation is built up by cascading of sections. The terminations were therefore required to be image impedances. Failure to obtain constancy of image impedance across the pass band, and particularly failure to provide a match to a constant-resistance load near cutoff, made the use of constant-k filters only a broad approximation.

Modern filter synthesis employs transfer functions which are selected as suitable approximations to the ideal filter characteristic. The pole locations in the left half plane are determined, and circuit parameters are matched to tabulated functions which have been determined to provide the selected pole locations. There is no concern for image impedances and no cascading of sections as a rule; n-pole filters can be designed as units with terminations included.

The order of the response equation denominator is equal to n, the number of energy-storage elements needed in the circuit. Response falls off at $-20n$ dBV/frequency decade at frequencies far in the stop band.

With ε chosen to make the gain variation of the Chebyshev response equal to 3 dBV, that filter gives a faster rolloff of response at cutoff than does the Butterworth circuit. The Butterworth circuit gives a phase shift which is nearly linear with frequency, reaching $-n\pi/4$ rad at cutoff.

The design of filters is usually done by first developing a low-pass model; then the element values and types are transformed to realize the desired form of filter response. A high-pass filter is designed by transforming each low-pass inductance to a capacitance of value inversely proportional to the original inductance, and each low-pass capacitance is transformed to an inductance of value inversely proportional to the original capacitance. To transform frequency from the normalized value at $\omega_o = 1$ r/s to any other ω_o', each inductor is divided by the new cutoff frequency in r/s, and each capacitor is also divided by the same value.

PROBLEMS

10.1 In the circuit of Fig. 10.24(a), find the values of α and β and the dB loss between terminals.

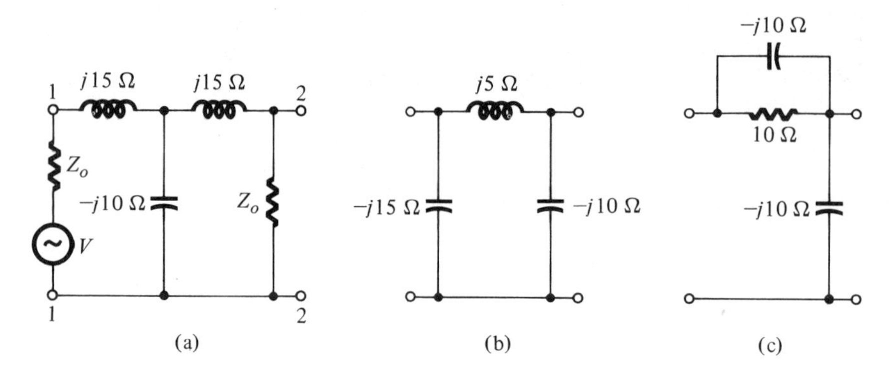

FIGURE 10.24

10.2 For the networks of Fig. 10.24(b) and (c), determine $\mathbf{Z}_o$.

10.3 (a) Design a constant-k low-pass T filter to cut off at 2500 Hz, with source and load of 10,000 Ω of resistance. (b) What is the attenuation at 3000 Hz?

10.4 Find the values of circuit elements needed for a constant-k high-pass T filter to cut off at 1500 Hz when working into a load of 1000 Ω. What is $\mathbf{Z}_{oT}$ at 1800 Hz?

10.5 Design a constant-k low-pass T filter to work into a load of 1250 Ω of resistance, with cutoff at 600 Hz and high attenuation at 750 Hz due to the shunt arms of the m-derived matching sections. Terminate it to give a power match across the pass band and draw the circuit. What is the attenuation at 900 Hz?

10.6 Connect two 0.05-H inductors and one 0.01-μF capacitor for a low-pass filter network. (a) What is the cutoff frequency? (b) What is R_k at zero frequency? (c) What is $\mathbf{Z}_{oT}$ at $0.5f_c$ and at $0.9f_c$?

10.7 For a 500-Ω resistance load, design a composite low-pass filter composed of the terminating half m-derived sections and one constant-k section, with a frequency of infinite attenuation of the m-derived sections at $1.1f_c$.

10.8 Develop the component values for a Butterworth network with $n = 3$, $\omega_c = 2000$ r/s, and $R_2 = R_1 = 1000\ \Omega$. Sketch the shape of the dB attenuation curve.

10.9 Develop the component values for a Chebyshev filter with $n = 2$, $\varepsilon = 0.663$, $\omega_c = 2000$ r/s, $R_1 = 1000\ \Omega$, and $R_2 = \infty$. Find the frequencies of peak gain.

10.10 Design a low-pass Chebyshev filter for $R_1 = 1000\,\Omega$, $R_2 = \infty$, and $\omega = 1000$ r/s, with a ± 0.75-dB attenuation ripple, to give -60-dB/decade attenuation at frequencies well above cutoff.

10.11 Design a low-pass constant-k filter with $f_c = 1000$ Hz and $R_k = 100\,\Omega$ at zero frequency. What element values are needed?

10.12 For the circuit of Fig. 10.25(a), the voltage transfer function, V_2/V_1, is to have the maximally flat form $1/\sqrt{1 + \omega^{2n}}$. (a) What should be the relation between R and L? (b) What values of R and L are needed for $\omega_c = 100$ r/s?

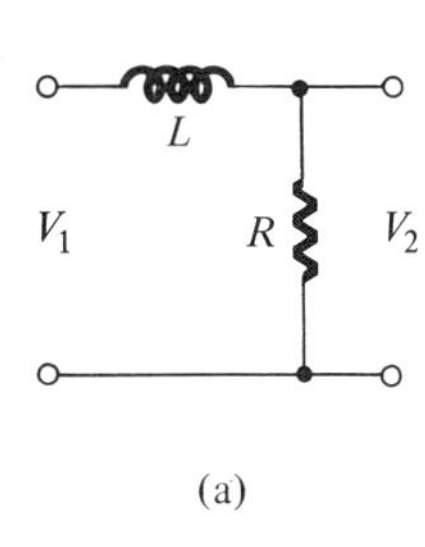

(a)

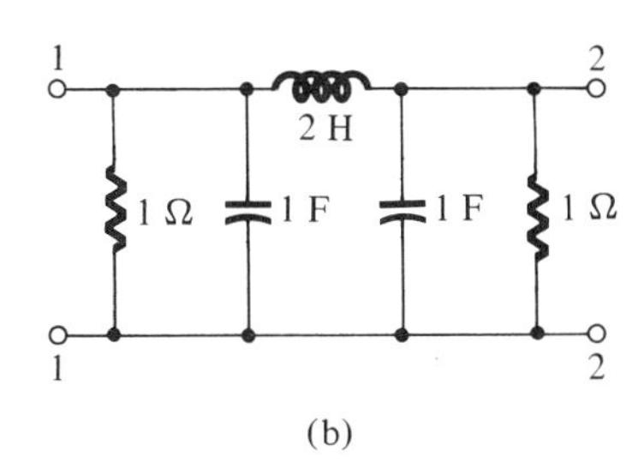

(b)

FIGURE 10.25

10.13 For the third-order network of Fig. 10.25(b), show that $G(s)$ has a maximally flat form and that

$$G(\omega) = \frac{1/2}{(s + 1)(s + 1/2 - j\sqrt{3}/2)(s + 1/2 + j\sqrt{3}/2)}$$

10.14 An equal-ripple response is given by

$$|G(\omega)| = \frac{1}{[1 + \varepsilon^2 C_n^2(\omega)]^{1/2}}$$

where $n = 5$ and $\varepsilon = 0.1$. Determine (a) the maximum value of $|G(\omega)|$, (b) the minimum value of $|G(\omega)|$ in the pass band, and (c) the cutoff frequency in terms of ω_o.

10.15 The half-power frequency of a Chebyshev network is to be 1.1 r/s + ω_o. The ripple magnitude is to be limited to ± 0.5 dB. What is the minimum value of n needed?

10.16 For $n = 6$, find the pole location coordinates for a maximally flat response for $f_c = 500$ kHz and $R_1 = R_2 = 500\,\Omega$.

10.17 Design a constant-k high-pass section with $\omega_c = 1500$ r/s and $R_L = 1000\,\Omega$.

10.18 Design a constant-k band-pass section with $f_H = 1000$ Hz, $f_L = 500$ Hz, and $R_L = 600\,\Omega$.

10.19 Determine the pass bands and stop bands of the symmetrical sections of Fig. 10.26.

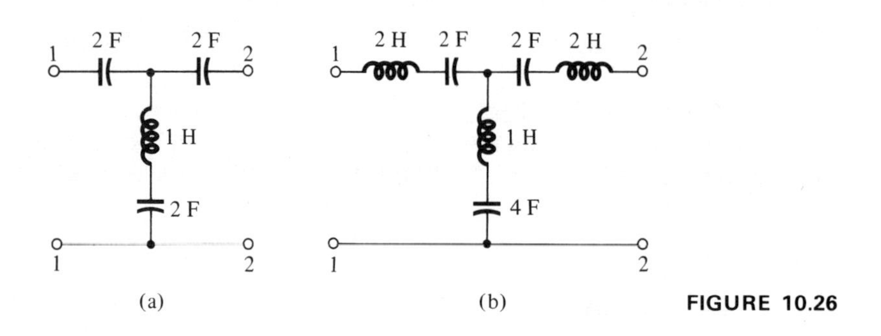

FIGURE 10.26

10.20 Figure 10.27(a) is a Chebyshev filter, normalized to $\omega_o = 1$ r/s and $R_L = 1\ \Omega$. Draw the circuit to be constructed if you desire a filter with cutoff at 1400 Hz and operating into a load of 600 Ω.

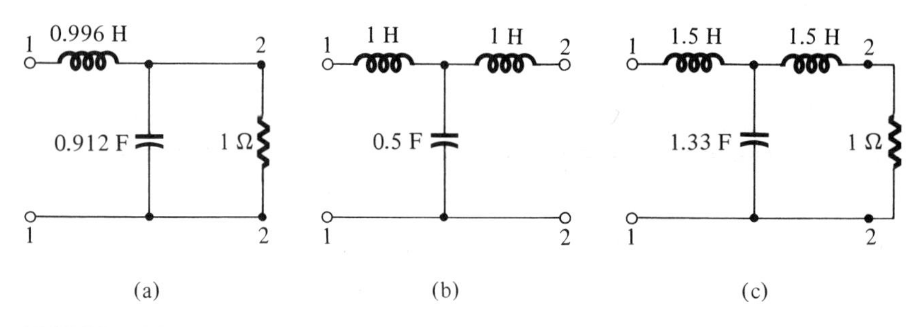

FIGURE 10.27

10.21 Plot Z_{oT} against ω for $\omega = 0$ to $2\omega_c$ for the network of Fig. 10.27(b).

10.22 Figure 10.27(c) is a Butterworth filter, $n = 3$, normalized to $\omega_c = 1$ r/s and $R_L = 1\ \Omega$. (a) Give the coordinates of the *s* plane poles. (b) Develop a high-pass model for $\omega_c = 5$ r/s and $R_L = 100\ \Omega$.

REFERENCES

1. J. D. Ryder, *Networks, Lines, and Fields*, 2d ed. Prentice-Hall, Inc., Englewood Cliffs, N.J., 1957.
2. M. E. Van Valkenburg, *Network Analysis*, 2d ed. Prentice-Hall, Inc., Englewood Cliffs, N.J., 1957.
3. W. L. Cassell, *Linear Electric Circuits*. John Wiley & Sons, Inc., New York, 1964.

active *RC* filters

eleven

Inductors and capacitors can be connected in T, π, or ladder networks to obtain low-pass, band-pass, or high-pass filter circuits. However, inductors are relatively large, and at low frequencies they are heavy and expensive because of their iron cores. Also, there is at present no satisfactory technique for producing an inductor for integrated microcircuits. Use of R and C elements as filters is not very satisfactory because of pass-band losses, and cutoff at the band edges is not sufficiently sharp for most applications.

Search for relief from these problems led to filter simulation with the analog computer, combining gain with the R and C elements. By inclusion of gain elements with the RC circuits, it is possible to restore pass-band losses and even to achieve a gain. Conjugate-pole pairs become possible, giving sharpness of cutoff equivalent to that of LC circuits, and we then have a capability for design of filter networks without inductors. The resultant networks are called *active RC filters* and, with the ready availability of the integrated circuit amplifier, are well suited to lightweight low-frequency filters.

The field is undergoing rapid development, and we shall merely introduce the topic here as a major application of the gain element and feedback.

11.1 Functions of the amplifier

The properties of the operational amplifier in analog computation provided the initial impetus for the use of similar gain elements with RC networks in filter applications. A high-gain amplifier combined with RC feedback created the conjugate-pole pairs needed for rapid frequency cutoff. The amplifier also provided isolation, as needed between the several RC networks and the circuit output loads. The possibilities can be shown by

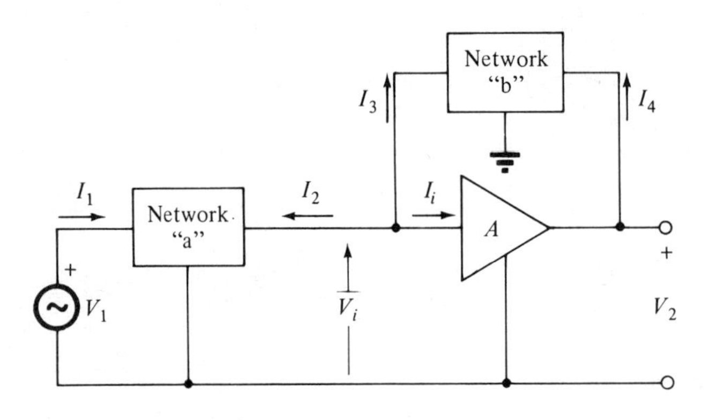

FIGURE 11.1
Gain element and feedback network.

placement of *RC* networks at *a* and *b* of the amplifier of Fig. 11.1, for which the following network admittance equations apply:

$$\begin{aligned} I_1(s) &= y_{ia}V_1(s) + y_{ra}V_i(s) \\ I_2(s) &= y_{fa}V_1(s) + y_{oa}V_i(s) \end{aligned} \tag{11.1}$$

$$\begin{aligned} I_3(s) &= y_{ib}V_i(s) + y_{rb}V_2(s) \\ I_4(s) &= y_{fb}V_i(s) + y_{ob}V_2(s) \end{aligned} \tag{11.2}$$

The amplifier is taken as an ideal gain element, and the input impedance and the internal gain are high so that

$$I_i(s) = 0$$

$$A \to \infty$$

Also we see that

$$V_2(s) = AV_i(s) \tag{11.3}$$

$$I_2(s) = -I_3(s) \tag{11.4}$$

Using this last equality with equations from 11.1 and 11.2 gives

$$y_{fa}V_1(s) + y_{oa}V_i(s) = -y_{ib}V_i(s) - y_{rb}V_2(s)$$

But $V_i(s) = V_2(s)/A$ and this substitution leads to

$$V_2(s)\left(\frac{1}{A} + \frac{y_{rb}}{y_{ib} + y_{oa}}\right) = \left(\frac{-y_{fa}}{y_{ib} + y_{oa}}\right)V_1(s) \tag{11.5}$$

With A very large the term $1/A$ may be dropped, giving

$$G(s) = \frac{V_2(s)}{V_1(s)} = -\frac{y_{fa}}{y_{rb}} \tag{11.6}$$

The input and output admittances of the two networks are not involved, leaving the gain function composed of transfer admittances. The zeros of y_{rb} represent the transmission zeros through network b, and the zeros may be chosen at will. However, these zeros of y_{rb} create the poles of $G(s)$, and so we have freedom in location of these poles. While we will restrict the pole locations to the left half of the s plane for reasons of stability, we can move the poles close to the $j\omega$ axis to increase the slope of the filter cutoff with frequency.

Because of the indicated freedom in pole locations, we choose desired frequency functions for Eq. 11.6 and obtain functions for RC networks which are equivalent in form and performance to the gain functions of LC filters. However, such circuits usually require a large number of passive circuit elements, and a high element count raises the cost; variation of element values to adjust the pole locations is also difficult in practice. In such filter simulations the amplifier gain figure A has no effect as long as it is large enough to be neglected in Eq. 11.5.

In another design method, somewhat simpler RC filter configurations are possible if we use an amplifier of low and controlled gain, but in these circuits the gain figure enters into the network design. The output impedance of the gain element will be very low because of the large feedback used to stabilize the gain, and the circuits may be used to drive other RC filter elements without further isolation.

A third method of active filter design employs the *negative impedance converter* (NIC) of Section 9.8. The NIC provides an impedance at one port which is the negative of the impedance at a second port and can be realized with a high-gain amplifier. It is possible to control the real parts of the pole locations and to vary the sharpness of filter cutoff. Unfortunately, the output impedance is not low and the gain figure enters into the transforming action; stability is not good, and the circuit is infrequently used, although much study is going on.

The *gyrator* is suited to transforming capacitance into inductance and can be used in the active filter. Again, circuit complexities and lack of stability seem to mitigate against practical application.

The amplifier forms of active RC filters use the readily available operational amplifier, manufactured in integrated circuit form. Because of their extensive application, we shall confine our study to the several amplifier forms of filters.

11.2 The filter functions

It is possible to achieve high-order filter performance by factoring of the high-order transfer functions into a product of first- and second-order terms and cascading of active filter networks designed to perform in accordance with those terms. That is,

$$\frac{1}{s^3 + as^2 + bs + 1} = \left[\frac{1}{(s+1)}\right]\left[\frac{1}{s^2 + ds + 1}\right] \tag{11.7}$$

Because of the low output impedances characteristic of the operational amplifier, first- and second-order circuits can be cascaded without interaction, resulting in the desired third-order performance. For this reason we need only to learn how to simulate first- and second-order filter forms.

The active *RC* filter design procedure is to choose a suitable gain transfer function and then to find the parameters of the *RC* circuit by equating the coefficients of the transfer function to the coefficients obtained from the chosen *RC* active network. Representative transfer functions will be cataloged in the next few paragraphs.

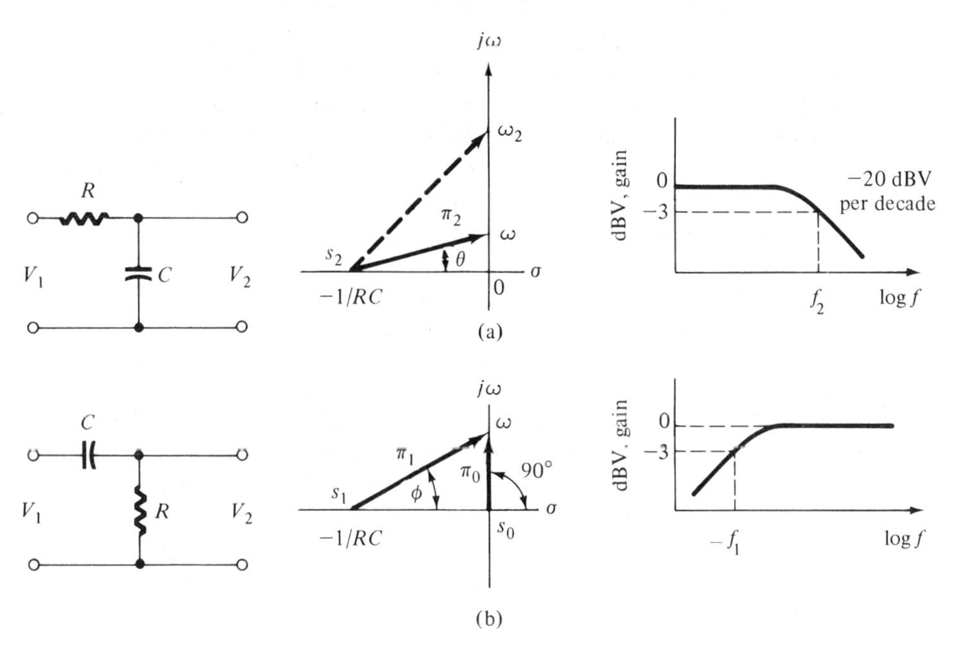

FIGURE 11.2
(a) Study of low-pass *RC* network; (b) high-pass *RC* network.

For first-order filters we have the circuits of Fig. 11.2. At (a) is a simple *low-pass network* with the transfer function

$$G(s) = \frac{V_2(s)}{V_1(s)} = \frac{1}{1 + sCR} \tag{11.8}$$

This function has a pole on the negative real axis at $s_2 = -1/CR$. The cutoff frequency is defined at the half-power bandwidth limit, where

$$\omega_2 = \frac{1}{CR}$$

At frequencies well above ω_2 the gain falls at -20 dBV/frequency decade on the plot of (c).

At (d) is the circuit of a first-order *high-pass* network with

$$G(s) = \frac{V_2(s)}{V_1(s)} = \frac{sCR}{1 + sCR} \tag{11.9}$$

having a pole on the negative-real axis at $s_1 = -1/CR$ and a zero at $s_o = 0$. The gain in (b) rises from zero at zero frequency at a rate of $+20$ dBV/frequency decade and reaches the defined -3-dBV cutoff at

$$\omega_1 = \frac{1}{CR}$$

The first-order high-pass result could also be written

$$G(s) = \frac{1}{1 + 1/sCR} \tag{11.10}$$

Comparison of this result with Eq. 11.8 shows that the low-pass function can be transformed into the high-pass function by substitution of $1/sT$ for sT. This is the general transformation with $T = CR$, and extended to the steady state by use of $\omega_0/j\omega$ for $j\omega/\omega_0$ as in Section 10.6.

In Section 10.10 we wrote the voltage-gain transfer function for a *second-order low-pass* filter as

$$G(s) = \frac{V_2(s)}{V_1(s)} = \frac{H\omega_o^2}{s^2 + s(\omega_o/Q) + \omega_o^2} \tag{11.11}$$

where H is the magnitude factor. The circuit appears in Fig. 11.3(a). There are two zeros at $s = \infty$, and the poles are given by

$$s_1, s_2 = -\frac{\omega_o}{2Q} \pm \omega_o\sqrt{\frac{1}{4Q^2} - 1}$$

To simulate a second-order LC filter, we need complex conjugate poles and so require underdamping, given by

$$1 > \frac{1}{4Q^2} \tag{11.12}$$

in the radical. The conjugate poles are then placed at

$$s_1, s_2 = -\frac{\omega_o}{2Q} \pm j\omega_o\sqrt{1 - \frac{1}{4Q^2}} \tag{11.13}$$

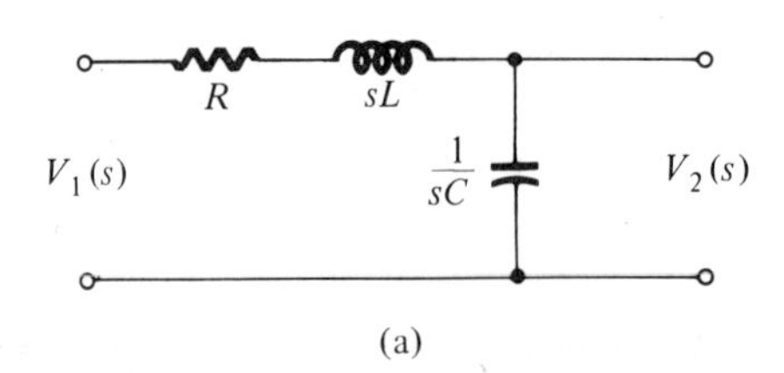

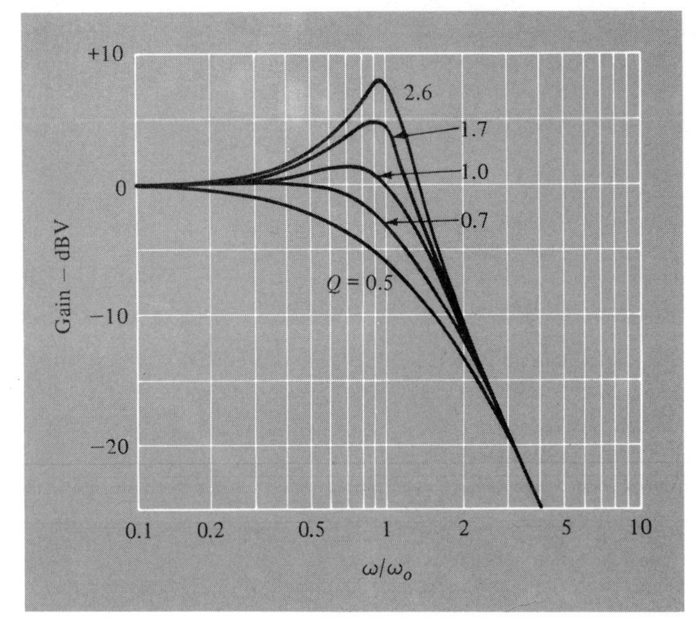

FIGURE 11.3
Second-order low-pass *LCR* filter.

The real part of the pole location is $\omega_o/2Q$, which is one-half the coefficient of the s^1 term of the denominator binomial. The steepness of the frequency response near cutoff is controlled by the nearness of the pole to the axis, that is, by the magnitude of $\omega_o/2Q$. The effect of Q variation on the shape of the response near cutoff at $\omega/\omega_o \cong 1$ is shown in Fig. 11.3(b). The response for $Q = 0.5$ is the limiting condition at critical damping, or $4Q^2 = 1$; this rate of falloff cannot be exceeded by *RC* passive filters. Larger values of Q are possible with active filters, resulting in conjugate pole pairs, paralleling the effect of inductance in *LC* filters and causing steeper response curves near cutoff. At frequencies well above cutoff all responses of second-order filters fall at -40 dBV/frequency decade.

With the *LCR* elements rearranged as in Fig. 11.4(a), the response for the *second-order high-pass filter* is

$$G(s) = \frac{V_2(s)}{V_1(s)} = \frac{Hs^2}{s^2 + s(\omega_o/Q) + \omega_o^2} \tag{11.14}$$

This function has a double zero at the origin, and a pair of complex conjugate poles, as in Fig. 11.4(b). The zeros control the response until ω is large, when

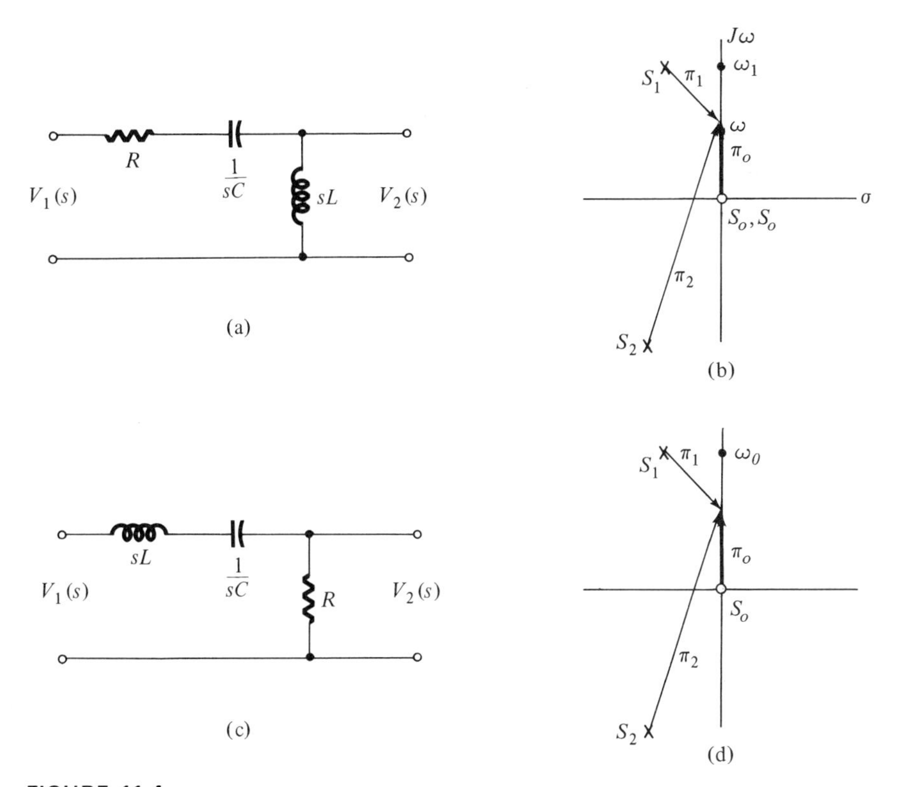

FIGURE 11.4

(a), (b) High-pass *LCR* network; (c), (d) band-pass *LCR* network.

one zero and one pole reduce to a multiplying constant. At frequencies above ω_o the other zero and second pole give a constant-pass action at high frequencies.

When the three circuit elements are connected as in Fig. 11.4(c), we have a typical *second-order band-pass filter* response:

$$G(s) = \frac{V_2(s)}{V_1(s)} = \frac{Hs\omega_o/Q}{s^2 + s(\omega_o/Q) + \omega_o^2} \tag{11.15}$$

A zero at $s = 0$ and a pair of complex conjugate poles determine the form of response. The response peaks at $\omega = \omega_o$, as for a resonant circuit, which (c) represents.

The three second-order response functions differ only in the numerator terms, and it is the zeros of transmission contributed by the numerators which determine that a given function is low pass, high pass, or band pass in response. Equations 11.11, 11.14, and 11.15 as well as the first-order Eqs. 11.8 and 11.9 provide a catalog of typical forms of first-order response functions. The gain functions obtained for *RC* active filters will be similarly modeled.

11.3

A high-pass active filter

As an illustration of the method by which the operational amplifier may be used with direct application of Eq. 11.6, suppose that wc need a high-pass filter with attenuation of -40 dBV/decade below cutoff at $\omega_o = 2$ r/s. This dictates a second-order filter acting in accordance with Eq. 11.14:

$$G(s) = \frac{Hs^2}{s^2 + s(\omega_o/Q) + \omega_o^2} \qquad (11.16)$$

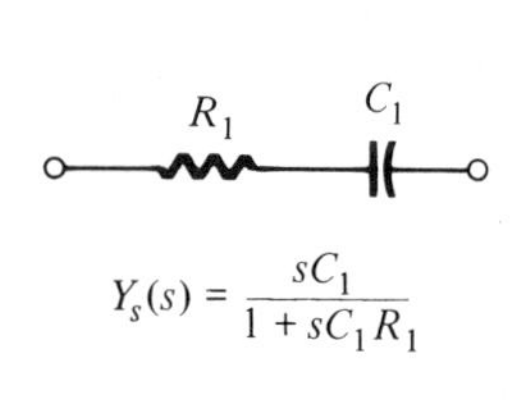

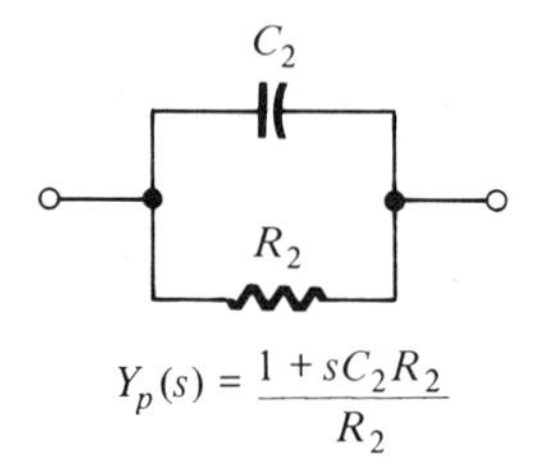

(a)

C_2
R_2
R_1
C_1
K
$V_1(s)$
$V_2(s)$

(b)

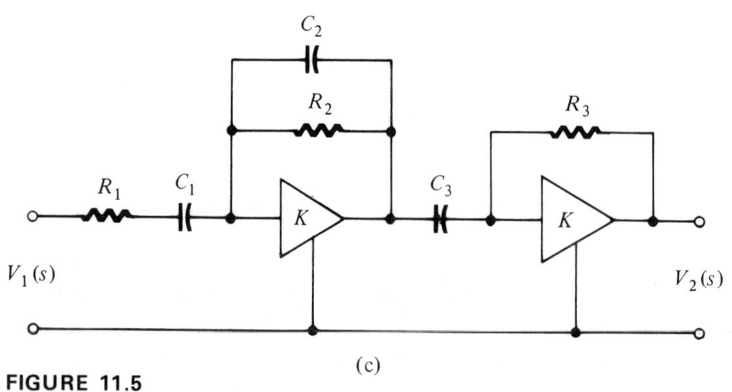

(c)

FIGURE 11.5
Evolution of an active *RC* filter.

Now note the admittances of the series and parallel RC circuits of (a) in Fig. 11.5:

$$Y_s(s) = \frac{sC_1}{1 + sC_1R_1} \qquad \text{and} \qquad Y_p(s) = \frac{1 + sC_2R_2}{R_2}$$

If we use $Y_p(s)$ as y_{rb} and $Y_s(s)$ as y_{fa} for the elements in Eq. 11.6, we require an amplifier gain

$$G_1(s) = \frac{-sC_1R_2}{(1 + sC_1R_1)(1 + sC_2R_2)} \tag{11.17}$$

and the amplifier with feedback appears in Fig. 11.5(b).

To complete the simulation of Eq. 11.16, we need another power of s in the numerator, without additive constants, and this can be obtained by cascading a second amplifier with

$$y_{rb} = \frac{1}{R_3} \qquad \text{and} \qquad y_{fa} = sC_3$$

which calls for a second amplifier gain,

$$G_2(s) = -sC_3R_3 \tag{11.18}$$

The overall gain function for the cascaded amplifiers of Fig. 11.5(c) is

$$\begin{aligned} G(s) = G_1(s)G_2(s) &= \frac{s^2C_1C_3R_2R_3}{(1 + sC_1R_1)(1 + sC_2R_2)} \\ &= \frac{s^2(C_3R_3/C_2R_1)}{s^2 + s[(C_1R_1 + C_2R_2)/C_1C_2R_1R_2] + (1/C_1C_2R_1R_2)} \end{aligned} \tag{11.19}$$

This is now of the form of Eq. 11.16, typical for a high-pass filter.

Setting like coefficients of Eq. 11.16 and 11.19 equal to each other, to adjust the pole positions, we can write

$$\frac{C_3R_3}{C_2R_1} = H \tag{11.20}$$

$$\frac{1}{C_1C_2R_1R_2} = \omega_o^2 \tag{11.21}$$

$$\frac{1}{C_1R_1} + \frac{1}{C_2R_2} = \frac{\omega_o}{Q} \tag{11.22}$$

There are more parameters than there are known relations, so we are free to choose some of the parameters. Since the second amplifier is quite independent, let

$$C_3R_3 = 1$$

Then

$$C_2 = \frac{1}{HR_1}$$

$$R_1 = \frac{1}{HC_2}$$

From Eqs. 11.21 and 11.22,

$$\omega_o = \frac{1}{Q(C_1R_1 + C_2R_2)}$$

and normalizing ω_o on 1 r/s,

$$1 = \frac{H}{Q[(C_1/C_2) + (R_2/R_1)]}$$

Choosing the capacitance and resistance ratios as unity, as easy values to maintain in integrated circuit production, we have

$$\frac{C_1}{C_2} = \frac{R_2}{R_1} = 1$$

$$H = 2Q$$

The gain is determined by the desired Q, and we are free to choose R_1 and C_1 or R_2 and C_2 to meet other circuit requirements.

11.4 Normalizing of parameter values

In the design of active filters, the cutoff angular frequency will usually be normalized to 1 r/s. Since certain of the circuit parameters are arbitrary, as just shown above, an appropriate resistor will be normalized to 1 Ω. An arbitrary capacitance can usually be made 1 F. We then decide on a suitable resistance level for the circuit, perhaps a resistor R shown as 1 Ω should have a real value of 1000 Ω; 100 or 10,000 Ω might be suitable in other applications. This fixes p, as used in previous normalizing operations. All resistance values of the model will be multiplied by p and all capacitances will be divided by p to maintain $\omega_o = 1$ as a design frequency.

To move ω_o to perhaps M r/s, all capacitances are divided by M, so that all capacitive reactances are kept invariant through the frequency translation.

11.5 The multiple-feedback low-pass filter

A more general approach to RC active filter design is through use of the circuit of Fig. 11.6. It can be made to provide a pair of complex conjugate poles, and zeros at the origin, as needed for high-pass and band-pass circuits. It employs two feedback paths with an amplifier of high internal gain A. Each of the impedances results from a single resistance or capacitance, with a maximum of five elements required.

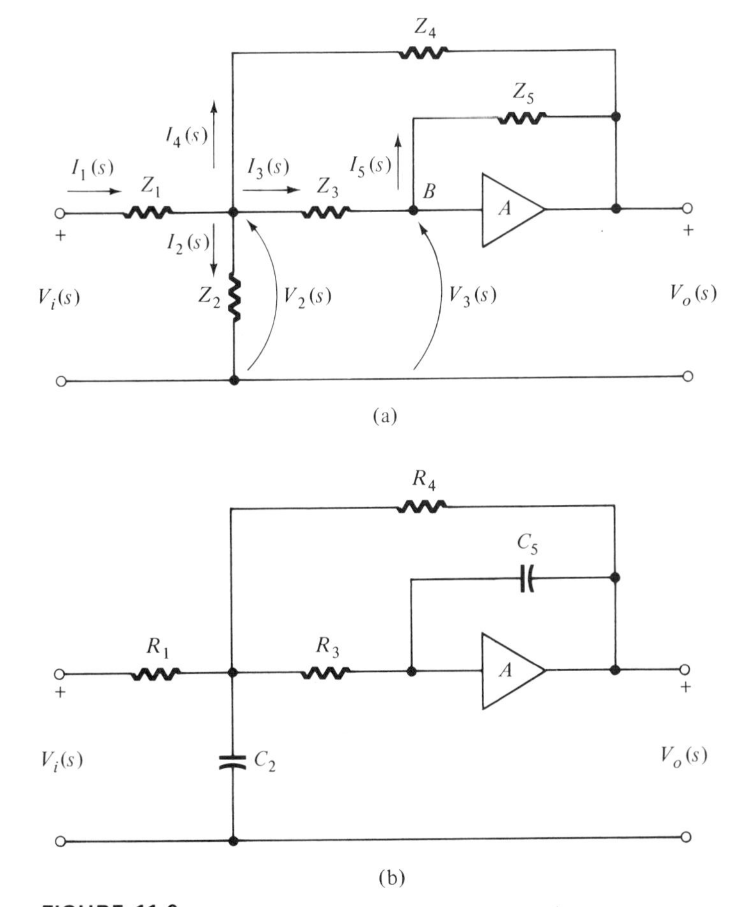

FIGURE 11.6

Multiple-feedback circuit for an *RC* active filter.

With $V_2(s)$ as shown, we write

$$I_1(s) = I_2(s) + I_3(s) + I_4(s) \tag{11.23}$$

Then

$$I_1(s) = \frac{V_i(s) - V_2(s)}{Z_1}$$

$$I_2(s) = \frac{V_2(s)}{Z_2}$$

With high gain A assumed, point B is essentially at 0 V or $V_3(s) = 0$. Then

$$I_3(s) = \frac{V_2(s)}{Z_3}$$

$$I_4(s) = \frac{V_2(s) - V_o(s)}{Z_4}$$

Also, $I_3(s) = I_5(s)$ because of the negligible amplifier input current, and

$$I_3(s) = I_5(s) = \frac{V_2(s)}{Z_3} = -\frac{V_o(s)}{Z_5}$$

$$V_2(s) = -V_o(s)\frac{Z_3}{Z_5} \tag{11.24}$$

Substitution in Eq. 11.23 leads to

$$V_i(s) = -V_o(s)\left[\frac{Z_1Z_3}{Z_5}\left(\frac{1}{Z_1} + \frac{1}{Z_2} + \frac{1}{Z_3} + \frac{1}{Z_4}\right) + \frac{Z_1}{Z_4}\right]$$

and consolidation of terms yields

$$\begin{aligned} G(s) &= \frac{V_o(s)}{V_i(s)} \\ &= \frac{-1}{(Z_1Z_3/Z_2Z_5) + (Z_1Z_3/Z_4Z_5) + [(Z_1 + Z_3)/Z_5] + (Z_1/Z_4)} \end{aligned} \tag{11.25}$$

The problem now becomes that of forcing an equality with one of the many filter transfer functions. We first consider the second-order *low-pass form* of Eq. 11.11:

$$G(s) = \frac{V_o(s)}{V_i(s)} = \frac{H\omega_o^2}{s^2 + (\omega_o/Q)s + \omega_o^2} \tag{11.26}$$

Since we are restricted to capacitors with impedance $1/sC$, the s^2 term must be derived from either or both of the first two terms of the denominator of

Eq. 11.25. That is, Z_5 and Z_2 or Z_4 must be capacitive; also if Z_5 is capacitive, the term in s will be derived from the third denominator term and then Z_1 and Z_3 must be resistive. The constant is derived from the last term, which requires Z_4 to be resistive since Z_1 has already been found as resistive. Therefore the circuit parameters will have the following impedances:

$$Z_1 = R_1 \qquad Z_4 = R_4$$

$$Z_2 = \frac{1}{sC_2} \qquad Z_5 = \frac{1}{sC_5}$$

$$Z_3 = R_3$$

and the circuit becomes that at (b) in Fig. 11.6.

Using these parameter values in Eq. 11.25, we have

$$G(s) = \frac{-1/(C_2C_5R_1R_3)}{s^2 + (s/C_2)[(1/R_1) + (1/R_3) + (1/R_4)] + (1/C_2C_5R_3R_4)} \tag{11.27}$$

We now require equality of the coefficients in Eqs. 11.26 and 11.27. It is apparent from the constant term in the denominator that

$$\omega_o = \frac{1}{\sqrt{C_2C_5R_3R_4}} \tag{11.28}$$

The numerator and the constant term in the denominator involve ω_o^2, and by dividing we obtain

$$-H = \frac{R_4}{R_1} \tag{11.29}$$

As an operational amplifier at zero frequency, the gain is

$$H = -\frac{R_f}{R_i} = -\frac{R_4}{R_1}$$

by circuit inspection; the result in Eq. 11.29 confirms that H is the zero-frequency or pass-band gain. The negative sign indicates the inherent phase reversal of such amplifiers.

The capacitances can be arbitrarily chosen, so we make

$$C_2 = C_5 = C \tag{11.30}$$

and choose any convenient capacitance value. From the coefficients of the s^1 term of Eq. 11.27 and from the typical equation for the low-pass filter, we have

$$\frac{\omega_o}{Q} = \frac{1}{C}\left(\frac{1}{R_1} + \frac{1}{R_3} + \frac{1}{R_4}\right) \tag{11.31}$$

With Eq. 11.28 we can write

$$\omega_o^2 C^2 = \frac{1}{R_3 R_4} \tag{11.32}$$

Then

$$R_3 = \frac{1}{\omega_o^2 C^2 R_4} \tag{11.33}$$

Rearranging Eq. 11.32 and using Eqs. 11.31 and 11.33 we have

$$\frac{\omega_o C}{Q} = \frac{H + 1}{R_4} + \omega_o^2 C^2 R_4$$

from which

$$\omega_o^2 C^2 R_4^2 - \frac{\omega_o C R_4}{Q} + H + 1 = 0 \tag{11.34}$$

This equation has a solution as a quadratic

$$\omega_o C R_4 = \frac{1}{2Q} \pm \sqrt{\frac{1}{4Q^2} - (H + 1)}$$

and so we have

$$R_4 = \frac{1}{2\omega_o C Q}[1 + \sqrt{1 - 4Q^2(H + 1)}] \tag{11.35}$$

choosing the positive sign to make R_4 a positive resistance. This result confirms that H must be negative for a real resistance. Equation 11.33 then is used for the other resistance parameter

$$R_3 = \frac{1}{\omega_o^2 C^2 R_4}$$

The quantities ω_o, Q, and H are usually specified; H may be less than 10 for $Q = 10$ and about 100 for $Q = 1$.

Assuming that $\omega_o = 1$ and that $Q = 0.707$ for a maximally flat response, the filter elements will have the normalized values of Fig. 11.7, with the response shown in (b).

This circuit utilizes a conventional high-gain amplifier with the internal gain reduced to the working level by the R_4/R_1 relation. Each element is a single resistance or capacitance, and circuit adjustment is simplified. However, it is not possible to obtain high Q values without careful adjustment of element values, a matter which is difficult with integrated circuits.

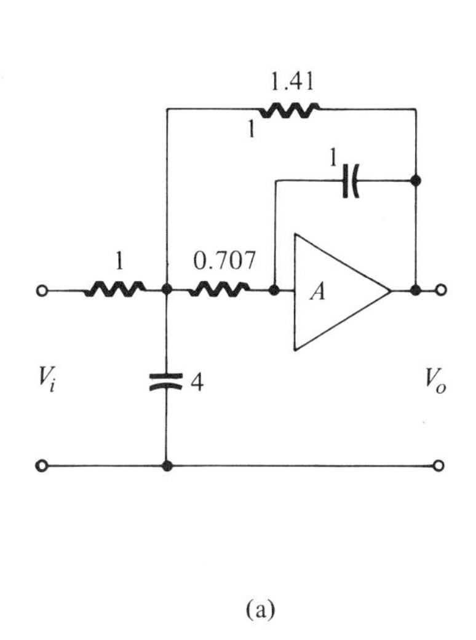

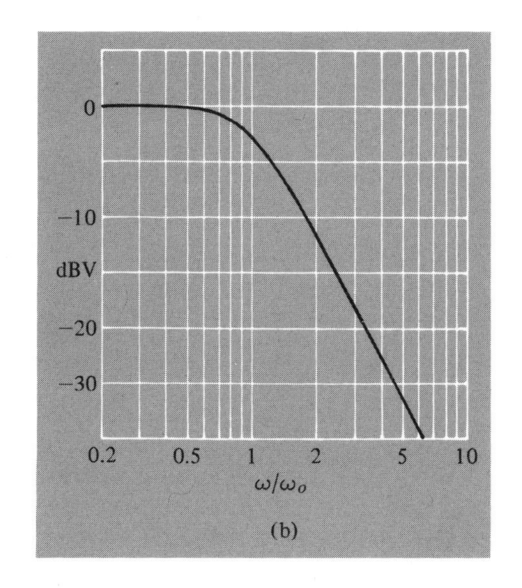

FIGURE 11.7

(a) Normalized low-pass circuit; $Q = 0.707$, $\omega_o = 1$ r/s, $H = 1$, $C_5 = 0.1\ \mu$F;
(b) measured response.

11.6

High-pass and band-pass circuits

The general network of Fig. 11.6(a) can be redesigned to provide a high-pass filter characteristic giving a response

$$G(s) = \frac{V_o(s)}{V_i(s)} = \frac{Hs^2}{s^2 + s(\omega_o/Q) + \omega_o^2} \tag{11.36}$$

We again choose the circuit elements to satisfy Eq. 11.25 by equating the coefficients to those of Eq. 11.36. A double zero is needed at zero frequency and it is evident that Z_1 and Z_3 could provide that as series capacitances. If Z_2 and Z_5 are made resistive, then the first term of the denominator yields the numerator zeros and the coefficient of the term in s^0. If we make Z_4 capacitive, the second and fourth denominator terms will contribute to the coefficient of the s^1 term.

The resultant circuit appears in Fig. 11.8, with the following parameter as impedances:

$$Z_1 = \frac{1}{sC_1} \qquad Z_4 = \frac{1}{sC_4}$$

$$Z_2 = R_2 \qquad Z_5 = R_5$$

$$Z_3 = \frac{1}{sC_3}$$

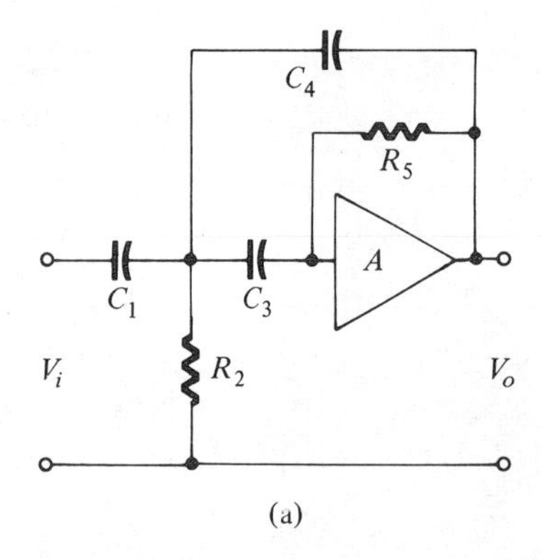

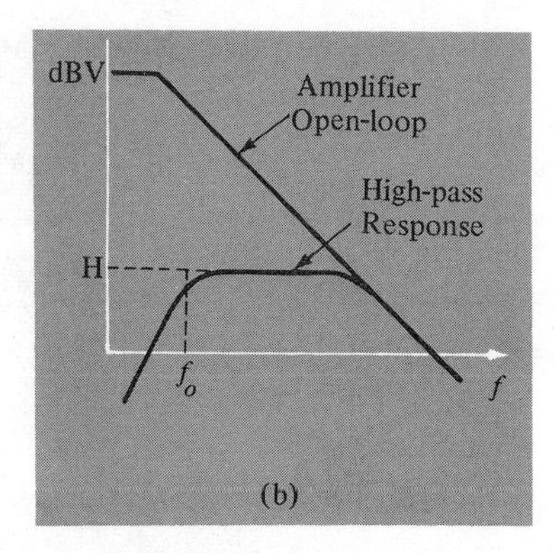

FIGURE 11.8

High-pass multiple-feedback circuit and response.

It should be no surprise that we find the R and C elements of the low-pass filter interchanged with those for the high-pass circuit.

Using these parameters, the transfer function can be written from Eq. 11.25 as

$$G(s) = \frac{V_o(s)}{V_i(s)}$$

$$= \frac{-(C_1/C_4)s^2}{s^2 + (s/R_5)[(1/C_3) + (1/C_4) + (C_1/C_3C_4)] + (1/C_3C_4R_2R_5)} \quad (11.37)$$

We recall that $A = z_f/z_i$ is the gain magnitude for the operational amplifier. At high frequencies in the pass band, the impedances of capacitors C_1 and C_4 control the feedback and the pass-band gain is

$$A = H = \frac{1/sC_4}{1/sC_1} = \frac{C_1}{C_4}$$

We may choose

$$C_1 = C_3 = C$$

and of any convenient value; accordingly

$$C_4 = \frac{C}{H} \quad (11.38)$$

Setting the constant term of the denominator of Eq. 11.37 equal to ω_o^2 from Eq. 11.36, we find

$$\omega_o = \frac{1}{\sqrt{C_3C_4R_2R_5}} \quad (11.39)$$

Equating the coefficient of the s^1 term in Eq. 11.37 to ω_o/Q,

$$\frac{\omega_o}{Q} = \frac{1}{R_5}\left(\frac{C_3 + C_4 + C_1}{C_3 C_4}\right) \tag{11.40}$$

and with $C_1 = C_3 = C$, we have

$$\frac{\omega_o}{Q} = \frac{1}{R_5}\left(\frac{2C + C_4}{CC_4}\right)$$

Using Eq. 11.38 we see that

$$\frac{\omega_o C}{Q} = \frac{1}{R_5}\left[\frac{2C + (C/H)}{(C/H)}\right]$$

$$= \frac{1}{R_5}(2H + 1) \tag{11.41}$$

Solving for the parameter R_5 gives

$$R_5 = \frac{Q}{\omega_o C}(2H + 1) \tag{11.42}$$

Again, using $C_1 = C_3 = C$ and $C_4 = C/H$ and combining Eqs. 11.39 and 11.42, we have

$$R_2 = \frac{H}{Q\omega_o C(2H + 1)} \tag{11.43}$$

Equations 11.38, 11.42, and 11.43 complete the circuit design, assuming that H, Q, and ω_o are specified. The response is shown in Fig. 11.8(b); at the high-frequency limit the gain rolloff of the operational amplifier controls the response.

To bring performance to the desired values, R_2 or R_5 may be adjusted to make Q have the correct value; then R_2 and R_5 should be adjusted, keeping their ratio constant to avoid changing Q, and bringing ω_o to specification.

The band-pass circuit has the general transfer function

$$G(s) = \frac{V_o(s)}{V_i(s)} = \frac{-Hs\omega_o/Q}{s^2 + s(\omega_o/Q) + \omega_0^2} \tag{11.44}$$

There is a single zero at the origin, so that one zero of the high-pass circuit can be eliminated by making Z_1 resistive; otherwise the circuit parameters can be identical in type to those of the high-pass circuit. That is,

$$Z_1 = R_1 \qquad Z_4 = \frac{1}{sC_4}$$

$$Z_2 = R_2 \qquad Z_5 = R_5$$

$$Z_3 = \frac{1}{sC_3}$$

The general relation of Eq. 11.25 then becomes

$$G(s) = \frac{V_o(s)}{V_i(s)}$$

$$= \frac{-(1/C_4R_1)s}{s^2 + (s/R_5)[(1/C_3) + (1/C_4)] + (1/C_3C_4R_5)[(1/R_1) + (1/R_2)]} \qquad (11.45)$$

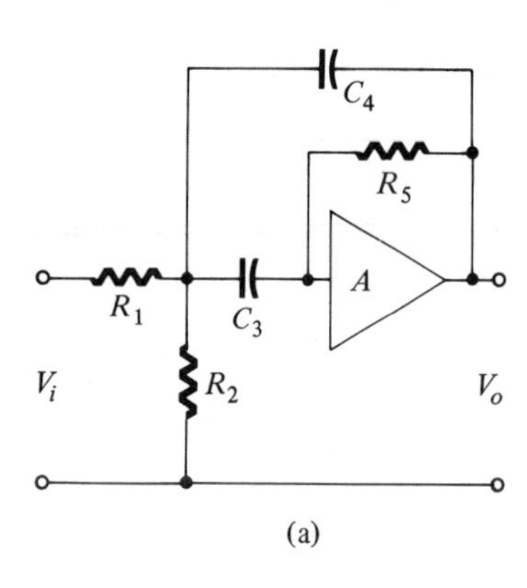

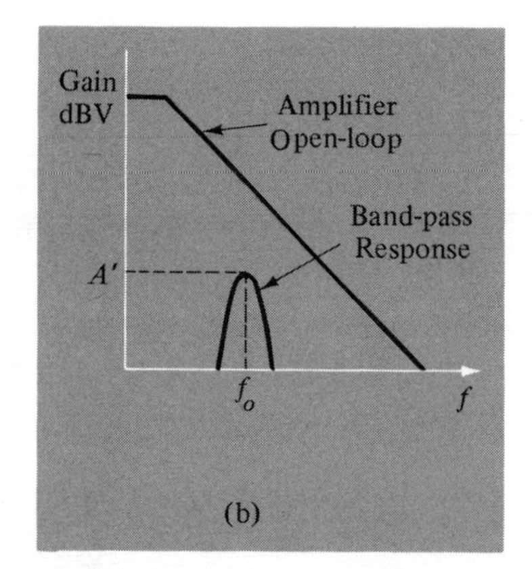

FIGURE 11.9
Band-pass multiple-feedback circuit and response.

The circuit is presented in Fig. 11.9(a). Equating the numerator coefficients for Eq. 11.45 and the type equation,

$$H\frac{\omega_o}{Q} = \frac{1}{C_4R_1}$$

but the coefficient of s^1 in the denominator of Eq. 11.45 must be ω_o/Q, so

$$H\left[\frac{1}{R_5}\left(\frac{1}{C_3} + \frac{1}{C_4}\right)\right] = \frac{1}{C_4R_1}$$

from which the band-pass amplifier gain is

$$H = \frac{R_5}{R_1[1 + (C_4/C_3)]} \tag{11.46}$$

From the coefficient of the s^0 term of Eq. 11.45,

$$\omega_o = \left[\frac{1}{C_3 C_4 R_5}\left(\frac{1}{R_1} + \frac{1}{R_2}\right)\right]^{1/2} \tag{11.47}$$

and since

$$\frac{\omega_o}{Q} = \frac{1}{R_5}\left(\frac{1}{C_3} + \frac{1}{C_4}\right) \tag{11.48}$$

then

$$Q = \frac{\sqrt{R_5(1/R_1 + 1/R_2)}}{\sqrt{C_3/C_4} + \sqrt{C_4/C_3}} \tag{11.49}$$

We can make the capacitors any convenient value so that

$$C_3 = C_4 = C \tag{11.50}$$

Using this relation in Eq. 11.48, we have

$$R_5 = \frac{2Q}{\omega_o C} \tag{11.51}$$

From Eq. 11.46, $R_1 = R_5/2H$ and so

$$R_1 = \frac{Q}{H\omega_o C} \tag{11.52}$$

and from Eq. 11.49,

$$R_2 = \frac{Q}{\omega_o C(2Q^2 - H)} \tag{11.53}$$

These are the design relations.

Resistance R_2 will be large; it can be varied to bring Q to the design value. Then R_2 and R_5 can be simultaneously varied to obtain ω_o without appreciably disturbing the value of Q. It is evident that internal gain changes can cause variation in circuit performance. Circuits to be studied in the next several sections divorce the gain adjustment from the frequency characteristics of the network.

However, the above should be useful in demonstrating the methods by which such circuits are designed.

11.7 Active filters using a controlled-gain amplifier

A class of circuits employing an amplifier of defined gain, K, was originated by Sallen and Key (Ref. 2). The R and C element values can be adjusted with the gain, but the gain itself is independent of filter element adjustment. This is important in circuit manufacture in integrated-circuit form. The circuit employs only four R or C elements in the low-pass version and can be designed for relatively high Q values if desired.

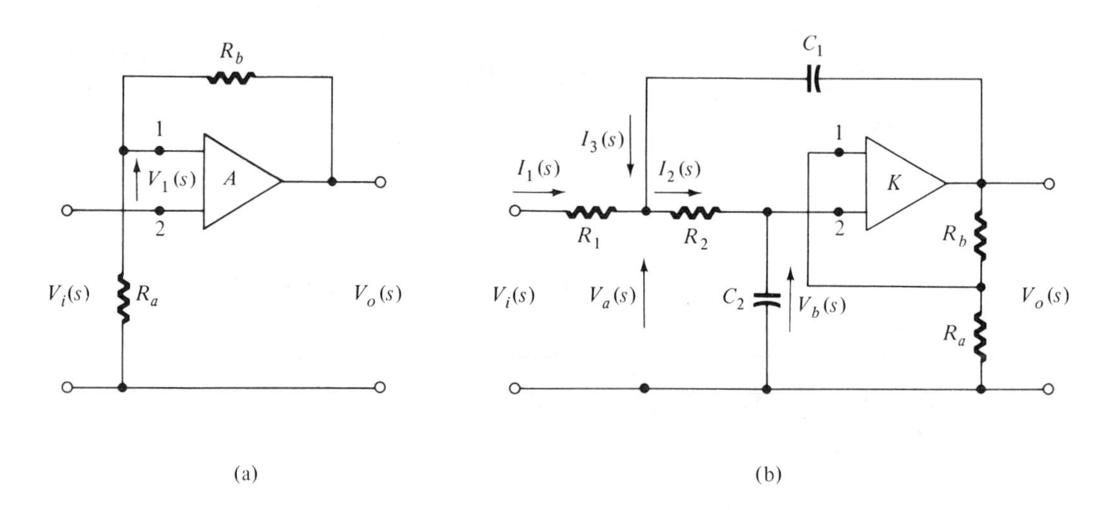

FIGURE 11.10

(a) Noninverting controlled-gain operational amplifier;
(b) gain element in low-pass filter.

The controlled-gain source is derived from the differential-input operational amplifier of Fig. 11.10, with internal gain A very large. We then have the overall gain K to input 1 obtainable from

$$V_1(s) = -\frac{R_a}{R_a + R_b} V_o(s)$$

$$K = -\frac{V_o(s)}{V_1(s)} = \left(1 + \frac{R_b}{R_a}\right) \tag{11.54}$$

This gain is independent of the other circuit parameters. The amplifier has the property of giving an equal positive gain magnitude at input 2. Normal filter input is applied to 2, as the *noninverting* or positive-gain terminal.

We again design the filter by comparison of the coefficients of the typical low-pass filter equation,

$$G(s) = \frac{H\omega_o^2}{s^2 + s(\omega_o/Q) + \omega_o^2} \tag{11.55}$$

with a transfer equation which we shall obtain for the network of (b) in Fig. 11.10. Summing the currents,

$$I_1(s) + I_3(s) = I_2(s)$$

$$\frac{V_i(s) - V_a(s)}{R_1} + \frac{V_o(s) - V_a(s)}{1/sC_1} = \frac{V_a(s) - V_b(s)}{R_2} \tag{11.56}$$

With positive gain, $V_o(s) = KV_b(s)$, and

$$\frac{V_b(s)}{V_a(s)} = \frac{1}{1 + sC_2R_2}$$

which is not surprising because C_2 and R_2 constitute a low-pass first-order network. We then have the result

$$G(s) = \frac{V_o(s)}{V_i(s)}$$

$$= \frac{K/(C_1C_2R_1R_2)}{s^2 + s\{(1/C_1)[(1/R_1) + (1/R_2)] + [(1-K)/C_2R_2]\} + (1/C_1C_2R_1R_2)} \tag{11.57}$$

It may be recognized that the gain value should be set to

$$H = K \tag{11.58}$$

usually chosen less than 10. Also,

$$\omega_o = \frac{1}{\sqrt{C_1C_2R_1R_2}} \tag{11.59}$$

From the coefficient of the s^1 term, and using ω_0

$$\frac{1}{Q} = \sqrt{\frac{C_2R_2}{C_1R_1}} + \sqrt{\frac{C_2R_1}{C_1R_2}} + (1 - K)\sqrt{\frac{C_1R_1}{C_2R_2}} \tag{11.60}$$

As before, choose

$$C_1 = C_2 = C \tag{11.61}$$

as any convenient capacitance. From Eq. 11.59,

$$R_1 = \frac{1}{\omega_o^2 C^2 R_2} \tag{11.62}$$

relating $R_1 R_2$ and C^2. Substitution of these equations into Eq. 11.60 leads to

$$\omega_o^2 C^2 R_2^2 - \frac{\omega_o C R_2}{Q} + 2 - K = 0$$

The value of R_2 follows as

$$R_2 = \frac{1}{2\omega_o C Q}\left[1 + \sqrt{1 + 4(K - 2)Q^2}\right] \tag{11.63}$$

If we make

$$C_1 = C_2 = C \qquad \text{and} \qquad R_1 = R_2 = R$$

then we have

$$R = \frac{1}{\omega_o C} \tag{11.64}$$

$$K = 3 - \frac{1}{Q} \tag{11.65}$$

and the filter parameters are determined. The value of $\omega_o = 1/\sqrt{C_1 C_2 R_1 R_2}$ is independent of gain, allowing the gain to be separately adjusted by R_a and R_b to meet the Q specification in Eq. 11.65.

11.8 Controlled-gain high-pass and band-pass filters

By interchange of the R and C elements, we form the high-pass circuit of Fig. 11.11(a) from the low-pass figure. The transfer function can be derived as

$$G(s) = \frac{V_o(s)}{V_i(s)}$$

$$= \frac{Ks^2}{s^2 + s\{(1/C_1 R_2) + (1/C_2 R_2) + [(1 - K)/C_1 R_1]\} + (1/C_1 C_2 R_1 R_2)} \tag{11.66}$$

The two capacitors, C_1 and C_2, create the needed double zero at the origin.

By comparison with the typical high-pass equation of Eq. 11.14 we see that

$$H = K \tag{11.67}$$

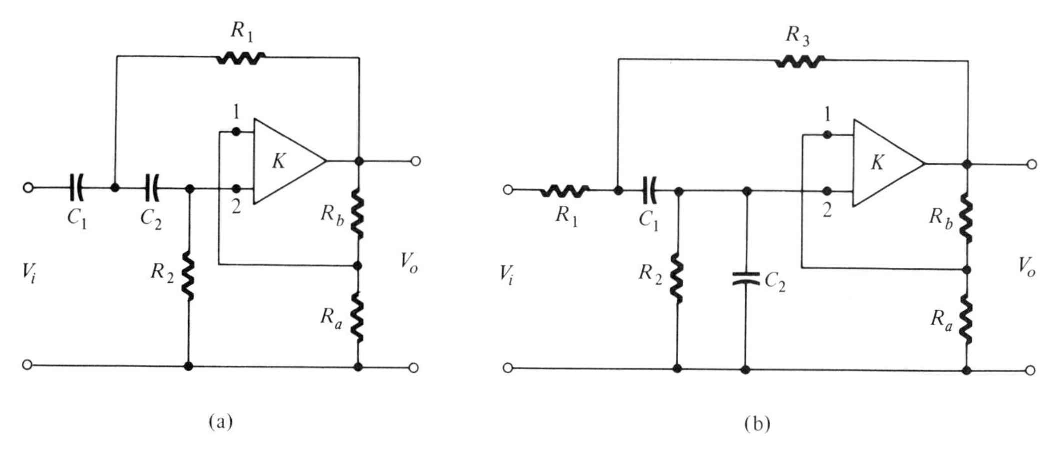

FIGURE 11.11

(a) Controlled-gain high-pass filter; (b) band-pass action.

where K is the gain of the amplifier, as established by R_a and R_b at the inverting input of the amplifier. The signal is applied to the noninverting input.

We also see that

$$\omega_o = \frac{1}{\sqrt{C_1 C_2 R_1 R_2}} \tag{11.68}$$

and equating coefficients of the s^1 terms in the denominator, we have

$$\frac{\omega_o}{Q} = \frac{1}{Q\sqrt{C_1 C_2 R_1 R_2}} = \frac{1}{C_1 R_2} + \frac{1}{C_2 R_2} + \frac{1-K}{C_1 R_1}$$

$$\frac{1}{Q} = \sqrt{\frac{C_2 R_1}{C_1 R_2}} + \sqrt{\frac{C_1 R_1}{C_2 R_2}} + (1-K)\sqrt{\frac{C_2 R_2}{C_1 R_1}} \tag{11.69}$$

Given H, Q, and ω_o, we can choose to make

$$C_1 = C_2 = C \tag{11.70}$$

of any convenient value. Inserting this condition into Eq. 11.69 gives

$$\frac{1}{Q} = 2\sqrt{\frac{R_1}{R_2}} + (1-K)\sqrt{\frac{R_2}{R_1}}$$

Using Eq. 11.68,

$$R_2 = \frac{1}{\omega_o^2 C^2 R_1} \tag{11.71}$$

Combining these expressions leads to

$$\omega_o^2 C^2 R_1^2 - \frac{\omega_o C R_1}{2Q} + \frac{1-K}{2} = 0$$

from which

$$R_1 = \frac{1}{4\omega_o C Q}[1 + \sqrt{1 + 8(K-1)Q^2}] \tag{11.72}$$

Then

$$R_2 = \frac{4Q}{\omega_o C[1 + \sqrt{1 + 8(K-1)Q^2}]} \tag{11.73}$$

The gain $H = K$ must be greater than unity to ensure that R_1 and R_2 are positive resistances. As for the low-pass circuit, we can choose $C_1 = C_2 = C$ and $R_1 = R_2 = R$; then $R = 1/\omega_o C$ and $K = 3 - 1/Q$.

The gain may be independently adjusted and used to raise the Q value without affecting the cutoff frequency; such adjustment can be made by change of R_a or R_b.

The band-pass circuit requires only a single zero at the origin to lower the low-frequency end of the response, but must also have a zero at infinity to reduce the response above the pass band. These zeros may be created from the high-pass circuit by replacing C_1 with R_1, thereby taking out one of the zeros at the origin, and by placing a capacitance C_2 in parallel with R_2 to create the zero at infinity. The circuit becomes that of Fig. 11.11(b), with the following transfer function derived by the usual methods:

$$G(s) = \frac{V_o(s)}{V_i(s)}$$

$$= \frac{Ks/C_2R_1}{s^2 + s\left(\frac{1}{C_2R_2} + \frac{1}{C_2R_1} + \frac{1}{C_1R_1} + \frac{1}{C_1R_3} + \frac{1-K}{C_2R_3}\right) + \frac{R_1 + R_3}{C_1C_2R_1R_2R_3}} \tag{11.74}$$

By comparison with the parameters of the typical band-pass equation, Eq. 11.15, we find

$$H = \frac{K}{1 + (R_1/R_2) + (C_2/C_1)[1 + (R_1/R_3)] + (1-K)(R_1/R_3)} \tag{11.75}$$

$$\omega_o = \sqrt{\frac{1}{C_1C_2R_2}\left(\frac{1}{R_1} + \frac{1}{R_3}\right)} \tag{11.76}$$

$$\frac{1}{Q} = \sqrt{\frac{R_1R_2R_3}{R_1 + R_3}}\left[\sqrt{\frac{C_1}{C_2}}\left(\frac{1}{R_1} + \frac{1}{R_2} + \frac{1-K}{R_3}\right) + \sqrt{\frac{C_2}{C_1}}\left(\frac{1}{R_1} + \frac{1}{R_3}\right)\right] \tag{11.77}$$

Given the center frequency ω_o and Q, we choose convenient values for

$$C_1 = C_2 = C$$
$$R_1 = R_2 = R_3 = R$$

Then Eq. 11.76 relates R and C as

$$R = \frac{\sqrt{2}}{\omega_o C} \tag{11.78}$$

Using R and C in Eq. 11.77 gives

$$K = 5 - \frac{\sqrt{2}}{Q} \tag{11.79}$$

and therefore K is less than 5.

Again the parameters are independent of the gain, and K can be close to 5 for a large Q, without affecting the resonant frequency ω_o. However, $Q > 10$ requires precise adjustment of K and is not practical.

11.9 A third-order low-pass filter design

A third-order response might be

$$G(s) = \frac{1}{s^3 + 2s^2 + 2s + 1} = \frac{1}{(s^2 + s + 1)} \frac{1}{(s + 1)} \tag{11.80}$$

By factoring of the denominator of the function we see that the filter might be realized with a low-pass second-order network, followed in cascade by a first-order RC network. This arrangement is shown in Fig. 11.12.

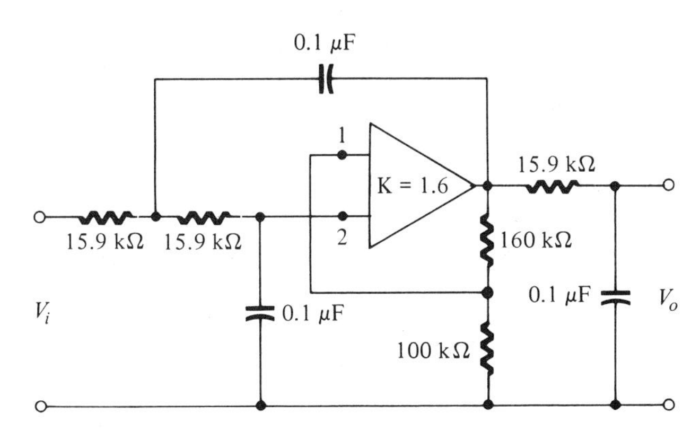

FIGURE 11.12
Third-order low-pass filter, $f_o = 100$ Hz.

The circuit may be designed for a cutoff frequency of $f_o = 100$ Hz. For a maximally flat response we know that $Q = 0.707$; then from Eq. 11.65,

$$K = 3 - \frac{1}{0.707} = 1.6$$

and this establishes R_a and R_b. Choosing $C_1 = C_2 = C = 0.1\ \mu$F, then from Eq. 11.64,

$$R_1 = R_2 = R = \frac{1}{\omega_o C} = \frac{1}{2\pi \times 100 \times 0.1 \times 10^{-6}}$$

$$= 15{,}900\ \Omega$$

This is a convenient value, and the circuit parameters for the second-order section have been obtained. This choice of parameters also makes $f_o = 100$ Hz for the first-order section and so we make $R_3 = 15{,}900\ \Omega$ and $C_3 = 0.1\ \mu$F. These values are indicated in Fig. 11.12. The response curve will roll off at $-20n$ dBV/frequency decade, or at a -60-dBV rate.

Designing with high-order transfer functions involves feedback loops over the entire network, and small variations in the feedback circuit become critical. Factoring of the higher-order response functions as shown, to obtain a cascade of first- and second-order networks, leads to more stable results. A factored network can be used in cascade because the gain elements have low output impedances and serve to isolate the several circuits.

11.10

The operational amplifier with parallel-*T* feedback

In Section 11.1 it was proposed that a frequency-selective network in the feedback branch of an operational amplifier could lead to a filter form of response. This situation is represented by Fig. 11.13(a), and the concept is carried out with a *parallel-T* feedback

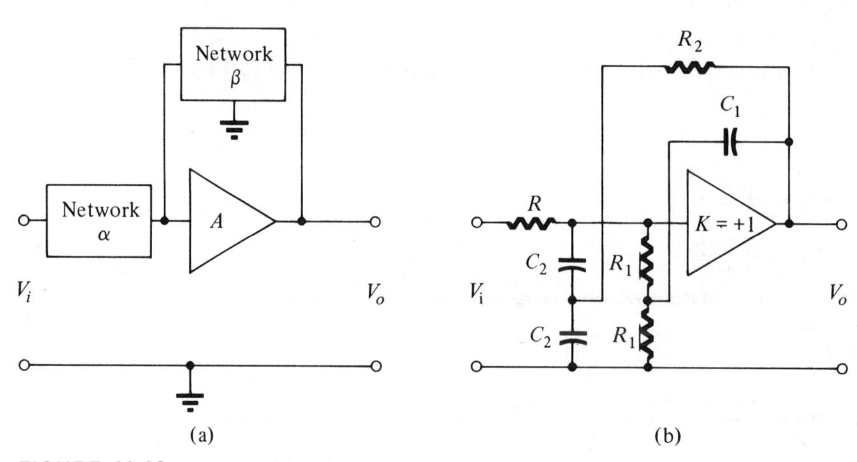

FIGURE 11.13

Active filter with parallel-*T* network for feedback.

network in (b). Such a network provides a pair of complex conjugate pole locations having small and controllable negative real parts. As connected, the circuit yields a filter with a band-pass response peaking at the chosen $\omega_o = 2\pi f_o$. Because of the active nature of the circuit, a negative resistance appears and is used to control Q, or the real part of the pole locations.

Consider the two feedback networks as in parallel and isolate the resistive branch as in Fig. 11.14(a). A current summation gives

$$I_1(s) + I_2(s) = I_3(s)$$

$$\frac{V_1(s) - V_a(s)}{R_1} + \frac{V_o(s) - V_a(s)}{1/sC_1} = \frac{V_a(s)}{R_1} \tag{11.81}$$

The circuit can easily be adapted to an amplifier in which the gain is made $K \cong +1$. Noting that

$$V_o(s) = KV_1(s) \cong V_1(s)$$

$$V_a(s) = V_1(s) - I_1(s)R_1$$

we can write the input impedance as

$$Z_R = \frac{V_1(s)}{I_1(s)} = 2R_1 + sC_1R_1^2$$

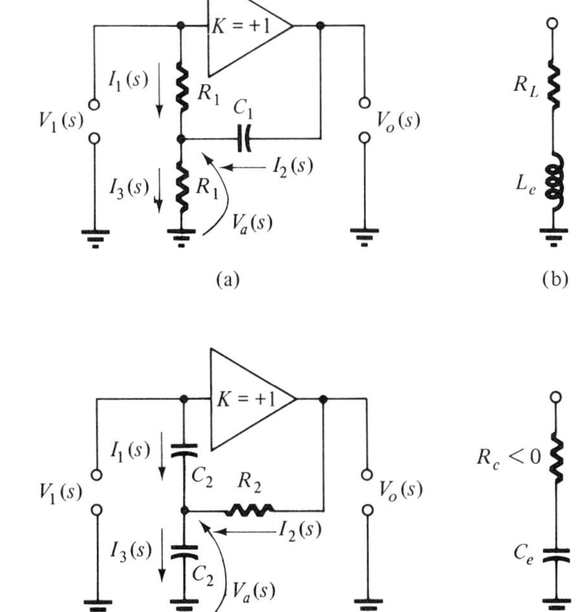

FIGURE 11.14

Separated branches of the parallel-*T* network.

and for the steady state with $s = j\omega$, this becomes

$$Z_R = 2R_1 + j\omega C_1 R_1^2 \tag{11.82}$$

The reactive term is positive and represents the reactance of an inductance, with

$$L_e = C_1 R_1^2 \tag{11.83}$$

and the circuit of (b) is equivalent to that at (a).

Repeating the analysis for the isolated capacitive feedback branch in (c), the input impedance is

$$Z_C = \frac{V_1(s)}{I_1(s)} = \frac{1}{s^2 C_2^2 R_2} + \frac{2}{sC_2}$$

With $s = j\omega$ this becomes

$$Z_C = -\frac{1}{\omega^2 C_2^2 R_2} - \frac{j2}{\omega C_2} \tag{11.84}$$

The reactive term represents an equivalent capacitance

$$C_e = \frac{C_2}{2} \tag{11.85}$$

but more importantly the real term is a negative resistance,

$$R_c = -\frac{1}{\omega^2 C_2^2 R_2} \tag{11.86}$$

The circuit is modeled in Fig. 11.14(d).

Placed in parallel, the equivalent circuits show a resonant frequency of

$$\omega_o = \sqrt{\frac{2}{C_1 C_2 R^2}} \tag{11.87}$$

with $R_1 = R_2 = R$. The effect is that of a parallel resonant circuit, with variable total R_e, which can be made negligibly small. The poles can therefore be moved close to the $j\omega$ axis and a sharp resonant response can be obtained. The equivalent Q is

$$Q = \frac{\omega_o L_e}{R_L - |R_c|} \tag{11.88}$$

and the bandwidth can be small. If the magnitude of R_c is made too great, the poles will move into the right half of the s plane and the circuit will become unstable. The circuit Q can be changed by varying the value of R_2.

The design is carried out by choice of a suitable L_e for the resonant circuit to simulate, and C_1 is chosen arbitrarily; Q is selected from the desired bandwidth, but this is somewhat degraded by the input resistor R. A response curve for a four-stage cascade appears in Fig. 11.15 for $f_o = 1000$ Hz.

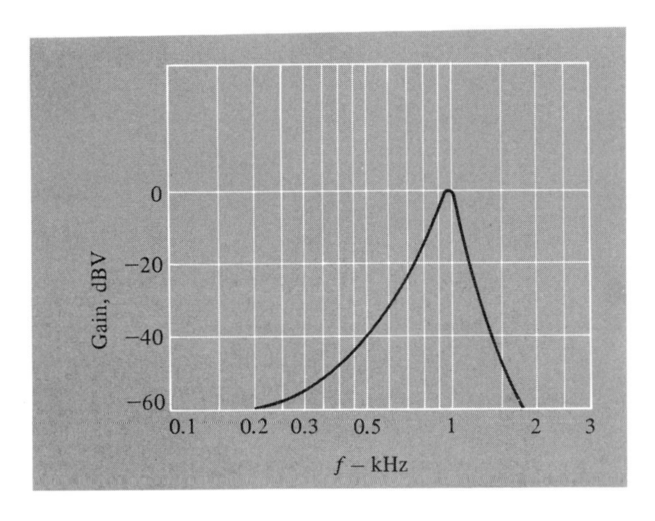

FIGURE 11.15

Response of four-section parallel-T filter.

The circuit employs seven elements per stage. The more general approach of Fig. 11.6 requires only five or six elements and represents a less complex design, but for high Q the parallel-T circuit is more easily constructed and aligned.

11.11 Sensitivity

The circuit elements of which the filter is composed will differ from the design values due to accidents of manufacture. We are concerned that the transfer function not be sensitive to such inevitable variations. *Sensitivity* to changes in a parameter M is defined as

$$S_M = \frac{\partial T/T}{\partial M/M} = \frac{M}{T}\frac{\partial T}{\partial M} \tag{11.89}$$

In this expression T is the transfer function under study and M is the element varied.

Many other definitions have been proposed, including one based on the shift of the poles by variation of an element. That is,

$$S_P = \frac{\partial \sigma/\sigma}{\partial M/M} + j\frac{\partial \omega/\omega}{\partial M/M} = \frac{M}{\sigma}\frac{\partial \sigma}{\partial M} + j\frac{M}{\omega}\frac{\partial \omega}{\partial M} \tag{11.90}$$

The real term shows the shift of the pole parallel to the real axis, and the j term shows the pole movement parallel to the $j\omega$ axis. The vector combination graphically depicts the pole motion as a circuit element or as the gain changes.

As an example, consider the transfer function of Eq. 11.57, making $C_1 = C_2 = 1.0$ F and $R_1 = R_2 = 1\ \Omega$ as normalized values. Make $Q = 1$; then

$$G(s) = \frac{K}{s^2 + (3 - K)s + 1} \tag{11.91}$$

This shows that $\omega_o/Q = 3 - K$, so that

$$Q = \frac{1}{3 - K} \tag{11.92}$$

since $\omega_o = 1$.

The sensitivity to changes in K can be found from the pole positions

$$s_1, s_2 = -\frac{3 - K}{2} \pm j\sqrt{1 - \frac{(3 - K)^2}{4}}$$

from which

$$\sigma = \frac{K - 3}{2}$$

$$\omega = \sqrt{1 - \frac{(3 - K)^2}{4}}$$

Taking the derivatives and substituting in Eq. 11.90 gives

$$S_K = \frac{K}{K - 3} + j\frac{K(3 - K)}{4 - (3 - K)^2} \tag{11.93}$$

From Eq. 11.92,

$$K = 3 - \frac{1}{Q}$$

and $K = 2.9$ for $Q = 10$. The gain sensitivity is then

$$S_K = -29 + j0.072$$

which has a large real part. Using Eq. 11.92 in the real part of the sensitivity S_K, we can write

$$S_K = 1 - 3Q \cong -3Q$$

and the sensitivity of the gain is primarily a function of Q. A reduction of Q to unity brings the sensitivity to approximately

$$S_K = -2 + j0.66$$

which is much more favorable. Similar sensitivities can be developed for the other circuit parameters.

This analysis points out the problem of obtaining large Q values for circuits which are subject to the vagaries of integrated circuit production.

11.12 General comments

We have here explored a new field which combines circuits and gain elements to synthesize a useful result. The general method is that of selection of a suitable filter circuit and the determination of its transfer function. We then equate the coefficients of the function with the coefficients derived from a general RC network, which will provide appropriate poles and zeros, and determine the parameters for the RC active filter. Forcing the equality of the coefficients ensures that the form of frequency response will be that desired. Use of inductances can be avoided, and smaller and more accurate filters can be produced.

For further study, the Sallen and Key paper (Ref. 2) is basic; Ref. 4 gives a rather complete bibliography.

PROBLEMS

11.1 Determine the voltage gain transfer functions, as functions of s, for the circuits of (a) and (b) in Fig. 11.16.

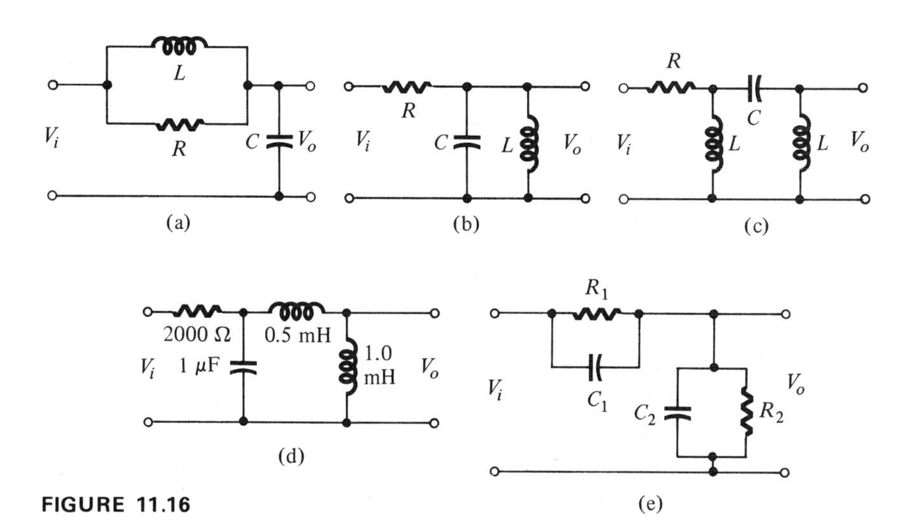

FIGURE 11.16

11.2 (a) Write the transfer function V_o/V_i for the circuit of Fig. 11.16(c) using s. (b) Determine V_0/V_i as a function of the resonant frequency ω_o for the circuit of (d) in Fig. 11.16.

11.3 (a) Find the transfer function V_o/V_i in terms of s for the circuit of Fig. 11.16(e). (b) Under what conditions of the parameters is this transfer function independent of frequency?

11.4 The transfer function of a network is

$$\frac{V_o}{V_i} = \frac{4s^2}{s^3 + 5s^2 + 7s + 2}$$

(a) Describe this network as a low-pass, band-pass, or high-pass network. (b) Propose an RC network with gain to simulate this function.

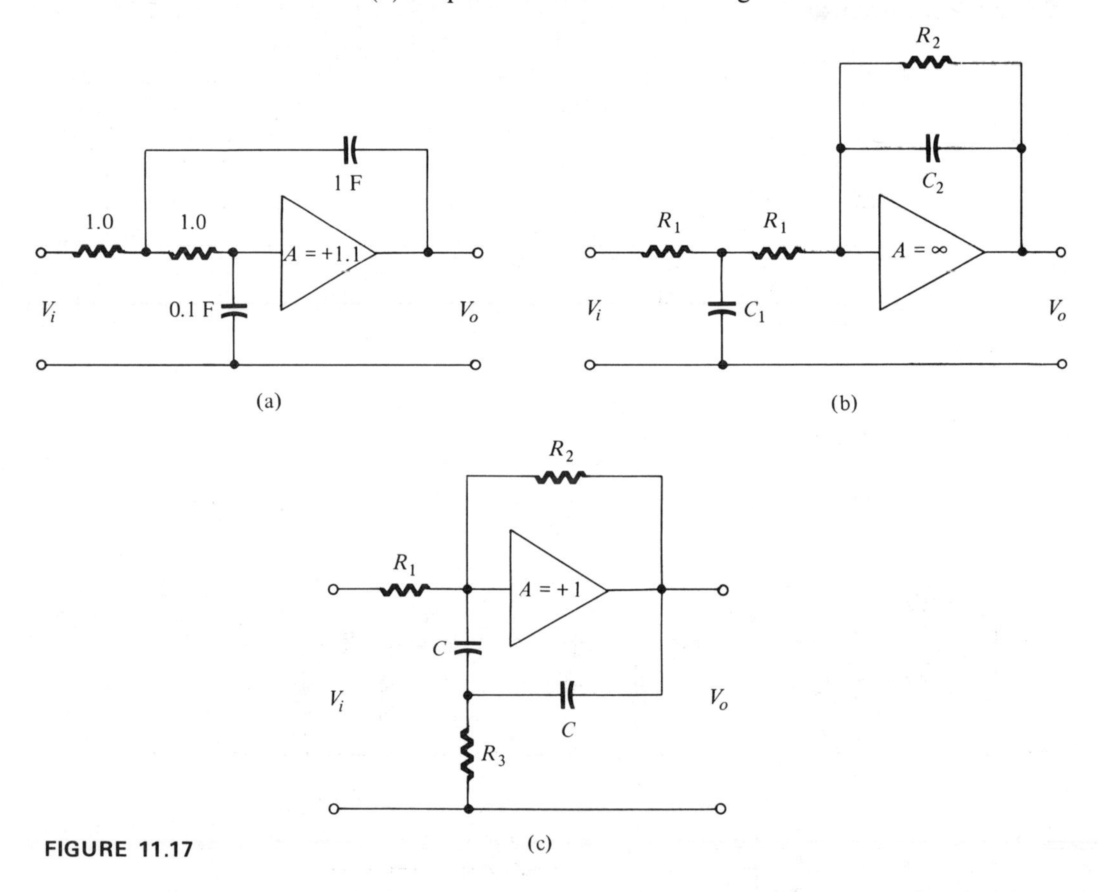

FIGURE 11.17

11.5 (a) Determine the transfer function for the circuit of Fig. 11.17(a). (b) What is the value of Q? (c) Scale the above parameters to a basic $R = 1000\ \Omega$, with $\omega_o = 200$ r/s.

11.6 (a) Find the gain at zero frequency in Fig. 11.17(b). (b) What is the transfer function in terms of s for that circuit?

11.7 (a) Determine the transfer function being simulated by the circuit of Fig. 11.17(c). (b) Find Q and the expression for ω_o.

11.8 For the circuit of Fig. 11.18(a), determine the transfer function in terms of s. Find the frequency at which oscillation will occur. Find the limit on Q if oscillation is to be avoided.

11.9 Determine the transfer function of (b) in Fig. 11.18. Is the performance low pass, band pass, or high pass? Find the frequency ω_o in terms of the gains. Find Q in terms of the gains.

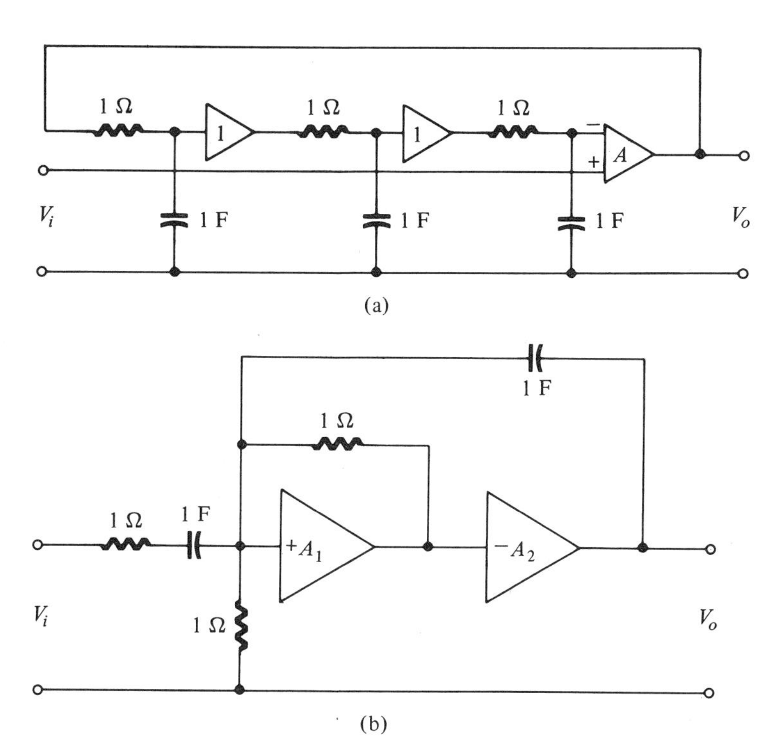

FIGURE 11.18

REFERENCES

1. J. G. Linvill, "Reactive Filters." *Proc. IRE*, 42, 555 (1954).
2. R. P. Sallen and E. L. Key, "A Practical Method of Designing RC Active Filters." *IRE Trans. Circuit Theory*, CT-2, 74 (March 1955).
3. S. S. Hakim, *Synthesis of RC Active Filter Networks*. McGraw-Hill Book Company, New York, 1969.
4. L. P. Huelsman, *Active Filters: Lumped, Distributed, Integrated, Digital, and Parametric*. McGraw-Hill Book Company, New York, 1970.
5. J. G. Graeme, G. E. Tobey, and L. P. Huelsman, *Operational Amplifiers*. McGraw-Hill Book Company, New York, 1971.

Index

C

D

L

M

N

O

S

T

U

V

W

Y

Z